Applications of Anionic Polymerization Research

ACS SYMPOSIUM SERIES **696**

Applications of Anionic Polymerization Research

Roderic P. Quirk, EDITOR
University of Akron

Developed from a symposium sponsored by the Division
of Polymer Chemistry at the 212th National Meeting
of the American Chemical Society,
Orlando, Florida,
August 25–29, 1996

American Chemical Society, Washington, DC

Library of Congress Cataloging-in-Publication Data

Applications of anionic polymerization research / Roderic P. Quirk, editor

p. cm.—(ACS symposium series, ISSN 0097–6156; 696)

"Developed from a symposium sponsored by the Division of Polymer Chemistry at the 212th National Meeting of the American Chemical Society, Orlando, Florida, August 25–29, 1996."

Includes bibliographical references and indexes.

ISBN 0–8412–3565–1

1. Addition polymerization—Congresses.

I. Quirk, Roderic P. II. American Chemical Society. Division of Polymer Chemistry. III. American Chemical Society. Meeting (212th : 1996: Orlando, Fla.) IV. Series.

QD281.P6A64   1998
547′.28—dc21   98–18647
                                                             CIP

This book is printed on acid-free paper.

Copyright © 1998 American Chemical Society

Distributed by Oxford University Press

All Rights Reserved. Reprographic copying beyond that permitted by Sections 107 or 108 of the U.S. Copyright Act is allowed for internal use only, provided that a per-chapter fee of $20.00 plus $0.25 per page is paid to the Copyright Clearance Center, Inc., 222 Rosewood Drive, Danvers, MA 01923, USA. Republication or reproduction for sale of pages in this book is permitted only under license from ACS. Direct these and other permissions requests to ACS Copyright Office, Publications Division, 1155 16th Street, N.W., Washington, DC 20036.

The citation of trade names and/or names of manufacturers in this publication is not to be construed as an endorsement or as approval by ACS of the commercial products or services referenced herein; nor should the mere reference herein to any drawing, specification, chemical process, or other data be regarded as a license or as a conveyance of any right or permission to the holder, reader, or any other person or corporation, to manufacture, reproduce, use, or sell any patented invention or copyrighted work that may in any way be related thereto. Registered names, trademarks, etc., used in this publication, even without specific indication thereof, are not to be considered unprotected by law.

PRINTED IN THE UNITED STATES OF AMERICA

Foreword

THE ACS SYMPOSIUM SERIES was first published in 1974 to provide a mechanism for publishing symposia quickly in book form. The purpose of the series is to publish timely, comprehensive books developed from ACS sponsored symposia based on current scientific research. Occasionally, books are developed from symposia sponsored by other organizations when the topic is of keen interest to the chemistry audience.

Before agreeing to publish a book, the proposed table of contents is reviewed for appropriate and comprehensive coverage and for interest to the audience. Some papers may be excluded in order to better focus the book; others may be added to provide comprehensiveness. When appropriate, overview or introductory chapters are added. Drafts of chapters are peer-reviewed prior to final acceptance or rejection, and manuscripts are prepared in camera-ready format.

As a rule, only original research papers and original review papers are included in the volumes. Verbatim reproductions of previously published papers are not accepted.

ACS BOOKS DEPARTMENT

Advisory Board

ACS Symposium Series

Mary E. Castellion
ChemEdit Company

Arthur B. Ellis
University of Wisconsin at Madison

Jeffrey S. Gaffney
Argonne National Laboratory

Gunda I. Georg
University of Kansas

Lawrence P. Klemann
Nabisco Foods Group

Richard N. Loeppky
University of Missouri

Cynthia A. Maryanoff
R. W. Johnson Pharmaceutical
  Research Institute

Roger A. Minear
University of Illinois
  at Urbana–Champaign

Omkaram Nalamasu
AT&T Bell Laboratories

Kinam Park
Purdue University

Katherine R. Porter
Duke University

Douglas A. Smith
The DAS Group, Inc.

Martin R. Tant
Eastman Chemical Co.

Michael D. Taylor
Parke-Davis Pharmaceutical
  Research

Leroy B. Townsend
University of Michigan

William C. Walker
DuPont Company

Contents

Preface..xi

INTRODUCTION

1. Principles of Anionic Polymerization: An Introduction....................................2
   Roderic P. Quirk, Qizhuo Zhuo, Sung H. Jang, Youngjoon Lee, and Gilda Lizarraga

2. Industrial Applications of Anionic Polymerization: An Introduction...........28
   Henry L. Hsieh

FUNDAMENTALS AND POLYMERZATION PROCESSES

3. Aggregation Behavior of Polymer Chains with Living Anionic Lipophobic Head-Groups..36
   L. J. Fetters, J. S. Huang, J. Sung, L. Willner, J. Stellbrink, D. Richter, and P. Lindner

4. Process Monitoring: UV Spectrophotometry as a Practical Tool....................50
   M. Bortolotti, G. T. Viola, and A. Gurnari

5. Polymerization of Butadiene with Novel *tert*-Amino Group Initiator............62
   Toshihiro Tadaki, Iwakazu Hattori, and Fumio Tsutsumi

6. Anionic Polymerization of Dienes Using Homogeneous Lithium Amide (N-Li) Initiators, and Determination of Polymer-Bound Amines.......77
   D. F. Lawson, D. R. Brumbaugh, M. L. Stayer, J. R. Schreffler, T. A. Antkowiak, D. Saffles, K. Morita, Y. Ozawa, and S. Nakayama

BLOCK COPOLYMERS

7. The Influence of Uncoupled Styrene–Butadiene Diblock Copolymer on the Physical Properties of SBS Thermoplastic Elastomers......................90
   H. R. Lovisi, L. F. Nicolini, A. A. Ferreira, and M. L. S. Martins

8. The Effect of SI Diblock in SIS Block Copolymer on Pressure-Sensitive Adhesive Properties..98
    Tetsuaki Matsubara and Minoru Ishiguro

9. Styroflex: A New Transparent Styrene–Butadiene Copolymer with High Flexibility Synthesis, Applications, and Synergism with Other Styrene Polymers...112
    Konrad Knoll and Norbert Nieβner

10. Tapered Block Copolymers of Styrene and Butadiene: Synthesis, Structure, and Properties...129
    Sergio A. Moctezuma, Enrico N. Martínez, Rodolfo Flores, and Enrique Fernández-Fassnacht

STAR POLYMERS

11. High-Temperature Anionic Polymerization Processes: Synthesis of Star-Branched Polymers..142
    F. Schué, R. Aznar, J. Couve, R. Dobreva-Schué, and P. Nicol

12. Asymmetric Star Block Copolymers: Anionic Synthesis, Characterization, and Pressure-Sensitive Adhesive Performance..............159
    J.-J. Ma, M. K. Nestegard, B. D Majumdar, and M. M. Sheridan

13. Preparation of Branched Polystyrenes with Known Architecture..............167
    J. L. Hahnfeld, W. C. Pike, D. E. Kirkpatrick, and T. G. Bee

DIENE POLYMERS

14. New Isoprene Polymers...186
    Makoto Nishikawa, Mizuho Maeda, Hiromichi Nakata, Hideo Takamatsu, and Masao Ishii

15. Commercial Production of 1,2-Polybutadiene...197
    A. A. Arest-Yakubovich, I. P. Golberg, V. L. Zolotarev, V. I. Aksenov, I. I. Ermakova, and V. S. Ryakhovsky

OTHER APPLICATIONS OF CONTROLLED ANIONIC POLYMERIZATION

16. **Practical Applications of Macromonomer Techniques for the Synthesis of Comb-Shaped Copolymers** 208
    Sebastian Roos, Axel H. E. Müller, Marita Kaufmann, Werner Siol, and Clemens Auschra

17. **Preparation of Lithographic Resist Polymers by Anionic Polymerization** 218
    Hiroshi Ito

POLYMERIZATION OF POLAR AND INORGANIC MONOMERS

18. **Poly(ethylene oxide) Homologs: From Oligomers to Polymer Networks** 236
    Christo B. Tsvetanov, Ivaylo Dimitrov, Maria Doytcheva, Elisaveta Petrova, Dobrinka Dotcheva, and Rayna Stamenova

19. **Anionic Polymerization of Lactams: Some Industrial Applications** 255
    K. Udipi, R. S. Dave, R. L. Kruse, and L. R. Stebbins

20. **Preparation of Poly(2,6-dimethyl-1,4-phenylene ether) (PPE) and Polyamide-6 Blends via Activated Anionic Polymerization of PPE and ε Caprolactam Solutions** 267
    I. Chorvath, M. D. M. Mertens, A. A. van Geenen, R. J. G. van Schijndel, P. J. Lemstra, and H. E. H. Meijer

21. **Nitroaryl and Aminoaryl End-Functionalized Polymethylmethacrylate via Living Anionic Polymerization: Synthesis and Stability** 291
    R. J. Southward, C. I. Lindsay, Y. Didier, P. T. McGrail, and D. J. Hourston

22. **Anionic Copolymerization of Dimethyl- and Diphenylcyclosiloxanes: Changes of Microstructure in the Course of Copolymerization** 304
    B. G. Zavin, A. Yu. Rabkina, I. A. Ronova, G. F. Sablina, and T. V. Strelkova

INDEXES

Author Index 319

Subject Index 321

Preface

This volume is based on an international symposium on "Anionic Polymerization. Practice and Applications" which was held during the National American Chemical Society Meeting in Orlando, Florida in the fall of 1996. The goal of the symposium was to provide state-of-the-art presentations by industrial research scientists practicing in the area of anionic polymerization. The symposium presentations [see *Polym. Prepr. Am. Chem. Soc. Div. Poly. Chem.* **1997**, *37(2)*, 626–733] and the corresponding papers included herein describe the usefulness of anionic polymerization research and emphasize the industrial research perspective on anionic processes for the preparation of polymeric materials.

Living anionic polymerization proceeds in the absence of the kinetic steps of chain transfer and chain termination and provides methodologies for the synthesis of polymers with controlled, well-defined structures. Polymers can be prepared with control of the major variables which affect polymer properties including molecular weight, molecular weight distribution, copolymer composition and microstructure, stereochemistry, chain-end functionality and molecular architecture.

This inherent aspect of control in living anionic polymerization stimulated tremendous industrial research activity which led to the development of numerous technologies for the preparation of important commodity and specialty materials. The papers collected in this monograph provide a unique glimpse into industrial anionic polymerization research and technology. To provide a self-contained and useful book, a general introduction to the principles of anionic polymerization (Quirk et al.) and the commercial applications of anionic polymerization (Hsieh) are included.

Four papers are grouped under the heading of fundamentals and polymerization processes: The subjects include chain-end aggregation (Fetters et al., Exxon), process monitoring by UV (Viola, Bortolotti et al., EniChem), tertiary amine-functionalized initiators (Tadaki et al., JSR) and soluble lithium amide initiators (Lawson et al., Bridgestone/Firestone). One of the unique and useful aspects of living polymerization is the ability to prepare block copolymers by sequential monomer addition or by direct tapered block copolymerization; the four papers on block copolymers include styrene–butadiene–styrene thermoplastic elastomers (Lovisi et al., Petroflex), styrene–isoprene–styrene triblocks for pressure-sensitive adhesives (Matsubara and Ishiguro, Nippon Zeon), packaging films based on styrene-butadiene block copolymers (Knoll and Niessner, BASF), and tapered styrene–butadiene copolymers for use in asphalt modification and as impact modifiers (Moctezuma et al. and Martinex, Industrias Negromex).

Another unique aspect of controlled anionic polymerization is the ability to prepare star polymers which often exhibit unique processing and property advantages. Three star polymer papers describe high temperature processes (Schué et al., Université Montpellier), pressure-sensitive adhesives based on asymmetric star block copolymers (Ma et al., 3M) and randomly branched polystyrenes (Hahnfeld et al., Dow).

The advantage of controlled diene microstructure for elastomers is exemplified by descriptions of new isoprene polymers (Ishii et al., Kuraray) and 1,2-polybutadiene (Arest-Yakubovich et al., Efremov Synthetic Rubber and Karpov Institute). Other applications of controlled anionic polymerization include macromonomer techniques (Roos and Müller, Röhm GmbH and Universität Mainz) and preparation of lithographic resists polymers (Ito, IBM).

Many commercial processes have been developed for the polymerization of polar and inorganic monomers. This subject includes poly(ethylene oxide) (Tsvetanov et al., Bulgarian Academy of Sciences), industrial applications of lactam polymerization (Udipi et al., Monsanto), poly(ε-caprolactam) as solvent for blends (Chorvath et al., Eindhoven University), functionalized poly(methyl methacrylates) (Lindsay, ICI Materials), and siloxane copolymers (Zavin et al., Russian Academy of Science).

In summary, this monograph describes many exciting and useful aspects of anionic polymerization research. It is anticipated that students, faculty and other practitioners of anionic polymerization will benefit from and be stimulated by learning about industrial research in this area from this work.

Acknowledgments

The symposium was supported by the ACS Polymer Division, the ACS Corporation Associates, ACS Petroleum Research Fund, FMC (Lithium Division), Bridgestone, Bridgestone/Firestone Research, Revertex, 3M Company, ICI, BASF, General Electric, and Cyprus Foote. This support is gratefully acknowledged.

RODERIC P. QUIRK
The Maurice Morton Institute of Polymer Science
University of Akron
Akron, OH 44325–3909

INTRODUCTION

Chapter 1

Principles of Anionic Polymerization: An Introduction

Roderic P. Quirk, Qizhuo Zhuo, Sung H. Jang, Youngjoon Lee, and Gilda Lizarraga

Maurice Morton Institute of Polymer Science, University of Akron, Akron, OH 44325–3909

The general mechanistic aspects of living alkyllithium-initiated polymerization of styrenes and dienes are presented. The structural variables which can be controlled are discussed including molecular weight and molecular weight distribution. Methods for the synthesis of well-defined block copolymers, star polymers and chain-end functionalized polymers are described. New developments in polymer synthesis using functionalized alkyllithium initiators are presented. The controlled alkyllithium-initiated polymerization of alkyl methacrylates is considered in terms of proper choice of initiator, solvent and temperatures. New additives which permit anionic alkyl methacrylate and acrylate polymerization at higher temperatures are discussed. The anionic polymerization of alkylene oxides (epoxides) is discussed with emphasis on ethylene oxide and propylene oxide. The mechanism and characteristics of anionic polymerization of lactams are also described.

The almost coincident reports of the delineation of the characteristics of living anionic polymerization by Szwarc, Levy and Milkovich (*1,2*) and of the uniqueness of lithium metal-initiated polymerization of isoprene to form high cis-1,4-polyisoprene by Firestone researchers (*3*) heralded the advent of controlled polymerization to form a wide variety of polymers with low degrees of compositional heterogeneity. Shortly thereafter, Tobolsky and coworkers (*4,5*) reported that the same high cis-1,4-polyisoprene was obtained in hydrocarbon solutions using n-butyllithium as initiator. In addition, they reported that the high stereospecificity was lost in polar solvents such as tetrahydrofuran or diethyl ether and also using either sodium or potassium counterions in hydrocarbon solution (*6*).

A living polymerization is a chain polymerization that proceeds in the absence of the kinetic steps of termination and chain transfer (*7,8*). Such polymerizations provide versatile methodologies for the preparation of polymers with control of the major structural variables that affect polymer properties. The following sections will describe general mechanistic aspects of living alkyllithium-initiated polymerizations of styrenes and dienes and the applications of controlled anionic synthesis for a variety specific polymer structures. A brief discussion of anionic polymerization of alkyl methacrylates, epoxides and lactams is also included.

Mechanism of Alkyllithium-Initiated Polymerization.

The mechanism of alkyllithium-initiated polymerization is analogous to other chain reaction polymerizations, except that the kinetic steps of random termination and transfer are absent, as shown in Scheme 1 for sec-butyllithium-initiated polymerization of styrene (8). Although alkyllithium compounds are generally soluble in hydrocarbon media, they are associated into dimers, tetramers and hexamers (9). Polymeric organolithium compounds are also associated into dimers, tetramers and higher aggregates (8,10). As a consequence of this association behavior of the initiator and propagating polymeric organolithium chain ends, the actual kinetics of initiation and propagation are complicated in hydrocarbon media. A further complication results from cross-association of the unreacted initiator with the growing

Scheme 1. Mechanism of alkyllithium-initiated polymerization of styrene.
Initiation

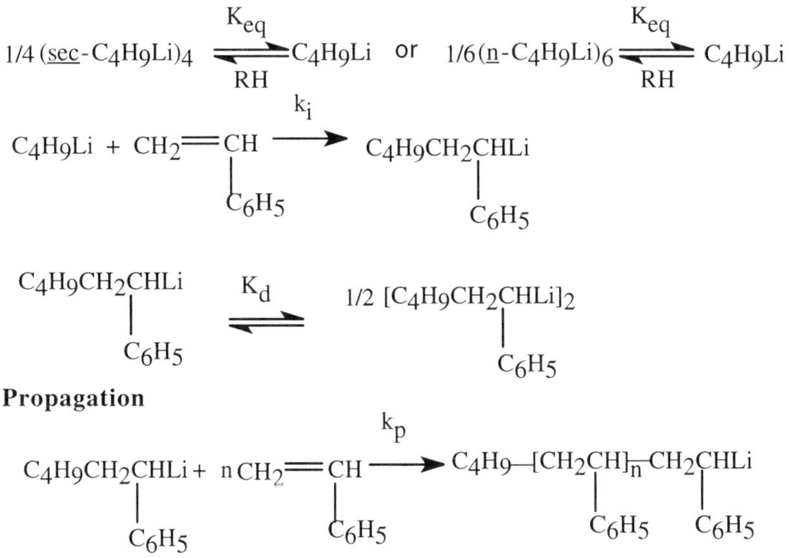

$$1/4\,(\underline{sec}\text{-}C_4H_9Li)_4 \xrightleftharpoons[RH]{K_{eq}} C_4H_9Li \quad \text{or} \quad 1/6\,(\underline{n}\text{-}C_4H_9Li)_6 \xrightleftharpoons[RH]{K_{eq}} C_4H_9Li$$

$$C_4H_9Li + CH_2{=}CH(C_6H_5) \xrightarrow{k_i} C_4H_9CH_2CHLi(C_6H_5)$$

$$C_4H_9CH_2CHLi(C_6H_5) \xrightleftharpoons{K_d} 1/2\,[C_4H_9CH_2CHLi(C_6H_5)]_2$$

Propagation

$$C_4H_9CH_2CHLi(C_6H_5) + n\,CH_2{=}CH(C_6H_5) \xrightarrow{k_p} C_4H_9{-}[CH_2CH(C_6H_5)]_n{-}CH_2CHLi(C_6H_5)$$

polymeric organolithium chain ends (eq. 1); because of this complication, reliable kinetic data for the initiation process can only be obtained at low conversions of the initiator.

$$(RLi)_n + (PLi)_m \rightleftharpoons (RLi)_x(PLi)_y \quad (1)$$

In general and especially in aromatic solvents, the kinetics of initiation exhibit a fractional order dependence on the concentration of alkyllithium initiator (8). For example, a one-sixth-order dependence on n-butyllithium concentration is observed for initiation of styrene polymerization in benzene solution (eq. 2). Since n-butyllithium is aggregated predominantly into hexamers in hydrocarbon solution, the fractional kinetic order dependence of the initiation process on total concentration of initiator has been rationalized on the basis that the species that reacts with styrene monomer must be the unassociated form of the initiator which is in equilibrium with the hexameric aggregate (see Scheme 1). Similarly, the analogous one-fourth order dependence observed for initiation using sec-butyllithium (eq. 3) is also consistent with reaction of the unassociated alkyllithium form, since sec-butyllithium is associated predominantly into tetramers in benzene solution (see Scheme 1).

$$R_i = k_i K_{eq}[\text{n-BuLi}]^{1/6}[S] \qquad (2)$$
$$R_i = k_i K_{eq}[\text{sec-BuLi}]^{1/4}[S] \qquad (3)$$

However, the observed inverse correlation between reaction order dependence for alkyllithium and degree of alkyllithium aggregation is not observed in aliphatic solvents. The use of aliphatic solvents leads to decreased rates of initiation and pronounced induction periods. In fact, a different reaction mechanism involving the direct addition of monomer with aggregated organolithium species has been proposed for aliphatic solvents (8).

The anionic propagation kinetics for styrene (S) polymerization with lithium as counterion are relatively unambiguous. The reaction order in monomer concentration is first order as it is for polymerization of all styrene and diene monomers in heptane, cyclohexane, benzene and toluene (1-3). The reaction order dependence on total chain end concentration, $[PSLi]_o$, is one-half as shown in eq. 4 (4).

$$R_p = -d[S]/dt = k_{obs}[PSLi]_o^{1/2}[S] \qquad (4)$$

Investigations of the kinetics of propagation for dienes have shown that although the rates exhibit first order dependence on monomer concentration, fractional order dependencies are generally observed for the concentration of active centers (8). Isoprene exhibited one-fourth order kinetics and butadiene exhibited one-fourth or one-sixth order dependencies. Recent kinetic studies for isoprene reported that the kinetic order dependence on active center concentration is 1/4 at $[PLi] > 10^{-4}$M and 1/2 at $[PLi] < 10^{-4}$M in benzene at 30°C (11). These results were explained in terms of a concentration-dependent change in degree of association from predominantly tetrameric at higher concentrations to predominantly dimeric at lower concentrations (see Scheme 2). Comparison of these kinetic orders with the degrees of association of the poly(dienyl)lithium chain ends is complicated by the lack of agreement regarding the predominant degree of association of these species in hydrocarbon solution. Predominant degrees of association of both two and four have been reported by different research groups using the same techniques, i.e. concentrated solution viscometry and light scattering. Furthermore, recent evaluation of the association states of poly(butadienyl)lithium chain ends in benzene by small-angle neutron scattering, as well as both dynamic and static light scattering, indicates that higher order aggregates (n > 100) exist in equilibrium with dimeric species (10). Several recent papers provide a critical evaluation of propagation kinetics with respect to the presumption of unreactivity of aggregated organolithium species (12,13).

Control Variables in Alkyllithium-Initiated Polymerizations.

Molecular weight. Molecular weight is one of the most important variables affecting polymer properties. The molecular weight in a living polymerization can be controlled

Scheme 2. Kinetic scheme for diene polymerization.

$$(PLi)_4 \overset{K_t}{\rightleftharpoons} 2(PLi)_2$$

$$(PLi)_2 \overset{K_d}{\rightleftharpoons} 2PLi$$

$$\text{Rate} = -d[I]/dt = k_{obs}[I]$$

$$k_{obs} = k_p K_t^{1/4} K_d^{1/2} [P_4]^{1/4} \text{ at high concentrations}$$

$$k_{obs} = k_p K_d^{1/2} [P_2]^{1/2} \text{ at low concentrations}$$

by the stoichiometry of the reaction and the degree of conversion (8). With a monofunctional initiator, one polymer chain will be formed for each initiator molecule or species. As a consequence, the expected number average molecular weight at complete conversion can be calculated using eq. 5. At intermediate degrees of

$$M_n = \text{grams of monomer/ moles of initiator} \qquad (5)$$

conversion or for equilibrium polymerization of monomers with low ceiling temperatures, the number average molecular weight is proportional to the grams of monomer consumed (eq. 6). Of course, it should be recognized that this ability to

$$M_n = \text{grams of monomer consumed/ moles of initiator} \qquad (6)$$

control molecular weight presupposes that the initiator is sufficiently reactive to be completely consumed prior to the consumption of monomer. For less reactive initiators such as n-butyllithium, incomplete consumption of the initiator can occur depending on the solvent, temperature, monomer and monomer concentration (14). From a practical point of view, it is possible to prepare polymers with predictable molecular weights ranging from $\approx 10^3$ g/mol to $> 10^6$ g/mol using living anionic polymerization methodology (15,16).

Molecular weight distribution. Synthetic polymers are mixtures of macromolecules with similar structures but with different sizes. Most chain reaction polymerizations produce polymers with a distribution of molecular weights which is governed by the random nature of the termination reaction, i.e., the molecular weight distribution corresponds to the so-called most probable distribution (M_w/M_n = 2.0). This distribution is a consequence of the simultaneous and random occurrence of initiation, propagation and termination. For a chain polymerization which proceeds in the absence of chain transfer and chain termination, it is possible to prepare a polymer with a narrow molecular weight distribution (Poisson distribution) (17-19). The essential requirements for formation of a polymer with a Poisson molecular weight distribution are as follows:

1. The growth of each polymer molecule must proceed exclusively by consecutive addition of monomer to an active terminal group.
2. All of these active termini, one for each molecule, must be equally susceptible to reaction with monomer, and this condition must prevail throughout the polymerization.
3. All active centers must be introduced at the onset of the polymerization.
4. There must be no chain transfer or termination (or interchange).
5. Propagation must be irreversible (i.e., the rate of depropagation should be vanishingly small).

From an operational point of view, it has been proposed that the characterization "narrow molecular weight distribution" be reserved for $M_w/M_n \leq 1.1$ (20). In general, it is possible to prepare a polymer with a narrow molecular weight distribution using a living anionic polymerization when the rate of initiation is competitive with the rate of propagation (21). These kinetics ensure that all of the chains grow uniformly for essentially the same period of time.

Flory (17,18) described the relationship between the polydispersity and the degrees of polymerization (number average, X_n; weight average, X_w) for a living polymerization as shown in eq. 7; the second approximation is valid for high

$$X_w/X_n = 1 + [X_n/(X_n + 1)^2] \cong 1 + [1/X_n] \quad (7)$$

molecular weights. From a practical point of view, it is easier to make higher molecular weight polymers with narrow molecular weight distributions; the preparation of lower molecular weight polymers with narrow molecular weight distributions generally requires careful attention to experimental details. Broader molecular weight distributions are obtained using less active initiators, with mixtures of initiators or with continuous addition of initiator as involved in a continuous flow, stirred tank reactor (8). With respect to processing characteristics, polymers, particularly elastomers, exhibit desirable non-Newtonian behavior when either their molecular weight distribution is broad or if they have significant amounts of long-chain branching (22,23). In contrast, linear polymers exhibit singularly Newtonian rheological behavior in that they exhibit constant bulk viscosity over a wide range of shear rates.

Synthesis of Block Copolymers.

One of the unique and important synthetic applications of living polymerizations is the synthesis of block copolymers by sequential monomer addition (24,25). The ability to prepare block copolymers is a direct consequence of the stability of the carbanionic chain ends on the experimental laboratory time scale when all of the monomer has been consumed. Since a living polymerization and the ability to prepare well-defined block copolymers requires the absence (or reduction to a negligible level) of chain termination and chain transfer reactions, monomer purity and the absence of side reactions with the monomer are necessary requirements (*26*).

A further consideration for successful design and synthesis of block copolymers is the order of monomer addition. In general, a carbanionic chain end formed from one monomer will crossover to form the chain end of another monomer and initiate polymerization of this monomer provided that the resulting carbanion is either of comparable stability or more stable than the original carbanion (*8*). The pK_a values of the conjugate acids of carbanions provide a valuable guide to the relative stabilities of carbanions. Thus, crossover will generally occur to monomers which have conjugate acid pK_a values which are the same or smaller than the pK_a of the conjugate acids corresponding to the initiating carbanionic chain ends. For example, to prepare a block copolymer of methyl methacrylate and styrene it is necessary to first polymerize styrene (estimated pK_a of toluene = 43) (*27*) and then add methyl methacrylate (estimated pK_a of ethyl acetate = 30-31) (*27*) to form the second block; the ester enolate anion formed from methyl methacrylate cannot initiate polymerization of styrene. With these limitations in mind, living anionic polymerization provides a powerful synthetic method for preparing block copolymers with well-defined structures, including copolymer composition, block molecular weights, block molecular weight distributions, block sequence and with low degrees of compositional heterogeneity (*8,21,26,28*).

There are three general methods for anionic synthesis of triblock copolymers: (1) three-step sequential monomer addition; (2) two-step sequential addition followed by coupling reactions; and (3) difunctional initiation and two-step sequential monomer addition (*8,26*). Each of these methods has certain advantages and limitations.

Block copolymer synthesis by three-step sequential monomer addition. The preparation of block copolymers by sequential addition of monomers using living anionic polymerization and a monofunctional initiator is the most direct method for preparing well-defined block copolymers. Detailed laboratory procedures for anionic synthesis of block copolymers are available (*29-33*). Several important aspects of these syntheses can be illustrated by considering the preparation of an important class of block copolymers (see Scheme 3), the polystyrene-<u>block</u>-polydiene-<u>block</u>-polystyrene triblock copolymers.

The goal of each step in this sequence is to prepare a block segment with predictable, known molecular weight and narrow molecular weight distribution without incursion of chain termination or transfer. The first step is simple alkyllithium-initiated polymerization of styrene. In order to obtain the desired polystyrene block segment with relatively low molecular weight (12,000-15,000 g/mole) and with narrow molecular weight distribution ($M_w/M_n \leq 1.1$) (see eq. 7), it is necessary to use a reactive initiator which effects a rate of initiation which is competitive with or faster than propagation as discussed previously. Hsieh and McKinney (*34*) have shown that this condition is fulfilled for <u>sec</u>-butyllithium, but not for <u>n</u>-butyllithium, with styrene and diene monomers. In a three-stage block copolymer system, at the completion of each step a sample of polymer can be removed and characterized

Scheme 3. Triblock copolymer synthesis by 3-step sequential monomer addition.

$$C_4H_9Li + n\ CH_2{=}CH(C_6H_5) \xrightarrow{\text{hydrocarbon solvent}} C_4H_9{-}[CH_2CH(C_6H_5)]_{n-1}{-}CH_2CH(C_6H_5)Li$$

PSLi

$$PSLi + m\ CH_2{=}CH{-}CH{=}CH_2 \longrightarrow$$

$$PS{-}[CH_2CH{=}CHCH_2]_{m-1}{-}CH_2CH{=}CHCH_2Li$$

PS-b-PBDLi

$$PS\text{-}b\text{-}PBDLi + n\ CH_2{=}CH(C_6H_5) \xrightarrow{\text{(Lewis base)}} PS\text{-}b\text{-}PBD{-}[CH_2CH(C_6H_5)]_{n-1}{-}CH_2CH(C_6H_5)Li$$

PS-b-PBD-b-PSLi

$$PS\text{-}b\text{-}PBD\text{-}b\text{-}PSLi + ROH \longrightarrow PS\text{-}b\text{-}PBD\text{-}b\text{-}PS$$

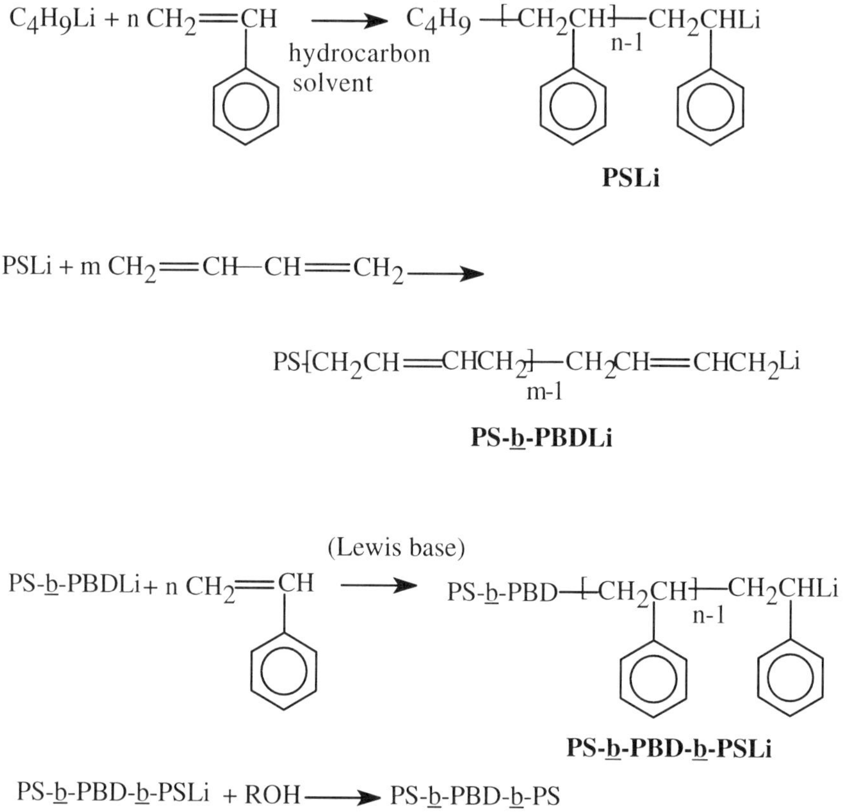

independently with respect to molecular weight, molecular weight distribution and composition.

  The second step in the three-step synthesis of a triblock copolymer by sequential monomer addition requires that the carbanionic chain end of the first block initiate polymerization of the second (diene) monomer. A hydrocarbon solvent is required to obtain polydiene block segments with high 1,4-micro-structure and low glass transition temperatures. Obviously, the monomer added at this step must be very pure to prevent significant termination of the active poly(styryl)lithium chain ends. To the extent that this occurs, the final product will be contaminated with polystyrene homopolymer and the molecular weight of the second block will be increased because of a decrease in chain end concentration (see eq. 5). The carbanion formed from addition to the second monomer must be either more stable or of comparable stability relative to the propagating carbanionic chain end corresponding to the first block segment. Thus, styrene and diene monomers can crossover to each other and also to a variety of polar vinyl and heterocyclic monomers. In order to obtain a narrow molecular weight distribution for the second block, the rate of crossover to the second monomer, i.e. the initiation reaction for the second block, must be competitive with or faster than propagation. This can be a problem for polar monomers which exhibit rapid rates of propagation. The crossover reaction of poly(styryl)lithium to butadiene to form a poly(butadienyl)lithium chain end is a very fast reaction compared to the rate of propagation for butadiene monomer. The order of reaction rate constants is $k_{SB}>k_{SS}>k_{BB}>k_{BS}$ *(35)*. An analogous kinetic situation exists for the order of reaction rate constants for styrene and isoprene *(36)*. Thus, it is relatively easy to obtain polystyrene-block-poly(butadienyl)lithium diblock segments with controlled, predictable block molecular weights and narrow molecular weight distributions.

In the third step, styrene monomer is added, usually in an amount which corresponds to formation of an end block with the same molecular weight as the first polystyrene block. The purity of styrene monomer is critical at this step also, since if impurities are introduced, living diblock chains will be terminated and the final triblock copolymer will be contaminated with the diblock copolymer, polystyrene-block-polybutadiene. In addition, impurities will lead to an increase in the molecular weight of the third block because of a decrease in chain end concentration in accord with eq. 5. Unfortunately, the rate of the crossover reaction of poly(butadienyl)lithium to styrene monomer to form a poly(styryl)lithium chain end is slow compared to the rate of styrene propagation. Because of the slow rate of styrene initiation relative to propagation, a broader molecular weight distribution would be expected for the final polystyrene block segment. In order to obtain a polystyrene end block with narrow molecular weight distribution, a Lewis base such as an ether or amine is often added before styrene monomer addition in this third stage of the triblock copolymer synthesis (37-41).

Block copolymer synthesis by two-step sequential monomer addition and coupling. Another general method for synthesis of A-B-A triblock copolymers involves a two step sequential monomer addition sequence followed by a coupling reaction with a difunctional electrophilic reagent as illustrated in Scheme 4. For the two-step synthesis of a polystyrene-block-polybutadiene-block-polystyrene triblock copolymer analogous to the polymer prepared by the three-step sequential monomer addition process, the first polystyrene block synthesis and the crossover reaction with butadiene monomer would be analogous to the three-step process. However, the amount of butadiene monomer added in the second step would correspond to only one-half of the amount required for the three-step process. The coupling reaction can be effected efficiently with a variety of difunctional linking agents such as α,α'-dichloro-p-xylene which is illustrated in Scheme 4. A variety of difunctional coupling agents have been investigated and useful, i.e., efficient, coupling agents include, dimethyldichlorosilane, 1,4-dihalo-2-butenes and esters such as ethyl benzoate *(42)*.

This two-step method with coupling offers many advantages over the three-step sequential monomer addition method. From a practical point of view, the polymerization time is reduced to one-half of that which is required for the three-step synthesis of a triblock copolymer with the same molecular weight and composition. One problem which is avoided in the two-step process is that the final crossover step from poly(butadienyl)lithium to styrene is eliminated. The elimination of the third

Scheme 4. Triblock copolymer synthesis by 2-step sequential monomer addition followed by coupling.

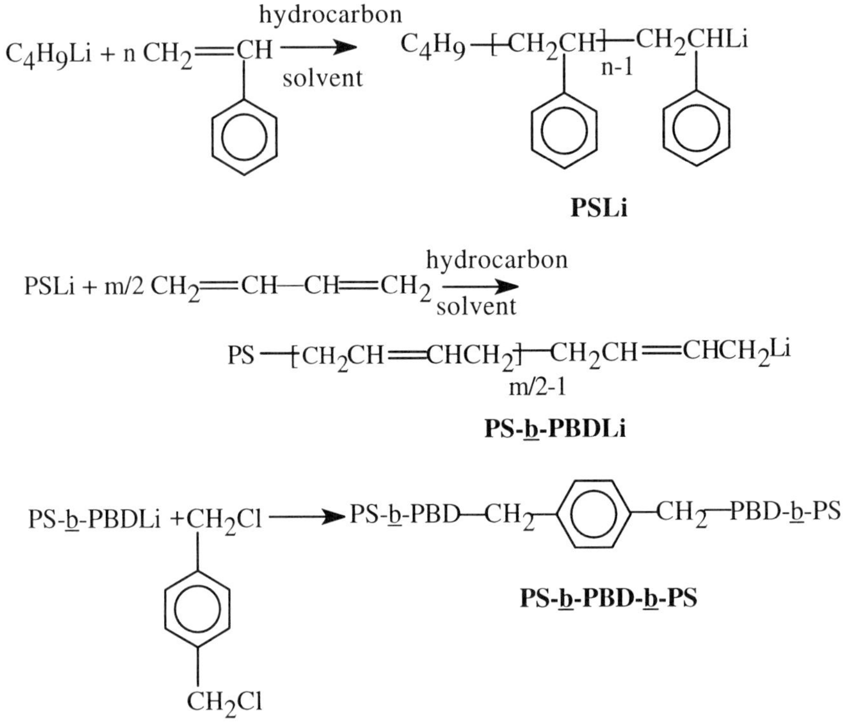

$$C_4H_9Li + n\ CH_2=CH\text{-(Ph)} \xrightarrow{\text{hydrocarbon solvent}} C_4H_9\text{-}[CH_2CH(Ph)]_{n-1}\text{-}CH_2CH(Ph)Li$$

PSLi

$$PSLi + m/2\ CH_2=CH\text{-}CH=CH_2 \xrightarrow{\text{hydrocarbon solvent}} PS\text{-}[CH_2CH=CHCH_2]_{m/2-1}\text{-}CH_2CH=CHCH_2Li$$

PS-b-PBDLi

$$PS\text{-}b\text{-}PBDLi + CH_2Cl\text{-}C_6H_4\text{-}CH_2Cl \longrightarrow PS\text{-}b\text{-}PBD\text{-}CH_2\text{-}C_6H_4\text{-}CH_2\text{-}PBD\text{-}b\text{-}PS$$

PS-b-PBD-b-PS

monomer addition step also decreases the possibility of termination by impurities in a third monomer addition.

Another advantage of the two-step process is that it is more versatile with respect to the chemical composition of the center block. With the two-step method, the center block can be a more reactive monomer which would not be capable of reinitiating polymerization of styrene because of the increased stability of the chain end. For example, a polystyrene-block-poly(methyl methacryloyl)-lithium diblock could be coupled with α,α'-dibromo-p-xylene to form a polystyrene-block-poly(methyl methacrylate)-block-polystyrene triblock; this triblock copolymer could not be synthesized directly by a three-step sequential monomer addition sequence.

The limitations of the two-step monomer addition plus coupling procedure are primarily associated with the coupling reaction. The efficacy of the coupling step requires both an efficient coupling reaction and high precision in controlling the stoichiometry of this reaction (43). In practice, it is difficult to control the stoichiometry of the coupling reaction and many two-step syntheses yield triblock copolymers with significant amounts of uncoupled diblock contaminants (43,44). It is reported that commercial Kraton™ triblock copolymers possess bimodal molecular weight distributions with 15-20 wt% of apparently diblock polymers with one-half of the molecular weight of the triblock copolymer (43,44). In general, the presence of diblock material affects the triblock copolymer morphology (*44*), which has a detrimental effect on the physical properties of triblock copolymers (*40,43*). However, the presence of diblock material reduces the viscosity of the polymer which facilitates processing.

Block copolymers by difunctional initiation and two-step sequential monomer addition. In principle, one of the most versatile methods for the synthesis of A-B-A triblock copolymers is the use of a difunctional initiator with a two-step sequential monomer addition sequence as illustrated in Scheme 5. The main limitation with respect to the synthesis of triblock copolymers using dilithium initiators is the fact that most dilithium initiators are not soluble in hydrocarbon media, as required for polymerization of dienes, because of the association of the chain ends to form insoluble network-like structures. Many successful dilithium initiators require the addition of polar additives such as ethers and amines for solubilization. Another problem with respect to dilithium initiators is due to the association of the polymeric organolithium chain ends which can result in higher viscosities of the dilithium polymer solutions compared to the monolithium analogs. Furthermore, any loss of difunctionality, either in the initiator or after addition of the diene, leads to formation of undesirable diblocks.

A practical two-step synthesis of a polystyrene-<u>block</u>-polybutadiene-<u>block</u>-polystyrene triblock copolymer requires the use of a hydrocarbon soluble, dilithium initiator such as the initiator shown in Scheme 5 which is formed by the addition of 2 moles of <u>sec</u>-butyllithium to 1,3-<u>bis</u>(1-phenylethenyl)benzene (DDPE) (*45-47*). Detailed experimental procedures have been reported recently for the preparation of a dilithium initiator based on the addition of two moles of <u>t</u>-butyllithium to meta-diisopropenylbenzene (*48*). A hydrocarbon soluble, dilithium initiator ensures that the polydiene center block will have a high 1,4-microstructure and a correspondingly low glass transition temperature. The number average molecular weight of the center diene block is uniquely determined by the stoichiometry of the reaction in accord with eq. 8. One of the advantages of using a difunctional initiator and a two-step

$$\overline{M_n} = \frac{\text{grams of monomer}}{1/2 \text{ moles of initiator}} \quad (8)$$

sequential monomer addition is the fact that, analogous to the two-step coupling method, the elimination of a third monomer addition step decreases the possibility of termination by impurities in a third monomer addition step. Also, the dilithium initiator is the most versatile method with respect to the chemical composition of the end blocks. The α,ω-dilithiumpolydiene center block can be used to initiate the polymerization of polar monomers to form both end blocks simultaneously. For example, this initiator has been utilized to prepare poly(<u>t</u>-butyl methacrylate)-<u>block</u>-polyisoprene-<u>block</u>-poly(<u>t</u>-butyl methacrylate) triblock copolymers (*45*). This type of triblock copolymer with polar end blocks cannot be prepared either by the three-step sequential monomer addition process or the two-step monomer addition with coupling method.

Synthesis of Star-Branched Polymers.

Molecular architecture, and particularly long-chain branching, can have a profound effect on the processing and properties of polymers. The methodology of

Scheme 5. Triblock copolymer synthesis by 2-step sequential monomer addition with dilithium initiator.

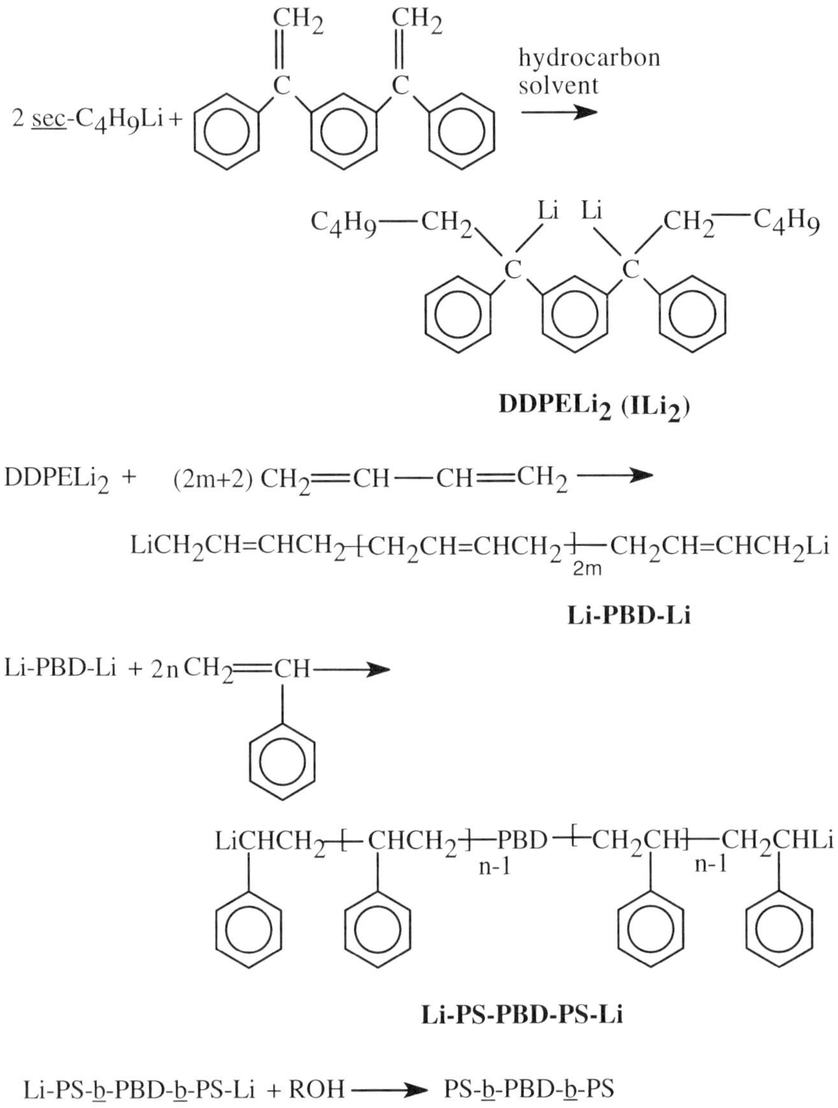

2 sec-C$_4$H$_9$Li + [DDPE] $\xrightarrow{\text{hydrocarbon solvent}}$

DDPELi$_2$ (ILi$_2$)

DDPELi$_2$ + (2m+2) CH$_2$=CH—CH=CH$_2$ ⟶

LiCH$_2$CH=CHCH$_2$[CH$_2$CH=CHCH$_2$]$_{2m}$CH$_2$CH=CHCH$_2$Li

Li-PBD-Li

Li-PBD-Li + 2n CH$_2$=CH—C$_6$H$_5$ ⟶

LiCHCH$_2$[CHCH$_2$]$_{n-1}$PBD[CH$_2$CH]$_{n-1}$CH$_2$CHLi

Li-PS-PBD-PS-Li

Li-PS-b-PBD-b-PS-Li + ROH ⟶ PS-b-PBD-b-PS

living anionic polymerization provides a variety of procedures for the synthesis of both compositionally homogeneous and compositionally heterogeneous branched

the three-armed star product (*52*). However, high linking efficiency to form the 3-armed star is obtained with methyltrichlorosilane (*52*).

There have been two general approaches which have been used to increase the efficiency of linking reactions of polymeric organolithium compounds with multifunctional silyl halides. The first procedure is to add a few units of butadiene to either the poly(styryl)lithium or poly(isoprenyl)lithium chain ends to effectively convert them to the corresponding less sterically-hindered poly(butadienyl)lithium chain ends. For example, after crossover to butadienyllithium chain ends, the yield of four-armed star polyisoprene with silicon tetrachloride was essentially quantitative in cyclohexane (*52*). The second method is to utilize a polychlorosilane compound in which the silyl halide units are more separated to reduce the steric repulsions in the linked product.

The efficiency of the linking reactions of polychlorosilanes with poly(dienyl)lithium compounds has been documented by synthesis of well-defined, narrow molecular weight distribution, 18-armed star-branched polyisoprenes, poly-butadienes, and butadiene end-capped polystyrenes by linking reactions with a decaoctachlorosilane [(SiCl)$_{18}$](*53,54*). The linking reactions of poly(butadienyl)-lithium (M_n = 5.3-89.6 x 10^3 g/mol) with carbosilane dendrimers with up to128 Si-Cl bonds have been reported to proceed smoothly at room temperature but requiring periods of up to 3 weeks (*55*).

From a practical point of view, the effective molecular weight distribution of an anionically prepared polymer can be broadened by reaction with less than a stoichiometric amount of a linking agent such as silicon tetrachloride. This results in a product mixture composed of unlinked arm, coupled product, three-arm and four-arm star-branched polymers. Heteroarm star branched polymers can be formed by coupling of a mixture of polymeric organolithium chains which have different compositions and molecular weights. This mixture can be produced by the sequential addition of initiator as well as monomers.

Divinylbenzene linking reactions. The linking reactions of polymeric organolithium compounds with divinylbenzenes (DVB) provides a very versatile, technologically important, but less precise method of preparing star-branched polymers. The linking reactions with DVB can be conceptually divided into three consecutive and/or concurrent reactions: (a) crossover to DVB; (b) block polymerization of DVB; and (c) linking reactions of carbanionic chain ends with pendant vinyl groups in the DVB block [poly(4-vinylstyrene)] (8,49,56,57). These reactions are illustrated in Scheme 6. The uniformity of the lengths of the DVB blocks depends on the relative rate of the crossover reaction (a) compared to the block polymerization of DVB (b) and the linking reactions (c). This block copolymerization-linking process has been described as formation of a DVB microgel nodule which serves as the branch point for the star-shaped polymer (*49*). In principle, **j** molecules of divinylbenzene could link together (**j**+1) polymer chains (*49*). Although the number of arms in the star depends on the ratio of DVB to polymeric organolithium compound, the degree of linking obtained for this reaction is a complex function of reaction variables (*49, 56-59*). It should be noted that these linking reactions are effected with technical grades of DVB which have been variously reported to consist of (a) 33 % DVB (11% p-divinylbenzene, 22% m-divinylbenzene) and 66% o-, m- and p-ethylvinylbenzene (EVB) (*49*); (b) 78

polymers (*8,25*). A star-branched polymer is a particularly important type of anionically prepared branched polymer that consists of several linear chains linked together at one end of each chain by a single branch or junction point. Thus, after anionic polymerization of a given monomer is completed, the resulting living polymer with a reactive carbanionic chain-end can be reacted with a variety of linking reagents to generate the corresponding star-branched polymer with uniform arm lengths. After analogous sequential polymerization of two or more monomers, linking reactions with the resulting living diblock, or multiblock, copolymers will generate star-branched block copolymers with uniform block arm length (*28*). It is noteworthy that, in principle, each arm is of uniform block copolymer composition with precise block molecular weights and narrow molecular weight distributions, i.e. with low degrees of compositional heterogeneity. Although various types of linking agents have been used to prepare star-branched copolymers, two of the most useful and important types of linking agents are the multifunctional silyl chlorides and divinylbenzenes.

Linking reactions with silyl halides. The most general methods for the preparation of regular star polymers are based on linking reactions of polymeric organolithium compounds with multifunctional electrophilic species such as silicon tetrachloride as shown in eq. 9. Analogous linking reactions with multifunctional halogenated hydro-

$$4 \text{ PLi} + \text{SiCl}_4 \longrightarrow \text{P}_4\text{Si} + 4 \text{ LiCl} \qquad (9)$$

carbons are complicated by side reactions such as lithium-halogen exchange, Wurtz coupling and elimination reactions (*9*). In contrast to these complexities for organolithium reactions with polyfunctional halogenated hydrocarbon-type linking agents, the reactions with chlorosilane compounds are very efficient and uncomplicated by analogous side reactions. However, the extents and efficiencies of these linking reactions are dependent upon the steric requirements of the carbanionic chain end (*49*).

In general, for a given multifunctional silicon halide, the efficiency of the linking reaction decreases in the order poly(butadienyl)lithium > poly(isoprenyl)-lithium > poly(styryl)lithium. In pioneering work by Morton, Helminiak, Gadkary and Bueche (*50*) in 1962, the reaction of poly(styryl)lithium (M_n = 60.6 x 10^3 g/mol; M_w/M_n =1.06) with a less than stoichiometric amount of silicon tetrachloride in benzene at 50°C for 48 hours produced a polymer product with M_w = 1.93 x 10^5 g/mol. After fractionation of this product, a four-armed star polymer (M_w = 2.57 x 10^5 g/mol; M_w/M_n = 1.09) and a three-armed star polymer (M_w = 1.70 x 10^5 g/mol; $M_w/M_n \approx$ 1.0) were isolated in weight fraction amounts corresponding to 0.252 and 0.349, respectively.

In contrast to the results of inefficient linking for poly(styryl)lithiums, the linking reactions of poly(butadienyl)lithiums with methyltrichlorosilane and silicon tetrachloride in cyclohexane at 50°C for 3 hours were reported to proceed efficiently to form the corresponding 3-arm and 4-arm stars, respectively (*51*). However, the linking efficiency of poly(isoprenyl)lithium with silicon tetrachloride is not high, analogous to poly(styryl)lithium. The stoichiometric reaction of poly(isoprenyl)lithium with silicon tetrachloride is reported to form predominantly

Scheme 6. Branching chemistry of polymeric organolithium compounds with divinylbenzenes.

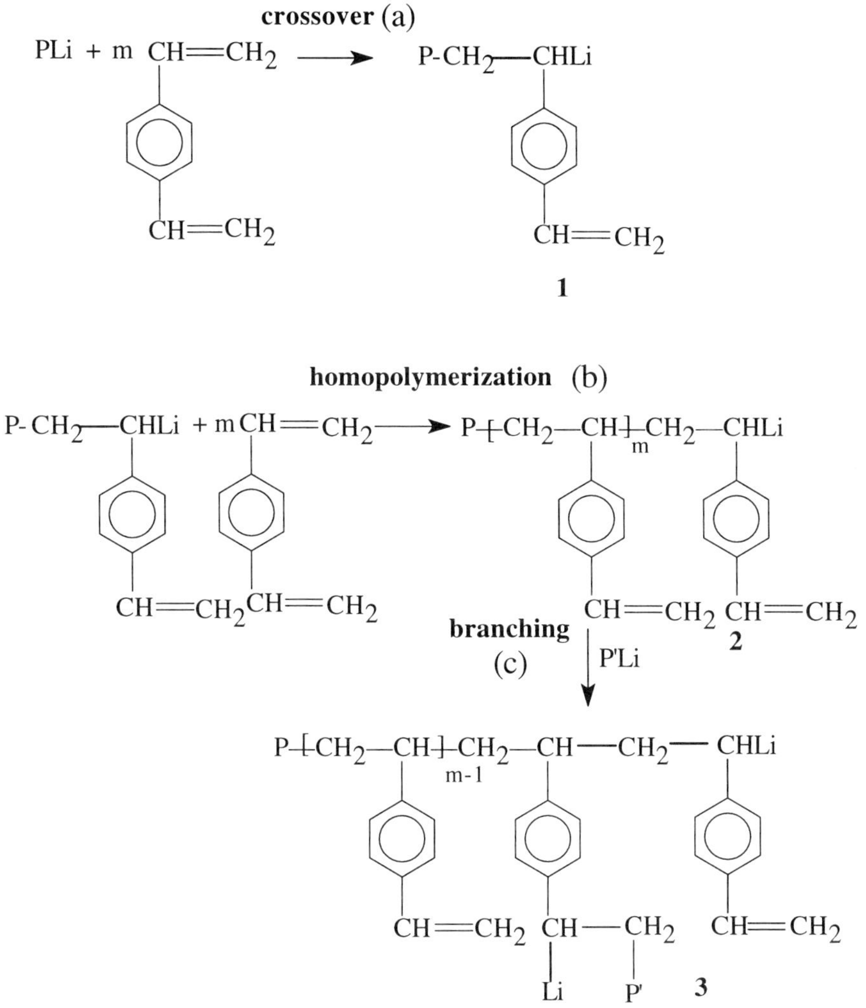

crossover (a)

PLi + m CH=CH₂(C₆H₄)CH=CH₂ ⟶ P-CH₂—CHLi(C₆H₄)CH=CH₂ **1**

homopolymerization (b)

P-CH₂—CHLi(C₆H₄-CH=CH₂) + m CH=CH₂(C₆H₄)CH=CH₂ ⟶ P-[CH₂—CH(C₆H₄-CH=CH₂)]ₘ—CH₂—CHLi(C₆H₄-CH=CH₂) **2**

branching (c) | P'Li

P-[CH₂—CH(C₆H₄-CH=CH₂)]ₘ₋₁—CH₂—CH(C₆H₄-CH(Li)-CH₂-P')—CH₂—CHLi(C₆H₄-CH=CH₂) **3**

% DVB (meta/para=2.6), 22% ethylvinylbenzene isomers (58);(c) 56% DVB, 44% EVB isomers (59); d) 18 mole % p-divinylbenzene, 39 mole% m-divinylbenzene, 10 mole % p-ethylvinylbenzene, and 33 mole % m-ethylvinylbenzene (49,60). The purity of DVB is critical, because impurities can terminate the active chain ends and this will also change the effective molar ratio of DVB to active chain end concentration.

    For poly(styryl)lithium chains, the rate of crossover to DVB is comparable to the rate of DVB homopolymerization and both of these rates are faster than the rate of the linking reaction of poly(styryl)lithium with the pendant double bonds in the poly(vinylstyrene) block formed from DVB. Therefore, it would be expected that the DVB block formed by crossover from poly(styryl)lithium would be relatively uniform and that the linking reaction would generally occur after the formation of the DVB block. In general, the linking efficiency of poly(styryl)lithiums by DVB is quite high except for very low ratios of DVB/PSLi (*8*).

For poly(dienyl)lithium chain ends, the rate of crossover to DVB is much slower than the rate of DVB homopolymerization. Therefore, it would be expected that the DVB block length would be longer for poly(dienyl)lithiums compared to poly(styryl)lithium. As a consequence, although one obtains a higher degree of branching, the linking efficiency is lower for diene versus styrene stars. However, it is possible to obtain good linking efficiencies for dienyllithium chains using ratios of DVB/PLi ≥ 3 (*8*).

When small amounts of DVB are added as a comonomer during polymerization of styrenes and dienes, it is incorporated into the chain in a statistical distribution dictated by the comonomer reactivity ratios and the feed composition. Subsequent to DVB incorporation, growing chains can add to the pendant vinylstyrene units followed by further monomer addition to generate long-chain branching sites (*8*). Although the resulting branch structure is not well-defined, it is very effective in modifying the low and high shear viscosity characteristics of the resulting polymers.

Synthesis of Chain-End Functionalized Polymers.

Chain-end functionalization by termination with electrophilic reagents.
Another unique facet of living anionic polymerization is the ability to tailor-make well-defined polymers with low degrees of compositional heterogeneity and with functional chain-end groups (*8,61*). The products of living anionic polymerization are polymer chains with stable carbanionic chain ends. In principle, these reactive anionic end groups can be readily converted into a diverse array of functional end groups by reaction with a variety of electrophilic species as shown in eq. 10, where PLi is a polymeric organolithium chain, E is an electrophilic reagent and X represents a chain-end functional group. Unfortunately, many of these functionalization reactions have

$$PLi + E \rightarrow P\text{---}X + LiY \qquad (10)$$

not been well characterized (*8,61*). Thus, it is generally necessary to carefully develop, analyze and optimize new procedures for each different functional group introduced by a specific post-polymerization functionalization reaction. Fortunately, several important functionalization reactions have been carefully investigated.

Carbonation. The carbonation of polymeric carbanions using carbon dioxide is one of the most useful and widely used functionalization reactions. However, there are special problems associated with the simple carbonation of polymeric organolithium compounds. Even when carbonations with high-purity, gaseous carbon dioxide are carried out in benzene solution at room temperature using standard high vacuum techniques, the carboxylated polymers are obtained in only 27-66% yields for poly(styryl)lithium, poly(isoprenyl)lithium and poly(styrene-b-isoprenyl)lithium (*62*). The functionalized polymer is contaminated with dimeric ketone (23-27%) and trimeric alcohol (7-50%) as shown in eq. 11 where P represents a polymer chain. It was proposed that the formation of these side-products is favored relative to the desired carboxylated polymer by aggregation of the chain ends in hydrocarbon

$$PLi \xrightarrow{CO_2} \xrightarrow{H_3O^{\oplus}} PCO_2H + P_2CO + P_3COH \qquad (11)$$

solution (*63*). Furthermore, it has been reported that the addition of sufficient quantities of Lewis bases such as tetrahydrofuran (THF) and N,N,N',N'-tetramethylethylenediamine (TMEDA) can reduce or even eliminate the association of polymeric organolithium chain ends (*64-66*). In accord with these considerations, it was found that addition of large amounts of either THF (25 vol%) or TMEDA ([TMEDA]/[PLi] = 1-46) was effective in favoring the carbonation reaction to the extent that the carboxylated polymer was obtained in yields >99% for poly(styryl)lithium, poly(isoprenyl)lithium and poly(butadienyl)lithium (*62*).

Hydroxylation. The preparation of hydroxyl-terminated polymers from polymeric organolithium compounds by reaction with ethylene oxide is one of the few simple, efficient functionalization reactions. The reaction of poly(styryl)lithium with excess ethylene oxide in benzene solution produces the corresponding hydroxyethylated polymer in quantitative yield without formation of detectable amounts of oligomeric ethylene oxide blocks (eq. 12) (*67*). For example, ^{13}C-NMR analysis of a hydroxyethylated polystyrene (M_n=1.3x10^{3}g/mol, M_w/M_n=1.08) showed no evidence for the formation of any ether linkages expected for oligomerization of ethylene oxide. This result is surprising in view of the steric strain

$$PSLi + \overset{O}{\triangle} \xrightarrow{H_3O^+} PSCH_2CH_2OH \qquad (12)$$

and intrinsic reactivity of ethylene oxide toward nucleophiles. Apparently the high degree of aggregation of lithium alkoxides and the strength of this association even in polar solvents renders them unreactive (*68*).

Amination. Richards, Service and Stewart (*69*) have reported that poly(butadienyl)lithium can be functionalized efficiently by reaction with N-(3-chloropropyl)dialkylamines as shown in eq. 13. The reported yields were as high as 95%. Recent work has confirmed the efficiency of this functionalization

$$PBdLi + Cl\text{-}(CH_2)_3\text{-}NR_2 \text{-------} > PBd\text{-}CH_2\text{-}CH_2\text{-}CH_2\text{-}NR_2 \qquad (13)$$

reaction for dienyllithiums in hydrocarbon solution [85 % functionalization for poly(isoprenyl)lithium and 90% functionalization for poly(butadienyl)lithium]. However, the corresponding functionalization reaction of poly(styryl)lithium forms the corresponding tertiary amine-functionalized polymer in only 65 % for direct addition and 77 % by inverse addition of PSLi to the alkyl chloride (*70*). This reinforces the fact that these functionalization reactions are sensitive not only to the experimental conditions, but also to the nature of the polymeric chain end.

Functionalized initiators. An alternative procedure for the preparation of end-functionalized polymers is to use alkyllithium initiators which contain functional groups as shown in Scheme 7 (*71*). Because most functional groups of interest (e.g., hydroxyl, carboxyl, primary amino) are not stable in the presence of either simple or

Scheme 7. Synthesis of linear and branched polymers using functionalized initiators.

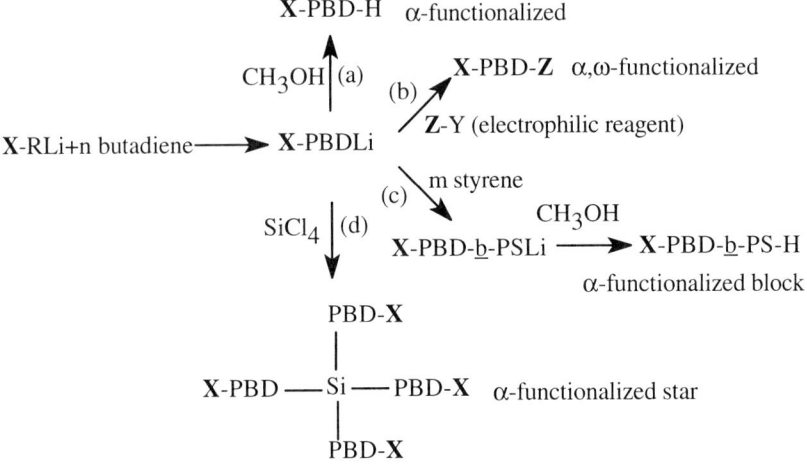

polymeric organolithium reagents, it is generally necessary to use suitable protecting groups in the initiator (71). A suitable protecting group is one that is not only stable to the anionic chain ends but is also readily removed upon completion of the polymerization. There are several distinct advantages in the use of a functionalized initiator. For a living alkyllithium-initiated polymerization, each functionalized initiator molecule will produce one macromolecule with a functional group from the initiator residue at the initiating chain end (α, alpha) and with the active carbanionic propagating species at the other chain end (ω, omega) regardless of molecular weight. Advantages relative to functionalization by electrophilic termination are that it is not necessary to be concerned about efficient and rapid mixing of reagents with viscous polymers or with the stability of the chain end, which is of concern at the elevated temperature polymerization conditions often employed (8). This methodology of producing α-functionalized living polymers which retain the carbanionic chain end provides the added advantages of the ability to prepare telechelic (α,ω–functionalized) polymers [see (b) in Scheme 7], functionalized block copolymers [see (c) in Scheme 7] and star-branched polymers with functional groups at the initiating ends of each branch [see (d) in Scheme 7] .

Shepherd and Stewart (72,73) have recently described a series of protected hydroxyl initiators based on the trialkylsilyloxy-protecting group prepared in hexane from the corresponding halides and lithium metal. Thus, 3-(t-butyldimethylsilyloxy)-1-hexyllithium initiated the polymerization of butadiene at 5°C followed by polymerization at room temperature; the propyllithium analog was also described. After termination with methanol, the resulting silyloxy-functionalized polybutadiene exhibited excellent correspondence between the calculated (M_n = 2x10^3 g/mole) and observed molecular weight (M_n = 2.1x10^3 g/mol), narrow molecular weight distribution (M_w/M_n = 1.1) and high 1,4-microstructure (89%). Furthermore, telechelic polybutadienes were prepared by either termination with ethylene oxide or by coupling with dimethyldichlorosilane. The t-butyl-dimethylsilyloxy-protecting group was removed by stirring the polymers in tetrahydrofuran at room temperature in the presence of tetra(n-butyl)ammonium fluoride and then precipitation into methanol. This is a milder method of removing the protecting group than that apparently required for the acetal-protected polymers.

Schwindeman, Granger and Engel (74) have described analogous w-alkoxy-protected 1-alkyllithium initiators for polymerization of dienes (see Scheme 8). Using

Scheme 8. Synthesis of α,ω–difunctional polymers using functionalized initiators.

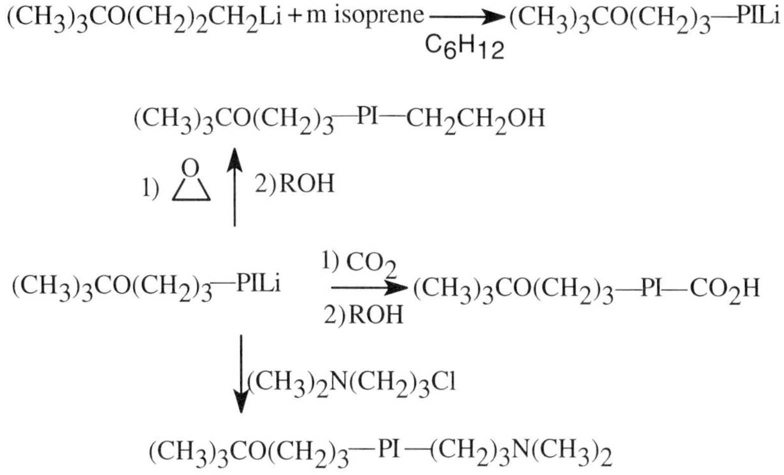

$$(CH_3)_3CO(CH_2)_2CH_2Li + m \text{ isoprene} \xrightarrow{C_6H_{12}} (CH_3)_3CO(CH_2)_3\text{—PILi}$$

$$(CH_3)_3CO(CH_2)_3\text{—PI—CH}_2CH_2OH$$

1) epoxide 2) ROH

$$(CH_3)_3CO(CH_2)_3\text{—PILi} \xrightarrow[2)ROH]{1) CO_2} (CH_3)_3CO(CH_2)_3\text{—PI—CO}_2H$$

↓ $(CH_3)_2N(CH_2)_3Cl$

$$(CH_3)_3CO(CH_2)_3\text{—PI—}(CH_2)_3N(CH_3)_2$$

these hydrocarbon-soluble initiators such as 3-(t-butoxy)-1-propyllithium or the corresponding isoprene chain-extended initiator, **1**, polydienes with high 1,4-microstructures and well-defined structures can be prepared. For example, initiation

$$(CH_3)_3CO(CH_2)_3\text{—}[CH_2CH=C(CH_3)CH_2]_2Li$$
1

of polymerization of butadiene with **1** at 30°C in cyclohexane produced the corresponding a-hydroxy-functionalized polybutadiene [M_n (calc) = 4.0×10^3 g/mole; M_n (obs) = 4.35×10^3 g/mole; M_w/M_n = 1.09; 81 % 1,4-microstructure] after deprotection with an acidic catalyst.

Anionic Polymerization of Alkyl Methacrylates

The proper choice of initiator and rection conditions is essential for controlled anionic polymerization of alkyl methacrylates. In general, a less reactive initiator such as 1,1-diphenylhexyllithium, formed by the addition of butyllithium to 1,1-diphenylethylene (eq. 14), is an effective initiator (75-80). More reactive initiators such as butyllithium react with the ester carbonyl group in competition with the Michael addition to the conjugated double bond; less than half of the initiator molecules initiate chain growth 81). These polymerizations must be carried out at low temperatures, e.g. -78°C,

$$C_4H_9Li + CH_2{=}C(C_6H_5)_2 \longrightarrow C_4H_9CH_2\underset{\underset{C_6H_5}{|}}{\overset{\overset{C_6H_5}{|}}{C}}Li \qquad (14)$$

(although it has been reported that polymerizations can be carried out at higher temperatures in the presence of lithium chloride (*82,83*). It is even possible to effect controlled anionic polymerization of t-butyl acrylate in the presence of lithium chloride (*84-86*) or lithium 2-(2-methoxyethoxy)ethoxide (*87*) in THF at low temperatures. In addition, the anionic polymerization of t-butyl methacrylate can be carried out at room temperature (*88*). The solvent has a dramatic effect on the stereochemistry of the polymerization. Highly isotactic polymer can be formed in hydrocarbon solvents such as toluene, whereas highly syndiotactic polymer is formed in tetrahydrofuran (*89*). However, polymerizations in toluene are characterized by broad or multimodal molecular weight distributions and termination reactions even at low temperature.

Anionic Polymerization Of Alkylene Oxides

Comprehensive studies on the mechanism and kinetics of the anionic ring-opening polymerization of epoxides have been reported in the last three decades and several reviews have been published on the subject (*91-94*). Ethylene oxide polymerizes readily to form (-O-CH$_2$-CH$_2$-)$_n$ in the presence of alkali metals (*95,96*), classical bases (carboxylates, alkoxides, hydroxides) (92), radical anions (*97*) and anions such as those derived from triphenylmethane, fluoradene, 9-methylfluorene and carbazole (98-100). The behavior of ethylene oxide in anionic polymerization is complicated by the association of ion pairs which subsist down to low chain end concentrations and occur even in high polarity solvents such as hexamethylphosphoramide (HMPA) (*101*) (ε =30 at 25°C). Moreover, ion pair to free ions dissociation constants are very low in solvents of medium polarity such as tetrahydrofuran (ε =7.3 at 25° C) (*102*), which results in low propagation rate constants. The stability of these polymeric alkoxide aggregates decreases with increasing cation size from Na$^+$ to Cs$^+$. The total inactivity of the Li$^+$ counterion in ethylene oxide polymerization has been verified numerous times (*95,103,104*). The association of alkali metal alkoxide end-groups of living poly(ethylene oxide) can be avoided by complexing the cations with suitable cryptands (*105,106*). When chain-end aggregation is avoided by the use of suitable conditions (i.e. choice of counterion, solvent, additives) and provided that protic substances that might act as chain transfer agents are absent, ethylene oxide polymerizes in a living fashion such that control of molecular weight and narrow polydispersities (M$_w$/M$_n$ $\leq$ 1.1) have been achieved (*106,107a,108*).

Propylene oxide can also be polymerized via anionic techniques (*92,94*). However, the anionic polymerization of propylene oxide is complicated by a chain transfer reaction of the alkoxide chain ends by hydrogen abstraction from the methyl group of propylene oxide to form allyl alkoxide that, subsequently, initiates new chains. This chain transfer reaction limits the attainable molecular weight and broadens the molecular weight distribution of the resulting polymer. It also reduces the hydroxyl functionality of polyols obtained by hydroxide-initiated polymerization.

Anionic Polymerization of Lactams

Strong bases such as alkali metals, metal hydrides, metal amides and metal alkoxides initiate the polymerization of lactam monomer (*107b*) according to Scheme 9. The imide dimer has been identified to be the actual initiating species. Slow initial reaction rates can be observed as the concentration of imide dimer builds up, which has prompted the use of very high reaction temperatures (255-285°C) in order to minimize this problem. Slow initial reaction rates are a consequence of the formation of the unstable nitranion as shown in step 1 in Scheme 9. In contrast, the imide dimer forms a more stable amide anion (see step 3, Scheme 9). As a further consequence of this, only the more reactive lactams such as caprolactam undergo polymerization by strong bases alone. This limitation has been overcome by forming an imide by reacting the monomer with an acyl chloride or anhydride to form the more reactive, or 'activated', N-acyllactam. The reaction temperature can then be lowered considerably. This acyllactam-promoted anionic ring-opening polymerization allows for fast cycle times, which makes it amenable to reaction injection molding (RIM) processing.

Lactam polymerizations are seldom living because of side reactions which destroy propagating centers and result in branching and chain cleavage (*109*). Controlled polymerization can be achieved with four-membered ring lactams. Thus, monodisperse polyamides have been prepared through the living anionic polymerization of a β-lactam having a bulky substituent, 3-butyl-3-methyl-2-azetidinone, in tetrahydrofuran at 27°C using butyllithium as initiator and the corresponding N-benzoyllactam as the activated monomer (*110*). Polymerization under mild conditions is desirable to suppress side reactions. The molecular weights obtained were as high as 2,000 g/mol and polydispersities were apparently narrow based on the published SEC chromatograms. Living anionic polymerization of 3,3-dimethyl- and 4,4-dimethyl-2-azetidinones in N,N-dimethylacetamide containing lithium chloride was achieved (*111*), despite the fact that these lactams contain active hydrogens at the α-position of the lactam carbonyl group. These polymerizations were initiated with potassium pyrrolidonate using the corresponding N-pivaloyl-lactams as the activated monomers. The molecular weights obtained were as high as 50,000 g/mol and $M_w M_n \leq 1.1$.

Summary and Conclusions

Alkyllithium-initiated anionic polymerization of styrene, dienes and alkyl methacrylates is one of the most versatile methodologies for the preparation of macromolecules with well-defined structures and low degrees of compositional heterogeneity. With this technique it is possible to synthesize macromolecular compounds with control of a wide range of compositional and structural parameters including molecular weight, molecular weight distributions, composition and

Scheme 9. Activated monomer mechanism for anionic ring-opening polymerization of ε-caprolactam.

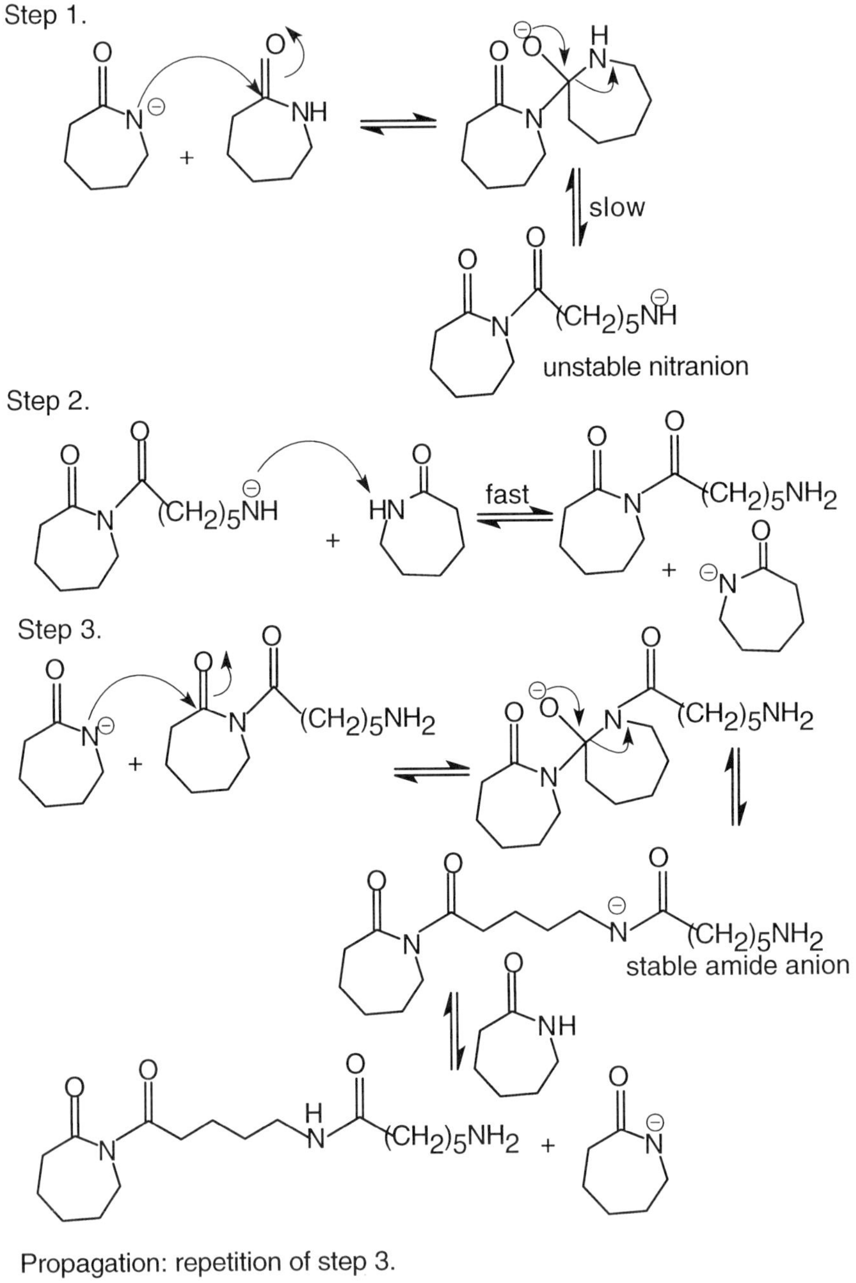
Step 1.
slow
$(CH_2)_5NH$
unstable nitranion
Step 2.
$(CH_2)_5NH$
fast
$(CH_2)_5NH_2$
Step 3.
$(CH_2)_5NH_2$
stable amide anion
Propagation: repetition of step 3.

microstructure, stereochemistry, branching and chain-end functionality. Anionic ring-opening polymerization of alkylene oxides provides controlled polymerization to form useful, hydrophilic polymers, macromonomers and amphiphilic block copolymers; the rapid ring-opening polymerization of lactams provides a route to nylons via reactive extrusion technology.

References

1. M. Szwarc, *Nature*, **178**, 1168(1956)
2. M. Szwarc, M. Levy and R. Milkovich, *J. Am. Chem. Soc.*, **78**, 2656(1956).
3. F. E. Stavely, et al., Paper Presented to the ACS Rubber Division, Philadelphia PA, November 1955; *Ind. Eng. Chem.*, **48**, 778(1956).
4. H. Hsieh and A. V. Tobolsky, *J. Polym. Sci.*, **25**, 245(1957).
5. H. Hsieh, D. J. Kelley and A. V. Tobolsky, *J. Polym. Sci.*, **26**, 240(1957).
6. H. Morita and A. V. Tobolsky, *J. Am. Chem. Soc.*, **79**, 5853(1957).
7. R. P. Quirk and B. Lee, *Polym. Int.*, **27**, 359(1992).
8. H. L. Hsieh and R. P. Quirk, *Anionic Polymerization: Principles and Practice*, Marcel Dekker, New York, 1996.
9. B. L. Wakefield, *The Chemistry of Organolithium Compounds*, Pergamon Press, Oxford, 1974.
10. L. J. Fetters, N. P. Balsara, J. S. Huang, H. S. Jeon, K. Almdal and M. Y. Lin, Macromolecules, **28**, 4996(1995).
11. S. Bywater and D. J. Worsfold in *Recent Advances in Anionic Polymerization*, T. E. Hogen-Esch and J. Smid, Eds., Elsevier, New York, 1987, p. 109.
12. R. N. Young, L. J. Fetters, J. S. Huang and R. Krishnamoorti, *Polym. Int.*, **33**, 217 (1994).
13. L. J. Fetters, J. S. Huang and R. N. Young, *J. Polym. Sci.: Part A: Polym. Chem.*, **34**, 1517(1996).
14. H. L. Hsieh, *J. Polym. Sci.*, **A3**, 163(1965).
15. E. L. Slagowski, L. J. Fetters and D. McIntyre, *Macromolecules*, **7**, 394(1974).
16. L. J. Fetters and M. Morton, *Macromolecular Synthesis, Collective Volume 1*, J. A. Moore, Ed., Wiley, New York, 1977, p. 463.
17. P. J. Flory, *Principles of Polymer Chemistry*, Cornell University Press, Ithaca, New York, 1953.
18. P. J. Flory, *J. Am. Chem. Soc.*, **62**, 1561(1940).
19. F. J. Henderson and M. Szwarc, *J. Polym. Sci., Macromol. Rev.*, **3**, 317(1968).
20. L J. Fetters in *Encyclopedia of Polymer Science and Engineering*, 2nd ed., J. I. Kroschwitz, Ed., Wiley-Interscience, New York, 1987, Vol. 10, p. 19.
21. P. Rempp, E. Franta and J.-E. Herz, *Adv. Polym. Sci.*, **86**, 145(1988).
22. F. C. Weissert and B. L. Johnson, *Rubber Chem. Tech.*, **40**, 590(1967).
23. J. T. Gruver and G. Kraus, *J. Polym. Sci.*, **A2**, 797(1964); *J. Polym. Sci.*, **A3**, 105(1965); *J. Appl. Polym. Sci.*, **9**, 739(1965).
24. O. W. Webster, *Science*, **251**, 887(1991).
25. L. J. Fetters and E. L. Thomas in *Material Science & Technology*, Vol. 12, VCH Verlagsgesellschaft, Weinheim, Germany, 1993, p. 1.

26. R. P. Quirk and M. Morton in *Thermoplastic Elastomers*, 2nd ed., G. Holden, N. R. Legge, R. P. Quirk and H. E. Schroeder, Eds., Hanser Pub., Munchen, 1996, p. 71.
27. F. G. Bordwell, *Acc. Chem. Res.*, **21**, 456(1988).
28. R. P. Quirk, D. J. Kinning and L. J. Fetters in *Comprehensive Polymer Science*, Vol. 7, *Specialty Polymers and Polymer Processing*, S. L. Aggarwal, Ed., Pergamon, Oxford, 1989, p. 1.
29. M. Morton and L. J. Fetters, *Rubber Chem. Tech.*, **48**, 359(1975).
30. R. W. Richards and J. L. Thomason, *Polymer*, **23**, 1988(1982).
31. J. C. Falk, M. A. Benedetto, J. Van Fleet and L. Ciaglia, *Macromol. Syn.*, **8**, 61(1982).
32. J. J. O'Malley and R. H. Marchessault, *Macromolecular Synthesis*, *Collective Volume 1*, J. A. Moore, Ed., Wiley, New York, 1977, p. 419.
33. D. Braun, H. Cherdron and W. Kern, *Practical Macromolecular Organic Chemistry*, K. J. Ivin (translator), Harwood Acad. Pub., New York, 1984, p. 232.
34. H. L. Hsieh and O. F. McKinney, *J. Polym. Sci., Part B, Polym. Lett.*, **4**, 843(1966).
35. R. Ohlinger and F. Bandermann, *Makromol. Chem.*, **181**, 1935(1980).
36. D. J. Worsfold, *J. Polym. Sci., A-1, Polym. Chem.*, **5**, 2783(1967).
37. M. Morton, J. E. McGrath and P. C. Juliano, *J. Polym. Sci., Part C, Polym. Symp.*, **26**, 99(1969).
38. L. J. Fetters, *J. Elastoplastics*, **4**, 34(1972).
39. S. Bywater in *Encyclopedia of Polymer Science and Engineering*, 2nd ed., Vol. 2, J. I. Kroschwitz, Ed., Wiley-Interscience, New York, 1985, p. 1.
40. M. Morton in *Block Polymers*, S. L. Aggarwal, Ed., Plenum Press, New York, 1970, p. 1.
41. N. R. Legge, *Rubber Chem. Tech.*, **60**, G83(1987).
42. H. L. Hsieh, *Rubber Chem. Tech.*, **49**, 1305(1976).
43. L. J. Fetters in *Block Copolymers: Science and Technology*, D. J. Meier, Ed., Harwood Acad. Pub., New York, 1983, p. 17.
44. L. J. Fetters, B. H. Meyer and D. McIntyre, *J. Appl. Polym. Sci.*, **16**, 2079(1972).
45. T. E. Long, A. D. Broske, D. J. Bradley and J. E. McGrath, *J. Polym. Sci., Part A, Polym. Chem.*, **27**, 4001(1989).
46. R. P. Quirk and J.-J. Ma, *Polym. Int.*, **24**, 197(1991).
47. L. H. Tung and G. Y. Lo in *Advances in Elastomers and Rubber Elasticity*, J. Lal and J. E. Mark, Eds., Plenum Press, New York, 1986, p. 129.
48. Y. S. Yu, Ph. Dubois, R. Jerome and Ph. Teyssie, *Macromolecules*, **29**, 2738(1996).
49. B. J. Bauer and L. J. Fetters, *Rubber Chem. Tech.*, **51**, 406(1978).
50. M. Morton, T. E. Helminiak, S. D. Gadkary and F. Bueche, *J. Polym. Sci.*, **57**, 471(1962).
51. R. P. Zelinski and C. F. Wofford, *J. Polym. Sci.*, A**3**, 93(1965).
52. L. J. Fetters and M. Morton, *Macromolecules*, **7**, 552(1974).
53. J. Roovers, N. Hadjichristidis and L. J. Fetters, *Macromolecules*, **16**, 214(1983).

54. P. M. Toporowski and J. Roovers, *J. Polym. Sci., Part A, Polym. Chem.* **24**, 3009(1986).
55. J. Roovers, L.-L. Zhou, P. M. Toporowski, M. van der Zwan, H. Iatrou and N. Hadjichristidis, *Macromolecules*, **26**, 4324(1993).
56. S. Bywater, *Adv. Polym. Sci.*, **30**, 89(1979).
57. R. N. Young and L. J. Fetters, *Macromolecules*, **11**, 899(1978).
58. G. Quack, L. J. Fetters, N. Hadjichristidis, and R. N Young, *Ind. Eng. Chem. Prod. Res. Dev.*, **19**, 587(1980).
59. M. K. Martin, T. C. Ward and J. E. McGrath in *Anionic Polymerization. Kinetics, Mechanisms, and Synthesis*, J. E. McGrath, Ed., ACS Symposium Series No. 166, American Chemical Society, Washington, D.C., 1981, p. 558.
60. J. W. Mays, N. Hadjichristidis and L. J. Fetters, *Polymer*, **29**, 680(1988).
61. R.P. Quirk in *Comprehensive Polymer Science*, First Supplement, S. L. Aggarwal and S. Russo, Eds., Pergamon Press, Elmsford, New York,1992, p. 83.
62. R. P. Quirk and J. Yin, *J. Polym. Sci., Part A: Polym. Chem.*, **30**, 2349(1992).
63. R.P. Quirk and W.-C. Chen, *Makromol. Chem.*, **183**, 2071(1982).
64. M. Morton, L. J. Fetters, R. A. Pett and J. F. Meier, *Macromolecules*, **3**, 327(1970)
65. R. Milner, R. N. Young and A. R. Luxton, *Polymer*, **24**, 543(1983).
66. R. N. Young, R. P. Quirk and L. J. Fetters, *Adv. Polym. Sci.*, **56**, 1(1984).
67. R. P. Quirk and J.-J. Ma, *J. Polym. Sci. Part A: Polym. Chem.*, **26**, 2031(1988).
68. V. Halaska, L. Lochmann and D. Lim, *Coll. Czech. Chem. Commun.*, **33**, 245(1968).
69. D. H. Richards, D. M. Service and M. J. Stewart, *Brit. Polym. J.*, **16**, 117(1984).
70. K. Han, Ph.D. thesis, University of Akron, 1995.
71. R. P. Quirk, S. Jang and J. Kim, *Rubber Reviews, Rubber Chem.Tech.*, **69**, 444(1996).
72. N. Shepherd and M. J. Stewart (to Secretary of State for Defence in U.K.), U.S. 5,331,058 (July 19, 1994).
73. N. Shepherd and M. J. Stewart (to Secretary of State for Defence in U.K.), U.S. 5,362,699 (November 8, 1994).
74. J. A. Schwindeman, E. J. Granger and J. F. Engel (to FMC Corporation), WO 95/22566 (August 24, 1995).
75. G. D. Andrews and L. R. Melby in *New Monomers and Polymers*, B.M. Culbertson and C. U. Pittman, Eds., Plenum, New York, 1984, p. 357.
76. D. Freyss, P. Rempp and H. Benoit, *J. Polym. Sci., Polym. Lett.*, **2**, 217(1964).
77. D. M. Wiles and S. Bywater, *J. Polym. Sci., Polym. Lett.*, **2**, 1175(1964).
78. D. M. Wiles and S. Bywater, *Trans. Faraday Soc.*, **61**, 150(1965).
79. R. P. Quirk and L. Zhu, *Brit. Polym. J.*, **23**, 47(1990).
80. B. C. Anderson, G. D. Andrews, P. Arthur, Jr., H. W. Jacobson, R. Melby, A. J. Playtis and W. H. Sharkey, *Macromolecules*, **14**, 1599(1981).

81. K. Hatada,T. Kitayama, K. Fujikawa, K. Ohta and H. Yuki in *Anionic Polymerization*, J. E. McGrath, Ed., ACS Symposium Series No. 166, 1981, p. 327.
82. S. K. Varshney, J. P. Hautekeer, R. Fayt, R. Jerome and Ph. Teyssie, *Macromolecules*, **23**, 2618(1990).
83. D. Kunkel, A. H. E. Müller, M. Janata and L. Lochmann, *Makromol. Chem., Macromol. Symp.*, **60**, 315(1992).
84. R. Fayt, R. Forte, C. Jacobs, R. Jerome, T. Ouhadi, Ph. Teyssie and S. K. Varshney, *Macromolecules*, **20**, 1442(1987).
85. J. P. Hautekeer, S. K. Varshney, R. Fayt, C. Jacobs, R. Jerome and Ph. Teyssie, *Macromolecules*, **23**, 3893(1990).
86. S. K. Varshney, C. Jacobs, J.P. Hautekeer, P. Bayard, R. Jerome, R. Fayt and Ph. Teyssie, *Macromolecules*, **24**, 4997(1991).
87. J.-S. Wang, R. Jerome, Ph. Bayard, M. Patin and Ph. Teyssie, *Macromolecules*, **27**, 4635(1994).
88. A. H. E. Müller, *Makromol. Chem.*, **182**, 2863(1981).
89. K. Hatada, T. Kitayama and K. Ute, *Prog. Polym. Sci.*, **13**, 189(1988).
90. A. H. D. Müller in *Comprehensive Polymer Science*, Vol. 3, *Chain Polymerization Part 1*, G. C. Eastmond, A. Ledwith, S. Russo and P. Sigwalt, Eds., Pergamon, Oxford, 1989, p. 386.
91. *Ring-Opening Polymerization*, D. J. Brunelle, Ed., Hanser Publishers, Munich, 1993.
92. S. D. Gagnon, in *Encyclopedia of Polymer Science and Engineering*; 2nd ed.; J. I. Kroschwitz, Ed., John Wiley & Sons, New York, 1986, Vol. 6, p. 225.
93. P. Dreyfuss and M. P. Dreyfuss in *Encyclopedia of Chemical Technology*; 3rd ed., R. E. Kirk and D. F Othmer, Eds., John Wiley & Sons, New York, 1982, Vol. 18, p. 616.
94. S. Boileau, in *Comprehensive Polymer Science*; 1st ed., G. C. Eastmond, A. Ledwith, S. Russo and P. Sigwalt, Eds., Pergamon Press, Oxford, 1989, Vol. 3, p. 474.
95. K. S. Kazanskii; A. A. Solov'yanov and S. G. Entelis, *Eur. Polym. J.*, **7**, 1421(1971).
96. M. C. Barker and B. Vincent, *Colloid Surf.*, **8**, 289(1984).
97. D. H. Richards and M. Szwarc, *Trans. Faraday Soc.*, **55**, 1644(1959).
98. P. Sigwalt and S. Boileau, *J. Polym. Sci., Polym. Symp*, **62**, 51(1978).
99. P. Sigwalt, *J. Poly. Sci., Polym. Symp.*, **50**, 95(1975).
100. C. J. Chang, R. F. Kiesel and T. E. Hogen-Esch, *J. Am. Chem. Soc.*, **95**, 8446(1973).
101. J. E. Figueruelo and D. J. Worsfold, *Eur. Polym. J.*, **4**, 439(1968).
102. F. E. Critchfield, J. A. Gibson, Jr. and J. L. Hall, *J. Am. chem. Soc.*, **75**, 6044(1953).
103. I. Cabasso and A. Zilkha, *J. Macromol. Sci. Chem.*, **A8**, 1313(1974).
104. J. Furukawa and T. Saegusa, *Makromol. Chem.*, **31**, 25(1959).
105. A. Deffieux, E. Graf and S. Boileau, *Polymer*, **22**, 549(1981).
106. S. Boileau, A. Deffieux, D. Lassalle, F. Menezes and B. Vidal, *Tetrahedron Lett.* 1767(1978).

107. G. Odian in *Principles of Polymerization;* 3rd ed., Wiley-Interscience, New York, 1991, (a) p. 536; (b) p. 562.
108. R. P. Quirk, J. Kim, C. Kausch and M. Chun, *Polym. Int.*, **39**, 3(1996).
109. J. Sebenda, in *Comprehensive Polymer Science*; 1st ed., G. C. Eastmond, A. Ledwith, S. Russo and P.Sigwalt, Eds., Pergamon Press, Oxford, 1989, Vol. 3, p. 511.
110. J. Sebenda and J. Hauer, *Polym. Bull.*, **5**, 529(1981).
111. K. Hashimoto, K. Hotta, M. Okada and S. Nagata, *J. Polym. Sci.: Part A: Polym. Chem.*, **33**, 1995(1995).

Chapter 2

Industrial Applications of Anionic Polymerization: An Introduction

Henry L. Hsieh

Phillips Fellow Emeritus, 700 Fearrington Post, Pittsboro, NC 27312

Anionic polymerization occupies a key position in the industrial production of polydiene rubbers (BR, IR), solution styrene/butadiene rubbers (SBR), thermoplastic elastomers of styrenic-type (SBS, SIS), and other non-rubber products, such as clear impact-resistant polystyrene resins, binders for solid rocket fuel, adhesives and others (1).

The polymerization of vinyl monomers with alkali metals was reported by several chemists as early as 1910. Ziegler studied the polymerization of butadiene with *n*-butyllithium (*n*-BuLi) in ether solution in the late 1920s and early 1930s. The German workers created methyl rubber in World War I (1914-1918) using 2,3-dimethylbutadiene, sodium metal, and carbon dioxide (a chain modifier). Later some other polydiene rubbers were produced also with sodium or potassium in World War I by the Germans. Russians continued to produce polybutadiene rubbers by using sodium or potassium metals in World War II (1939-1945). United States and others during the WWII period and early postwar period relied on the emulsion processes for the production of styrene/butadiene copolymers. These copolymers, SBRs, were the backbone of the synthetic rubber industry.

In the early 1950s, the synthesis of polyisoprene having high *cis*-configuration (*cis*-polyisoprene) was discovered by using either lithium metal or alkyllithium in hydrocarbon solution. Synthetic natural rubber was born. The stereospecific polymerization of isoprene stimulated and regenerated interest in alkyllithium-initiated polymerizations. Intensive research in this area has been done in industrial laboratories as well as in academic institutions. About the same time, the so-called living polymerization (no premature termination) by employing alkali metal types of initiator was recognized. While the stereospecificity created the initial excitement, it was the unprecedented control over polymer properties provided by anionic polymerization that led to the development of several important commercial products. New products and new applications continue to appear in the marketplace based on anionic polymerization. This control is most easily exerted when polymerization is initiated by hydrocarbon soluble organolithium and includes: polymer composition, microstructure, molecular weight, molecular weight distribution, monomer sequence distribution in copolymers, choice of functional end groups, and branching. The recently published book entitled Anionic Polymerization: Principles and Practical Applications by Hsieh and Quirk (2) provides an extensive discussion regarding structure and bonding in carbon ionic compounds, anionic and living polymerization, kinetics and mechanism in anionic polymerization, anionic synthesis of polymers with well-defined structures, anionic polymerization of polar monomers and commercial applications of anionically prepared polymers. The last section mentioned should be of particular interest to the readers of this symposium series book. It covered polydiene rubbers, styrene-diene rubbers, styrenic thermoplastic elastomers, applications of styrenic thermoplastic elastomers in plastics modifications, adhesives, and footwear, clear impact-resistant polystyrene, nonfunctional liquid polybutadienes and telechelic elastomers and prepolymers. A brief summary of each of these industrial developments will be presented here as an introduction to the chapters that follow in which the recent advances in industrial applications of anionic polymerization are discussed.

Firestone and Shell started commercial production of *cis*-polyisoprene by the anionic process in the 1950s, but these plants are no longer in operation. Firestone technology employed lithium metal, while Shell used an alkyllithium initiator. The anionic produced polyisoprene rubber has 92-94% *cis*-1,4 microstructure, while the other technology using tri-alkylaluminum/titanium tetrachloride as initiator produces polyisoprene rubber of 96-98% *cis*-1,4 microstructure, ensuring more rapid crystallization and hence higher tear strength in vulcanizates, which are closer in mechanical properties than those from natural rubber (3-5). About the same time, Phillips and Firestone started manufacturing polybutadiene rubber by the anionic route and ever since, their use has grown steadily, particulary in tire applications. These polybutadienes in general have low vinyl content (typically, 8-10%), but some interesting applications have been found for medium vinyl polybutadiene as well (6). Polybutadiene with 50-55% vinyl content behaves like emulsion SBR in tire-tread formulations and exhibit similar tread wear, wet skid resistance, low heat buildup, and other characteristics. Medium vinyl polybutadiene is also effective as a partial or complete replacement for SBR in a variety of routine applications.

Polybutadiene rubbers with molecular weight of 250,000 g/mole will have a desirable combination of vulcanizate properties, but will also have a high cold flow, high compounded Mooney viscosity, and poor processibility. These problems were solved by broadening the molecular weight distribution and/or introducing long-chain branching in the polymer molecule. Introduction of one or two long-chain branches into a polydiene molecule leads to profound changes in rheological behavior. For instance, in polybutadiene, at low molecular weight, the Newtonian viscosity is decreased relative to a linear polymer of the same molecular weight (7). At higher molecular weight, the Newtonian viscosity rises rapidly above the corresponding value for a linear polybutadiene. At a higher shear rate, the viscosity of the branched polymers is uniformly lower than that of the linear polymers of an identical molecular weight. In other words, the non-Newtonian behavior of the branched polymers become rapidly more pronounced at a higher molecular weight. A long-chain branched polymer has higher resistance to flow at low shear rate (i.e., low cold flow) and has more flow at moderate and high shear rate (i.e., better processing) than the corresponding linear polymer. Commercially, both the use of a difunctional monomer, such as divinylbenzene (8), and the use of a multi-functional coupling (linking) agent, such as silicon tetrachloride, are employed to produce long-chain branched polymers (9,10).

Nonfunctional liquid polybutadienes are produced frequently with organolithium organosodium, and organopotassium compounds or the metals themselves. A complex of RLi/ROM(M=Na, K, Rb, Cs) can also be used for this purpose (11-13). Often a promoter (e.g., THF, TMEDA) is used in conjunction with a chain transfer agent or telogen (e.g., toluene) (14,15). A wide range of microstructure varying from 10% vinyl to 90+% vinyl can be produced. Functional polybutadiene prepolymers (telechelic polybutadienes) were first commercially produced by Phillips in 1962 (16,17) These liquid polymers with either carboxy or hydroxy-end groups (carboxy-telechelic or hydroxy-telechelic) were used almost exclusively as solid rocket fuel binder for thirty years (18). Improved versions with lower vinyl content, thus lower Tg, were introduced in more recent years. Nippon Soda in Japan introduced a product with 90% vinyl. It was produced in THF at a very low temperature employing a sodium-based initiator. This product, in its applications, served as a prepolymer to produce a high-molecular weight product via multi-linking reactions with a di-or tri-coupling agent. At the same time, the active vinyl side chains served as sites for further chemical reactions, such as cyclization with the neighboring vinyl group.

Butadiene/styrene copolymers are the workhorses of the synthetic rubber industry. Anionic polymerization is the most versatile technology to produce the copolymers. The anionic process has the ability to control the monomer sequence distribution as no other known polymerization process is capable of accomplishing to a similar degree (19,20). In addition, the anionic process can conveniently regulate molecular weight, molecular weight distribution, vinyl content (short branching of the diene and long-chain branching. Thus, pure block (B→S), tapered block (B→B/S→S), random block (B/S→S) as well as random (B/S) copolymers can be produced on a commercial scale. In the copolymerization of butadiene (or isoprene) and styrene, the reactivity ratios are influenced by the solvent and the nature of the anionic initiator (i.e., the ionic character of the character of the carbon-metal bond). Pure block copolymers can be produced by incremental addition of monomers. Tapered (graded) block copolymers can be made by polymerizing the butadiene/styrene mixtures in hydrocarbon solution with n-BuLi initiator. A tapered block copolymer containing 75% butadiene and 25% styrene (about 18% styrene in block form) was the first solution copolymer produced commercially by Phillips in 1962 (19). This polymer has outstanding extrusion characteristics, low water absorption, low ash, and good electric properties. As a result, it is useful in such diverse applications as wire and cable coverings, shoe soles, and floor tiles. Random butadiene/styrene copolymers, solution SBRs, are produced commercially in several different technologies based on independent discoveries and developments. Onc can prepare random SBRs by using an alkyllithium initiator in the presence of small amounts of ether of tertiary amine (21). Use of these polar randomizers also increases the vinyl unsaturation in the copolymer. The random copolymers can also be prepared by slow and continuous addition of the two monomers or by incremental addition of butadiene to a styrene rich monomer mixture during polymerization. These methods would produce low vinyl unsaturation (10%). Thus, one can also produce constant composition copolymers by continuous polymerization. Another general method to prepare random copolymers is the use of a mixture of alkyllithium and alkali metal alkoxide (e.g., RONa, ROK, RORb, ROCs) as the initiator (22-24). Thus, one can prepare copolymers of various randomness and various vinyl unsaturation.

Solution SBRs have much narrower molecular weight distribution than those of emulsion-produced products. The low molecular weight component in the emulsion copolymer helps it to process better, but has an undesirable influence on hysteresis. The solution copolymer, on the other hand, is characterized by a sharp molecular weight distribution with only a small amount of lower and higher molecular weight fractions. This feature leads to improved hysteresis properties and better abrasion resistance, but also results in more difficult processing. In general, solution SBRs have performed well in tire-tread formulation. They require about 20% less accelerator than emulsion SBR and give higher compounded Mooney, lower heat buildup, increased resilience, and better retread abrasion indexes.

Intensive efforts were made to improve further the performance of the random copolymers in tire-tread applications. This effort was focused in particular upon increasing abrasion resistance or decreasing rolling resistance while maintaining or improving processing characteristics and wet traction. It was shown that abrasion resistance decreases and skid resistance increases as styrene or vinyl content is increased. Processability is also improved by the increase of styrene or vinyl content in the rubber. Rolling resistance is predominately related to the loss tangent (i.e., the tan δ at comparatively low frequency and at appropriate temperatures above the glass transition temperature, [Tg]). Rolling resistance is related to low tan δ. The sliding friction or traction on a wet pavement during braking involves a high-frequency deformation of the tire tread, and is predominantly dependent on the Tg.

Selectively hydrogenated copolymers of styrene and dienes (both block and random) are tailor made and used as a viscosity index (VI) improver in multi-grade lubricating oils (25,26). Hydrogenation is essential to impart resistance to oxidative degradation. Crystallinity is undesirable and by control of microstructure and distribution of randomly incorporated styrene, an outstanding property of these polymers is their shear stability.

One group of styrene/diene block copolymers is unique in that they are characterized as two phases and two Tgs and by high raw strength, complete solubility and reversible thermoplasticity. These copolymers are known as styrenic thermoplastic elastomers (27). The blocks are arranged with the rubbery (soft) block in the middle of two polystyrene (hard) blocks, S-B-S type. The glassy polystyrene block is associated in domains, but since there is one at each end of a long rubbery block, a single molecule may be connected to two different domains. As long as the polystyrenes are in their glassy, rigid state, the product will appear as a network structure with the rubbery chains anchored between two fixed domains. In this form, as long as the polystyrenes are below their glass transition temperature, the polymer has good tensile strength without chemical cross-linking or vulcanization. The unique behavior of these thermoplastic elastomers combines the features that were previously considered to be mutually exclusive, namely, thermoplasticity together with elastic behavior. The styrenic thermoplastic elastomers are used extensively in injection-molded rubber goods, footwear, pressure-sensitive and hot-melt adhesives, and in mechanical goods. These copolymers also find applications in blends with various thermoplastic resins. The commercial productions of these triblock copolymers utilize (1) the use of a dilithium initiator and sequential addition of monomers; (2) the use of an alkyllithium initiator and sequential addition of monomers; or (3) the use of an alkyllithium initiator and utilizing linking reactions. Chemically modified (e.g., hydrogenated) versions of these copolymers are also available commercially. Shell Chemical is the major producer of these styrenic thermoplastic elastomers.

Other thermoplastic elastomers include olefins, urethanes, copolyester ethers and others. In 1995, global demand for these polymers reached 2.14 billion pounds and is estimated to reach 2.87 billion pounds by the year 2000. Styrenics produced by anionic polymerization represent about half of the total demand.

Styrene-butadiene block copolymers containing over 60% styrene constitute a family of transparent resinous thermoplastics with good impact strength (28). The lower diene content makes these copolymers non-elastomeric. These resins are generally crystal clear, have moderate to good toughness, good impact strength, excellent to good rigidity, excellent printability and gloss, and excellent organoleptic properties. These resins supplied to free-flowing pellet form can be injection-molded, blow-molded, or thermoformed. Ease of processing allows for a variety of end uses including packaging, medical devices, toys, bottles, office articles, disposable cups, and films. These resins are basically a mixture of block copolymers of styrene and butadiene formed *in situ* by incremental additions of alkyllithium initiator and monomers followed by some form of chain linking. These clear resins are being produced in the United states, Europe and Asia. The estimated total capacity is around 500 million pounds and increasing. Phillips is the first and the major manufacturer. Others include BASF, Firestone, Fina, Denka-Kaguku, and Asahi Chemicals.

Commercial applications of anionic polymerization via graft copolymerizations, either graft from or graft on, or the use of macromonomers (Macromers) are receiving increasing attention. New products are being introduced in the marketplace.

In conclusion, anionic polymerization has long emerged from laboratory curiosity to an important industrial process. Millions of tons of elastomers, thermoplastics and speciality polymers are manufactured in plants around the world. Dedicated and innovative scientists, engineers and technologists are working diligently to develop new products and processes. Many of these developments are published in this book.

Literature Cited
1. Hsieh, H. L.; Farrar, R. C.; Udipi, K. *Chem. Tech.* **1981**, *11*, 626.
2. Hsieh, H. L.; Quirk, R. P. *Anionic Polymerization: Principles and Practical Applications*, Marcel Dekker, Inc., New York, NY, **1996**.
3. Scott, K. W.; Trick, G. S.; Mayer, R. H.; Saltman, W. M.; Pierson, R. M. *Rubber Plastic Age*, **1961**, *42*, 175.
4. Hackthorn, M. J.; Brock, M. J. *Rubber Chem. & Tech.* **1972**, *45*, 1295.

5. Brock, M. J.; Hackthorn, M. J. *Rubber Chem. & Tech.* **1972**, *45*, 1303.
6. Railsback, H. E.; Stumpe, N. A. *Rubber Age*, **1965**, *107*, 27.
7. Kraus, G.; Gruver, J. T. *J. Polym. Sci.* **1975**, *A3*, 93.
8. Zelinski, R. P.; Hsieh, H. L. U.S. Patent 3,280,084.
9. Zelinski, R. P.; Hsieh, H. L. U.S. Patent 3,280,084.
10. Uraneck, C. A.; Short, J. N., *Rubber Chem. & Tech.* **1968**, *41*, 1375.
11. Kamienski, C. W.; Merkley, U.S. Patent 3,751,501.
12. Wofford, C. F. U.S. Patent 3,356,754.
13. Hsieh, H. L.; Wofford, C. F. *J. Polym. Sci.* **1969**, *A-1*, 449.
14. MCEloroy, B. J.; Merkley, J. H. U.S. patent 3,678,121.
15. Langer, A. W., Jr. U.S. patent 3,451,988.
16. Crouch, W. W. Rubber Plastics Age, U.S. patent 42,276 (1961).
17. Wentz, C. A.; Hopper, E. E. *Ind. Eng. Chem. Res. Dev.* **1967**, *6*, 209.
18. Sayles, D. C. *Rubber World* **1965**, *153*, 89.
19. Hsieh, H. L. *Rubber Plastics Age* **1965**, *46*, 394; *Rubber Chem. & Tech.* **1966**, *39*, 491.
20. Hsieh, H. L.; Glaze, W. H. *Rubber Chem. & Tech.* **1970**, *43*, 22.
21. Zelinski, R. P. U.S. patent 2,975,160.
22. Wofford, C. F. U.S. patent 3,294,768.
23. Hsieh, H. L.; Wofford, C. F. *J. Polym. Sci.* **1969**, *A1*, 461.
24. Hsieh, H. L. *J. Polym. Sci.* **1970**, *A1*, 533.
25. Schiff, S.; Johnson, M. M.; Streets, W. L. U. S. patent, 3,554,911.
26. St Clair, D. J.; Evans, D. D. U. S. patent 3,772,196.
27. Chapter 18, *Styrenic Thermoplastic Elastomers* in reference 1 above.
28. Chapter 20, *Clear Impact-Resistant Polystyrene* in reference 1 above.

Recommended General Background Reading

1. J. P. Kennedy and E. G. Tornquist, Polymer Chemistry of Synthetic Elastomers, Part 1 and Part II, Interscience Publishers, New York, 1969.
2. W. M. Saltman, The Stereo Rubbers, John Wiley & Sons, New York, 1977.
3. G. Holden, N. R. Legge, R. P. Quirk and H. E. Schroeder, *Thermoplastic Elastomers, A Comprehensive Review*, 2nd edition, Hanser Publishers, Munich Vienna, New York, 1996.

Fundamentals and Polymerization Processes

Chapter 3

Aggregation Behavior of Polymer Chains with Living Anionic Lipophobic Head-Groups

L. J. Fetters[1], J. S. Huang[1], J. Sung[1], L. Willner[2], J. Stellbrink[2], D. Richter[2], and P. Lindner[3]

[1] Exxon Research and Engineering Company, Corporate Research Laboratories, Clinton Township, Route 22, East Annandale, NJ 08801–0998
[2] Institut für Festkorperforschung, Forschungszentrum, Jülich, Germany D-52425
[3] Institute Laue-Langevin, 156, Avenue des Martyrs, BP 156 F-38042 Grenoble, Cedex 9, France

> Small angle neutron scattering (SANS) has been used to evaluate the self-assembling behavior of the highly lipophobic styryl- and dienyllithium head-groups in benzene and cyclohexane. These monofunctional chains were found to self-organize into aggregates ranging from dimers to mass-fractal clusters.

This paper describes the self-assembling behavior in hydrocarbon solvents of living polymer chains having the lithium counter-ion at one end. These head-groups undergo self-assembly due to their considerable ionic character(**1**). The commonly invoked mechanism(**2-4**) uses the interdependent conjectures that the aggregated head-groups are dormant, that propagation ensues only when the head-group is momentarily unaggregated and that reaction orders directly yield structural information regarding the association state of the self-assembled aggregates. The advocates of that mechanism have not offered an explanation, other than the implied one of kinetic convenience, as to why and how the aggregated state purloins head-group activity.

For these polymerizations the generic propagation rate expression is:

$$R_p / [M] = k_{app} [PLi]_0^{1/n} \qquad (1)$$

where n was taken to denote the number-average aggregation state taken as 4 for dienyllithium and 2 for styryllithium head-groups, k_{app} the apparent propagation rate constant and $[PLi]_0$ the total concentration of organolithium head-groups. A survey(**5**) of the exponents of equation (1) for various diene systems shows that, in addition to the orthodox 1/4 order, other values ranging from ~ 0.1 to 1 have been observed. Styrenic monomers exhibit the more limited range of 0.48 to 0.87 for n^{-1}. Static light scattering (SLS) measurements yielded apparent aggregation values of ~ 4 for the butadienyl- and isoprenyllithium head-groups and 2 for styryllithium(**6**); values in felicitous accord with those based upon the measured reaction orders. Those authors(**6**) did note the presence of "anomalous scattering" which was perceived to occur only at angles below 90°. This scattering was ascribed to "fine glass particles" produced by the rupture of breakseals. That explanation is incorrect. Pyrex is a weak scattering material in hydrocarbon media as demonstrated by its use as sample cells in light scattering experiments. Furthermore the ~4 fold density difference between the solution and Pyrex particles effectively precludes their long term existence as suspended material. The SLS data of Figure 1 shows that the anomalous scattering was not caused by the presence of particulate Pyrex. The solid line represents the scattering signal from cyclohexane whilst the solid symbols denote the measured scattering with the fragmented Pyrex present. The scattering behavior for *both* systems is angle (Q) independent over the angular range of 6 to 160^0. Thus the cause of the anomalous scattering reported for the cyclohexane solutions in ref.(**6**) lies elsewhere and the presence of large scale aggregated structures is an obvious explanation. It was also observed (Fig. 1 of ref. 6) that the intensity (90°) did not vary linearly with concentration for polyisoprenyllithium in cyclohexane. That behavior can be caused by an association-dissociation process that could lead to changes in aggregate size as concentration is altered. Thus, measurements at different concentrations would observe dissimilar species. That problem is serious for experiments that rely on extrapolations to zero concentration in order to generate single particle data. An additional potential pitfall is the existence of non-trivial correlations between the associated ionic aggregates.

The claim has also been made(**6**) that SLS measurements demonstrated the dimeric aggregation state for the SLi head-group. That conclusion is invalid since quantitative SLS measurements on colored solutions will be vitiated by optical absorption, i.e., λ_{max} is 328 *nm* for SLi in cyclohexane(**7**). This problem is further complicated by the change in absorption as the head-group population varies. These SLS results (**6**) have been accepted as the definitive measurements regarding head-group aggregation states (see refs. **2, 3** and **4**). This generally universal and long-term acceptance was doubtlessly encouraged by the pleasing apparent confluence of association states from the SLS measurements and propagation reaction order results.

The essential premise of the conjectured mechanism for dienes invokes equilibria involving tetramers (**T**), dimers (**D**) and singlets (**S**) where the more highly aggregated structures are favored:

$$(PLi)_4 \xrightleftharpoons{K_{TD}} 2(PLi)_2 \quad\quad (2)$$

$$(PLi)_2 \overset{K_{DS}}{\rightleftharpoons} 2PLi \qquad (3)$$

For styrenic monomers equation (3) is taken to describe the mechanism involving chain propagation. For both types of monomers the aggregates are deemed unreactive while the sole active entity is conjectured to be the singlet head-group. This postulate (which dates back nearly four decades) has acquired the status of fact through frequent repetition. In conflict with that premise is the observation **(8)** that vitrified polystyryllithium will react with gaseous 1,3-butadiene at room temperature followed by propagation. This observation contradicts the above mechanism since the formation of free head-groups would involve the reptational diffusion of chain ends away from each other along the path of the tube; an impossible event in the glassy state over any laboratory time scale. In turn the reassembly process demands the statistically improbable near-simultaneous reversal of the reptational direction of movement of the free head-groups. It has also been shown**(9)** that isoprene and butadiene vapors will react with crystalline ethyllithium with propagation following. These dual observations yield prima facie support to the notion that the initiation and propagation events can involve aggregated structures.

Viscosity measurements**(10-12)** involving concentrated solutions of chains with the styryl- and dienyllithium head-groups have without exception delivered the weight-average aggregation state (N_W) of 2 where N_W denotes the measured viscometric association state as opposed to that of n which is adduced from propagation reaction orders. That commonality of findings was referred **(13)** to as "perturbing"; an assessment based on the conflict with the conjectured mechanism which does not tolerate identical primary aggregation states for the dienyl- and styryllithium head-groups. As an outcome the solution based viscometric data were dismissed as unreliable**(2,3,4,13)**.

A provocative finding regarding head-group aggregation is available via the 1966 report of Makowski and Lynn**(14)**. Their approach involved melt state measurements of viscosities of oligomeric butadienyllithium and their terminated counter-parts. Those results, at 20°C, are shown in Figure 2. It is evident that during the early stages of chain growth the viscosity of the butadienyllithium system *decreases*, even though polymer chain length is *increasing*. This attenuation of viscosity is reversed at a number-average degree of polymerization (DP_n) of about 15 and is replaced by an increase in viscosity for progressively higher DP_n values. For comparison sake, the viscosities of the terminated chains (measured) and the dimer counterpart (calculated) are given in Figure 2. That data is replotted in Figure 3 as η_{ag}/η_d vs. the DP_n of the parent chain. The subscripts refer to aggregate and dimer. This rendition of the Makowski-Lynn data shows the diminution of the η_{ag}/η_d ratio as the parent chain DP_n increases. The changes seen are compatible with the initial presence of highly extended large scale structures

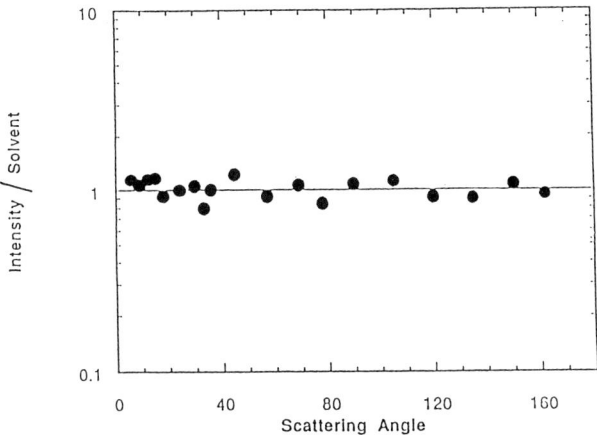

Figure 1. Scattering behavior of powdered Pyrex in cyclohexane at 20°C. The solid line denotes baseline scattering behavior of cyclohexane while the symbols denote the scattering with particulate Pyrex present.

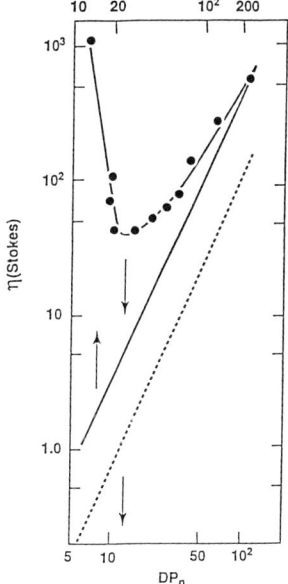

Figure 2. Viscosity behavior at 20°C of oligmeric butadienyllithium as a function of the measured DP_n of the parent terminated polymer. The dashed line represents singlet chain behavior while the solid line denotes dimer behavior; $\eta_d = 4.24 \times 10^{-3} DP_n^{2.20}$ (Stokes); cc: = 0.992.

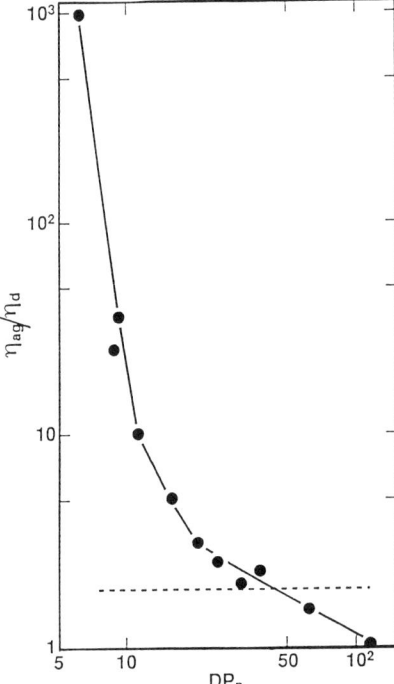

Figure 3. The ratio of η_{ag}/η_d as a function of the measured DP_n of the parent terminated polymer. The dashed line denotes the approximate behavior of tetrameric aggregates (η_t/η_d).

which undergo an attenuation in the extent of aggregation as chain length increases. For DP_n of $> 10^2$ the dimer structure seems to be favored in the melt state.

Such macrolattice formation in the melt for amorphous associating polymers has been found for polyisoprenes with the highly polar sulfo-zwitterion groups at one end **(15)**. For low DP_n values (ca. 32 to 68) high-resolution small angle X-ray scattering (SAXS) studies revealed a lattice with the symmetry of a triangular array of cylinders with the zwiterionic head-groups as the spine. The lattice ordering was strong with the nominal domain size in excess of 5×10^3 Å. At higher DP_n this distinctive spatial ordering underwent a structural phase transition to a bcc cubic lattice with long range order. Rheological measurements have also shown the existence in the melt of large scale structures**(16)**. Furthermore, in consonance with the SAXS results, the extent of aggregation was found**(16)** to diminish as the chain length increased. Thus, the behavior of the zwitterionic head-group is similar to that of its oligomeric butadienyllithium counterpart **(14)**.

The mechanism in question was subjected**(7,17)** to a computer based (Silicon Graphics 4/D 35) non-linear curve fitting program with the purpose of determining the singlet head-group concentrations, the respective equilibrium constants {equations (2) and (3)} and the corresponding propagation rate constants. This was done for the data sets involving the dienyl**(7)** and styryl**(17)** head-groups. Generally incoherent intramural behavior was found for both monomer families. An example of this is represented by styrene in cyclohexane **(17)** at 25 and 40°C. The lower temperature data set yielded $K_{DS} = 1.6 \times 10^{-4}$ M while the 40°C data set yielded a $K_{DS} = 4.5 \times 10^{-8}$ M ; see equation (3). Thus, the physically absurd picture emerges of a 15 degree temperature *increase* engendering a 3.6×10^3 fold *decrease* in K_{DS}. Since the formation of aggregates must necessitate a loss in translational freedom, and thus a decrease in entropy, the process must be exothermic in order to occur spontaneously. It follows that K_{DS} must *decrease* with *decreasing* temperature. This result serves as strong indirect evidence for the flawed nature of the mechanism in question. When the foregoing is coupled with other cited**(7,17)** examples a re-evaluation of the head-group aggregation behavior is mandated. This was done using SANS which allows the coverage of a large range of scattering vectors, Q, where:

$$Q = \frac{4\pi}{\lambda} \sin \frac{\theta}{2} \qquad (4)$$

with λ the wavelength of radiation and θ the scattering angle. The Q range covered via SANS was 0.13 to 1.3×10^{-3} Å^{-1}.

Experimental

The general experimental protocols have been given elsewhere**(5)** in detail. The following serves as a brief summary. Freshly prepared s-BuLi was used as the initiator. The solutions were clear and colorless and did not contain the decomposition product LiH. d_6-Benzene was used as the solvent while the base chain was made from h_8-styrene which was distilled directly into the reactor. This precludes the presence of a thermally prepared high molecular weight polymer. Such material will form at ca. -15°C in the neat monomer over storage time. The head-group change was done using deuterated dienes to form the dienyllithium head-groups while hydrogenous diphenylethylene and ethylene oxide were used to form their respective head-groups. These latter two species do not propagate. Thus only a single unit adds to each head-group. Active and terminated solutions were captured from the reactor into 5mm quartz cells. Termination was done by distilling CD_3OD into the reactor. The SANS scattering was done using the D-11 spectrometer at the Institute Laue-Langevin (ILL); Grenoble, France. The SLS measurements (λ= 6328 Å) were done with the Santa Barbara Light Scattering Instrument.

Standard vacuum line procedures[18] were utilized for solvent and monomer purifications. The presence of hydrogenous particulate materials has no influence on the SANS measurements since those length scales are outside our Q range. Silicate based particulate material lacks contrast. The combination of deuterated solvents and hydrogenous styrene was chosen since it has been long recognized[19] that the Q independent incoherent scattering of deuterated solvents is generally about a decade less intense than that found for their hydrogenous counterparts. In all cases, the base chain was polystyrene which ranged in molecular weight from 3.1×10^3 to 3.3×10^4. Concentrations were kept below the overlap concentration ($c^* = 0.75 \, M_w \, \pi^{-1} \, R_G^{-3}$) where the dimer M_w was used in the calculation. The calculated radii of gyration (R_G) came from the power laws given elsewhere[20]. The measured values where obtained via Guinier plots. Molecular weights were determined using a combination of SANS, SLS and SEC. The latter measurements revealed apparent M_z/M_w and M_w/M_n ratios no greater than 1.05. These values were not corrected for column broadening effects. Hence the authentic values are < 1.05. Sample identification follows the format[5] involving the head-group and molecular weight of the base polystyrene chain, e.g., SBLi(3.1) denotes the butadienyllithium head-group attached to a 3.1K polystyrene chain.

Results and Discussion

The initial SANS measurements[5] were limited to a relatively small Q range. This situation has been corrected by the reopening of the ILL following a four-year closure. The coherent scattering intensity I(Q) can be defined as the coherent cross section, $d\Sigma(Q)/d\Omega$;

$$I(Q) = (\Delta\rho^2)(N_A)^{-1} S(Q) = d\Sigma(Q)/d\Omega \qquad (6)$$

Here Ω is the solid angle into which the radiation is scattered, S(Q) is the static scattering factor (or structure factor) and $\Delta\rho^2$ denotes the average contrast factor between the polymer and the solvent. S(Q) contains the intraparticle parameter the radius of gyration (R_G). In effect the parameter S(Q) predicts the probability of neutrons being scattered as a function of Q. It should be recalled that for polymers SANS measurements are feasible since hydrogen and deuterium differ markedly in their neutron scattering lengths. That parameter is a measure of neutron-nucleus interactions and is independent of atomic number.

Figure 4 shows the SBLi(3.1) head-group over a two decade Q range at the respective volume fractions of 9.60 and 0.58%. The upward curvature at low Q in I(Q) signals the presence of aggregates larger than tetramers. In comparison the terminated counterparts yield intensities independent of Q which is the expected behavior of linear chains. The data shown for the polymeric systems are the measured intensities prior to subtraction of the background scattering of d_6-benzene. The d_6-benzene scattering is Q independent over the entire Q range. This demonstrates that no contribution to the aggregate scattering emerges from the solvent.

One of the more interesting phenomena in random growth (self-assembly) events is that systems with only short range forces and lacking long range order can aggregate into large scale structures that are distinctive and statistically well defined. Some of these structures are definable as mass fractals which are objects whose molecular weight (M) scales as their radius (R) to a power D which is less than the dimension of space d. Thus we have the familiar power law relation where $R \propto M^{-D}$ and $1 < D < 3$.

Beaucage(21,22) has developed the following to describe I(Q) of a mass fractal structure:

$$I(Q) = G \exp(-Q^2\{R_G\}^2/3) + B \exp(-Q^2\{R_S\}^2/3)(1/Q^*)^P$$
$$+ G_S \exp(-Q^2\{R_S\}^2/3) + B_S(1/Q_S^*)^{P_S} \qquad (7)$$

with $Q^* = Q/[\mathrm{erf}(QkR_G/\{6\}^{0.5})]^3$, $Q^*_s = Q/[\mathrm{erf}(QkR_S/\{6\}^{0.5})]^3$ and er**f** is the error function. G, G_S, B and B_S are amplitude factors and k is an empirical constant found(21,22) to be ≈ 1.06. The first term of (7) describes the sizes of the large scale aggregates (R_G) which in turn are composed of arbitrary subunits (S) whose size is given in the third term of the equation as R_S. The second term in this formula describes the mass-fractal regime with two structural limits. The final two terms capture the characteristics of the subunit.

The underlying picture of the aggregate is a supramolecular mass fractal structure(23,24) of overall size R_G and fractal dimension P which in turn is formed of subunits of size R_S with the subunit fractal dimension of P_S. The formula, equation (7), is a mathematical model designed to describe such a structure and thus must not be viewed as a comprehensive theory. Nonetheless the use of this expression yields, Figure 5, an excellent fit to the low concentration data of Figure 4 over the *entire* Q range. Also shown in Figure 5 is the linear fit (12 data points) for the low Q data which yields the slope, D, of 2.99. That value is a signature of a mass fractal structure. In Figure 6 the results are shown in the form of a Kratky

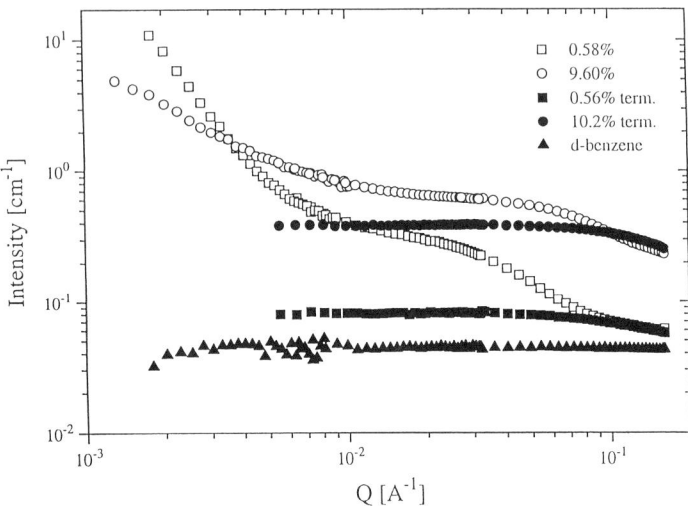

Figure 4. SANS intensities, I, vs. Q for d_6-benzene, the SBLi (3.1) head-group at two different concentrations and the terminated parent polystyrene chain.

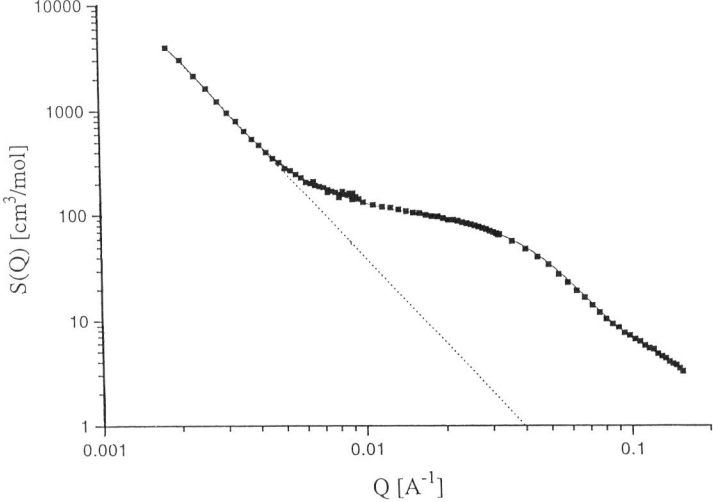

Figure 5. S(Q) vs Q for the low concentration SBLi(3.1) head-group. The fit is based upon equation (7).

plot. As can be seen the model yields a good description of the peak and the upturn at low Q. This latter feature is missing for linear chains. The parameters of the fit via equation (7) for the mass-fractal are R_G = 1260 Å, P = 2.96, R_s = 41Å and P_S = 1.67 which is the value expected for a swollen chain in a good solvent**(21,23,24)**. The mass fractal dimension of two is found for P for a chain that displays Gaussian statistics. Hence, the value of 2.96 can be taken to signal the presence of a very dense structure relative to its size. The chain dimension of the polystyrene tails in the aggregates based on the SBLi head-group is 21 Å. This value invites comparison with the calculated**(20)** value of R_G of 16 Å for the base polystyrene chain. This finding is evidence of some chain stretching in the micelles and is behavior consistent with that of polymer brushes **(25)**.

Figure 7 shows the SBLi(33) head-group at two concentrations. The Guinier plot for the high concentration run yields an R_G of 93 Å while the calculated value**(20)** is 89 Å for the dimer. Conversely, the molecular weight was found to be 8.0 x 10^4 g mol^{-1} which of course, is larger than the anticipated value of 6.6 x 10^4 g mol^{-1} for the dimer. Similar behavior was observed for the SILi(33) head-group. Dynamic light scattering on the SBLi(33) and SILi(33) solutions show the presence of aggregated species with a R_H = 260 Å**(26)**. Thus, this combination of measurements has shown the coexistence of dimers and moderate scale structures (multimers). The lower concentration is well fitted by equation (9). In this case the mass-fractal gradient D is 2.35. The extension of these trends to higher molecular weights and polymer concentrations leads to the conclusion that under those conditions the dominant species is the dimer. That result is in accord with the findings based upon the melt and concentrated viscosity measurements.

This is understandable on entropic grounds. The type of structures formed are the outcome of the interplay between the enthalpic contribution resulting from the act of aggregation and head-group packing preferences and the entropic loss undergone by the tails when in their tethered state. In other words the act of aggregation causes a loss in translational freedom of the tail and thus a loss in entropy. The spontaneity of the dimeric aggregation event dictates its exothermic (ΔH_{DS}) identity which appears to be about 33 kT per dienyllithium head-group**(10,27,28)**. The structures can be viewed as dense polymer brushes and thus the entropy loss incurred by the polystyrene tails involves contributions arising from the elastic energy of chain stretching. Increasing chain length will result in an increase in entropic loss that would be expected to lead to a decrease in the mean state of aggregation. This explains the decrease in aggregation states found by Makowski and Lynn**(14)** with increasing chain length of oligomeric butadienyllithium as well as the similar changes observed for the polyisoprenes tipped with zwitterions**(15,16)**. In an analogous fashion increasing the polymer concentration will cause an increase in chain interactions with again an entropic penalty resulting. Thus the dual increases in chain length and chain concentration during the polymerization event would be anticipated to cause a corresponding decrease in the level of aggregation. Hence these polymerizations can be pictured

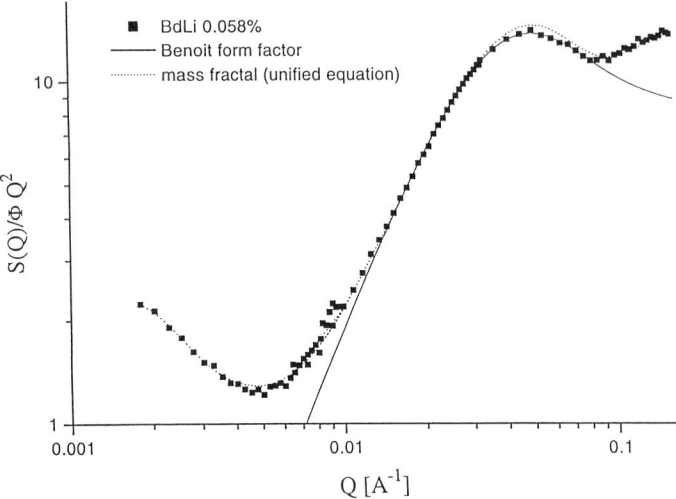

Figure 6. Kratky plot for the low-concentration SBLi (3.1) head-group. The dashed line is from equation (7) while the solid line is based on the Benoit form factor.

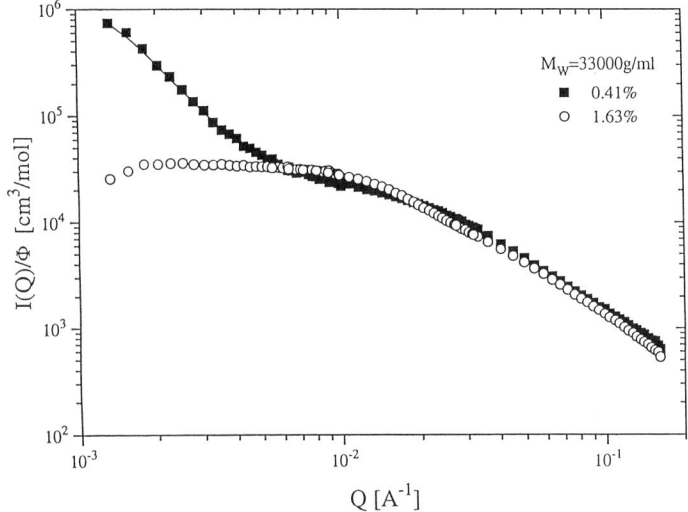

Figure 7. $I(Q)/\Phi$ vs. Q for the SBLi(33) head-group at two different concentrations. Φ denotes concentration.

as producing an ever changing continuum of structures during, at least, the earlier stages of chain growth.

SANS evaluations were done on the SLi(12.4), SBLi(12.4) and the SILi(12.4) aggregates in d_{12}-cyclohexane at 40^0C (the θ state). At present it is sufficient to mention that all three head groups form large scale structures. In addition to the head-groups discussed in this paper those derived from 1,1 diphenylethylene (DPE) and ethylene oxide have been evaluated in d_6-benzene. It can be noted at this juncture that both head-groups also form large scale aggregates. Thus this self-assembling behavior of the styrene and DPE based chain-ends signals the extinction of the reaction phenomenon anointed "the memory effect"(**29-31**). That kinetic phantasm was predicated on the notion of the sole existence of dimers for the styrenic based head-groups along with the compulsory dormancy of those aggregates.

Conclusions

The current findings have led to the following insights regarding the association capacities of these living anionic lipophobic head-groups. The co-existence of large scale aggregates with intermediate (multimers) and small scale species (dimers) is shown by SANS over the Q range of 1.3×10^{-3} to ca. 0.13 Å^{-1}. The population of large scale structures decreases as polymer concentration and molecular weight jointly increase with dimers then becoming the primary aggregate for the diene and styrenic head-groups. This commonality of association state is in agreement with the viscosity based findings(**10-12**). These results thus correct the assertion(**32**) that those data were "frequently incorrect and self-contradictory". The self-assembling behavior of 1,3 diphenylallyllithium, allyllithium and benzyllithium in the solid state is compatible with the large scale self-assembled SLi , SILi and SBLi structures in solution (see ref. 1 for sources). Also the tetrameric structure is *not* the primary aggregation state for the SILi and SBLi head-groups in solution. Within the framework of the self-assembling behavior of surfactants(**33**), ionomers(**34**) and block copolymers(**35-37**) the association behavior of these anionic lipophobic head-groups is understandable. These data thus correct the erroneous notions and conjectures(**2-4,6,13,31,38**) regarding aggregate reactivity and the self-assembling capacities of these anionic head-groups in hydrocarbon milieu.

Literature Cited

1. Weiss, E. *Angew. Chem.,* **1993**, *105*, 1565; *Angew. Chem., Int. Ed., Engl.,* **1993**, *32*, 1501.
2. van Beylen, M.; Bywater, S.; Smets, G.; Szwarc, M; Worsfold, D. J. *Adv. Polym. Sci.,* **1988**, *86,* 87.
3. Szwarc, M.; van Beylen, M. *Ionic Polymerization and Living Polymers,* Chapmann and Hall; London, **1993**.

4. Szwarc, M. *Ionic Polymerization Fundamentals*, Hanser Publishers, Munich, **1996**.
5. Fetters, L. J.; Balsara, N.P.; Huang, J.S.; Jeon, H.S.; Almdal, K.; Lin, M. Y. *Macromolecules*, **1995**, *28*, 4996.
6. Worsfold, D. J.; Bywater, S, *Macromolecules,* **1972**, *5,* 393.
7. Worsfold, D. J. *J. Polym. Sci.* **1967**, *A-1,* 2783: Johnson, A. F.; Worsfold, D. *J. Makromol. Chem.* **1965**, *85,* 273: *J. Polym Sci.* **1965**, *A-2,* 449. The asseveration has been made that for SLi "the absorption is weak in aliphatic hydrocarbons and λ_{max} is > 400 *nm*"; see ref. 4, p. 35. No explanation or literature citation was provided to support that declaration. In benzene λ_{max} = 335 *nm* and 343 *nm* in THF; see Bywater, S.; Worsfold, D. J. *Can. J. Chem.*, **1962**, *40,* 1564.
8. Young, R.N.; Fetters, L.J.; Huang, J.S.; Krishnamoorti, R. *Polym. Int.,* **1994**, *33,* 217.
9. Polykov, D.K.; Balashova, N.I.; Polykova, G.R.; Izyumnikov, A.L. *Doklady AN SSSR,* **1974**, *218,* 152.
10. Morton, M.; Fetters, L.J. *J. Polym. Sci.,* **1964**, *A-2,* 3331.
11. Morton, M.. Fetters, L. J. Pett, R. A. Meier, J. F. *Macromolecules,* **1970**, *3,* 3273.
12. Al-Jarrah, M. M.; Young, R. N. *Polymer*, **1980**, *21,*119.
13. Szwarc, M. in *Anionic Polymerization: Kinetics Mechanisms and Synthesis* McGrath, J. E. Ed., ACS Symp. Series **166**, ACS Washington, DC, **1981**; p. 1.
14. Makowski, H. S.; Lynn, M. *J. Macromol. Chem.,* **1966**, *1*, 443.
15. Shen, Y.; Safinya, C. R.; Fetters, L. J;. Adam, M.; Witten, T.; Hadjichristidis, N. *Phys. Rev. A,* **1991**, *43*, 1886.
16. Fetters, L. J.; Graessley, W. W.; Hadjichristidis, H.; Kiss, A. D.; Pearson, D.; S. Younghouse, L. B. *Macromolecules*, **1988**, *21*, 1644.
17. Fetters, L. J.; Huang, J. S.; Young, R. N. *J. Polym. Sci: Part A: Polym. Chem.,* **1996**, *34*, 1517.
18. Morton, M.; Fetters, L. J. *Rubb. Chem. Tech.*, **1975**, *48*, 359.
19. See e.g., Higgins, J. S.; Benoit, H. C. *Polymers and Neutron Scattering,* Oxford Science Publications, Oxford, England, **1994**.
20. Fetters, L. J.; Hadjichristidis, N.; Lindner, J. S.; Mays, J. W. *J. Phys. Chem. Ref. Data*, **1994**, *23,* 619.
21. Beaucage, G.; Schaefer, D. W. *J. Non-Cryst. Solids*, **1994**, *172-174*, 797.
22. Beaucage, G. *J. Appl. Cryst.* **1996**, *29,* 134.
23. Adam, M.; Lairez, D. *Fractals* **1993**, *1*, 149.
24. Daoud, M.; Stanley, H. E.; Stauffer, D. in *Physical Properties of Polymers Handbook,* Mark. J. E., Ed., America Institute of Physics, Woodbury, N.Y. **1996**; p. 71.
25. Halperin A.; Tirell, M.; Lodge, T. P. *Adv. Polym. Sci.* **1992**, *100*, 31.
26. Fetters, L. J.; Huang, J. S.; Lindner, P.; Willner, L.; Stellbrink, J.; Richter, D. in preparation.

27. Hommes, N. J. R.; Buhl, M.; Schleyer, P. v. R. *J. Organomet.Chem.,* **1991**, *409,* 307.
28. Young, R. N.; Quirk, R. P;. Fetters, L. J. *Adv. Polym. Sci.,* **1984**, *56,* 1; see Table 5, p. 10.
29. Laita, Z.; Szwarc, M. *Macromolecules,* **1969**, *2,* 412.
30. Yamagishi, A.; Szwarc, M. *Macromolecules*, **1978**, *11 ,* 504.
31. See ref. 3, pgs. 361 to 366 and ref. 4, pgs. 182 to 185.
32. See ref. 3, p. 162.
33. See e.g., *Structure and Dynamics of Strongly Interacting Colloids and Supramolecular Aggregates in Solution,* Vol. *369,* Chen, S.-H.; Huang, J. S.; Tartaglia, P. Eds., Series C:Mathematical and Physical Sciences, Kluwar Academic Publishers, Dordrecht, **1992**.
34. See e.g., *Developments in Ionic Polymers,* Vol. *1,*. Wilson, A. D.; Prosser, H. J. Eds., Applied Science, London, **1983**.
35. Balsara, N. P.; Tirrell, M.; Lodge, T. L. *Macromolecules,* **1991**, *24*, 1975.
36. Zhao, J. Q.; Pearce, E. M.; Kwei, T. W.; Jeon, H. S.; Kesoni, P. K.; Balsara, N. P. *Macromolecules,* **1995**, *28,* 1972.
37. Zhou, Z.; Yang, Y.-W.; Booth, C.; Chu, B. *Macromolecules,* **1996**, *29*, 8357.
38. Bywater, S. Polym. Inter. **1995**, *38*, 325.

Chapter 4

Process Monitoring: UV Spectrophotometry as a Practical Tool

M. Bortolotti, G. T. Viola, and A. Gurnari

EniChem Elastomeri, R&D Center, via Baiona 107/111, 48100 Ravenna, Italy

Real time process control techniques permit consistent product quality improvement. It is therefore important to take advantage of on-line analytical facilities even for consolidated production process like living anionic polymerization of SBS and SIS. Although UV spectroscopy was extensively used in the laboratory in early studies on anionic polymerization, UV monitoring of industrial processes was postulated but no further communications appeared about the feasibility of such an analytical tool.

We present in this communication results of work covering the monitoring of pilot plant production of typical block copolymers (styrene-butadiene and styrene-isoprene) under conventional solution polymerization conditions (10-15% solids).

Information obtained obviously permits control of the most important molecular parameters related to product performance (basically MW of arm for linear/radial SBS and linear SIS); furthermore this gives on-line data that is relevant to process parameters, i.e. solvent or monomer purity, abnormal kinetics, etc. Another useful feature offered by on-line UV spectroscopy is the detection of abnormal concentrations of batches. This gives warning of high concentration which may lead to dramatic run-away reactions.

On line measurements take an important role in the improvement of quality of production and in the increase of economics of operations. It is therefore of primary interest, even for consolidated production processes like styrene-diene block copolymerization through living anionic polymerization, to take advantage of on line UV spectroscopy, a powerful tool, for monitoring anionic polymerization.

Currently, on line measurements carried out in the polymerization step are mainly physical process relevant parameters (i.e. flow rates, reactor temperature and pressure, agitator power). Finished product analysis significant to product relevant parameters (i.e. rheological measurements like Melt Indexes) are possible. The only drawback is that these analysis can be done some time after the material has come out from the reactor. Possible correction are not taken in real time. Moreover, the complexity of the relationship between viscoelastic response of elastomers in respect of their molecular parameters can not lead to straightforward correction steps.

UV spectroscopy has been extensively used since '60s in many laboratories[1]. This has been done mainly on high vacuum equipped glass reactors. We took the challenge to bring this powerful tool also to common 1 l glass reactor and finally to the pilot plant.
UV spectroscopy for polymerization monitoring in industrial processes was postulated[2] many years ago, but no further communication has appeared about the feasibility of this analytical tool.
In this paper monitoring of pilot plant production of different kinds of polymers is reviewed and advantages and limitations discussed.
Potential data gained from on line UV spectroscopy are really impressive and are summarized here. It is not necessary to emphasize their relevant importance for industrial process control:
 - molecular weight of base (before any coupling reaction) copolymer
- molecular weight of first block and therefore efficiency of propagation from first to second block
- conversion during styrene polymerization
- direct evaluation of efficiency of linking reactions (reacting polydienyllithium with polyfunctional electrophiles, i.e. $SiCl_4$, $PhSiCl_3$, etc.)[3]
- direct monitoring of deactivation of living polymers during addition of terminating agents. (i.e. ethanol or water).
This information obviously permits control of the most important molecular parameters related to product performance (basically MW of arm for linear/radial SBS and linear SIS), because of the close correlation existing between molecular weights and the architecture of polymers and the technological performances[4]. Furthermore, this gives on-line data that are relevant to process parameters, i.e. solvent or monomer purity, abnormal kinetics, etc. With these data any possible process problem can be correctly addressed.
Another useful feature offered by on-line UV spectroscopy is the detection of abnormal concentrations of batches. This can give warning of high concentrations potentially leading to dramatic run-away reactions.
Also impressive is the versatility of UV spectroscopy in terms of monitoring batch reaction of different products (only considering EniChem Elastomeri products by anionic route):
-Thermoplastic elastomers (SBS, SIS)
-Multiblock copolymer (SBS, tapered-block)
-Diblocks (SB) copolymers, pure and tapered
-Random styrene-butadiene copolymers (S-SBR)
-Low cis polybutadiene (PBDE).
We present in this communication results of work covering the monitoring of pilot plant production of typical block copolymers (styrene-butadiene and styrene-isoprene), tapered diblocks and random styrene-butadiene rubbers under conventional solution polymerization conditions (10-15% solids).

EXPERIMENTAL

The system utilized in the pilot plant consists of a photometer equipped with a flow through-cell in which a solution of living polymer in hydrocarbon solvent is continuously pumped from and returned to the reactor.
For process control, it was decided to abdicate to sensitivity and selectivity offered by typical laboratory photometer, for such technical requirements as long term stability, ruggedness, ease of maintenance and use.
Therefore a process photometer was used (Perkin Elmer Fagos 100: technical most relevant data are collected in Table I) that enables continuous monitoring at different wavelengths. Cells (optical length ranging from 0.5-1.5 mm) are connected via fiber optics to the

photometer which is located in a flame-proof cabinet. Signals from the photometer are sent to a PC which is configured for alarms, possibly giving on-line active lithium concentrations for process control.

Table I: Perkin Elmer Fagos 100, technical relevant data

Light Source	Xenon Flashlamp
Interference filters	1 to 18 (spectral width <10 nm)
Range of measurement	2×10^{-4} A up to approx. 3 A
Detector	Si
Liquid Probe	SS, temp. -20 to 120°C

Cromophores relevant to process monitoring appear at different wavelengths. Polystyryllithium anions absorb at 330 nm peak maximum; the shoulder of the absorbance bands of polydienyllithium can be located around 290 nm. Maximum peaks related to polydienyllithium are not easily recognized due to a superimposition with a strong absorption band of phenyl ring (if present).

Also styrene concentration can be monitored during polymerization using the absorption band at 290 nm. Diene concentration can be monitored at 250 nm in absence of styrene, whose high extinction in the considered spectrum region raises the absorbance up to 4 U.A .

Sequential styrene-butadiene-styrene copolymers of different M.W.s were synthesized and analyzed. Diblock tapered styrene-butadiene and random styrene-butadiene copolymers were also monitored and some considerations regarding active chain ends at total conversion are also possible. Coupling reaction of living styrene-butadienyllithium with $SiCl_4$ at different ratios of Li/Cl were also monitored. It was then possible to get different coupling efficiencies of branched radial structures.

Before and after every polymerization, the line and cell in which the living polymer solution was pumped were washed with hydrocarbon solvent thus permitting registration of blank signals. After polymerization, washing operations were conducted till the UV absorbance between 250-270 nm (polystyrene absorption) decreased to initial values.

Samples for M.W. analysis were collected directly from the reactor through bottle under nitrogen atmosphere and shortstopped, if necessary, by adding a small amount of isopropyl alcohol. Samples were redissolved in THF and filtered. An HP 1090 gel permeation chromatography (GPC) was used to determine the molecular weight averages. PL columns (2×10^5, 10^4, 10^3, 10^6, 500 A) were installed in series.

RESULTS AND DISCUSSION

A typical polymerization reaction (Styrene reaction, isoprene reaction, coupling to linear SIS) is shown in Figure 1. Two signals related to styryllithium (330 nm) and to isoprenyllithium (290 nm) are displayed with reaction time.

In order to establish a correlation for practical usefulness several SBS's of different M.W. were sinthetized. A very good relationship is obtained (Figure 2, 3) between molecular weights of copolymers (GPC analysis) and their absorbances registered during their synthesis, following both the 330 nm styryllithium absorption band and the 290 nm polydienyllithium absorption band. There is no need[2] to calibrate by taking spectra of deactivated polymer because there are no other cromophores absorbing at the wavelengths considered.

Also, for diblock tapered copolymers, it is possible to achieve a good relationship between molecular weight of copolymer and absorption signal registered at total conversion as shown

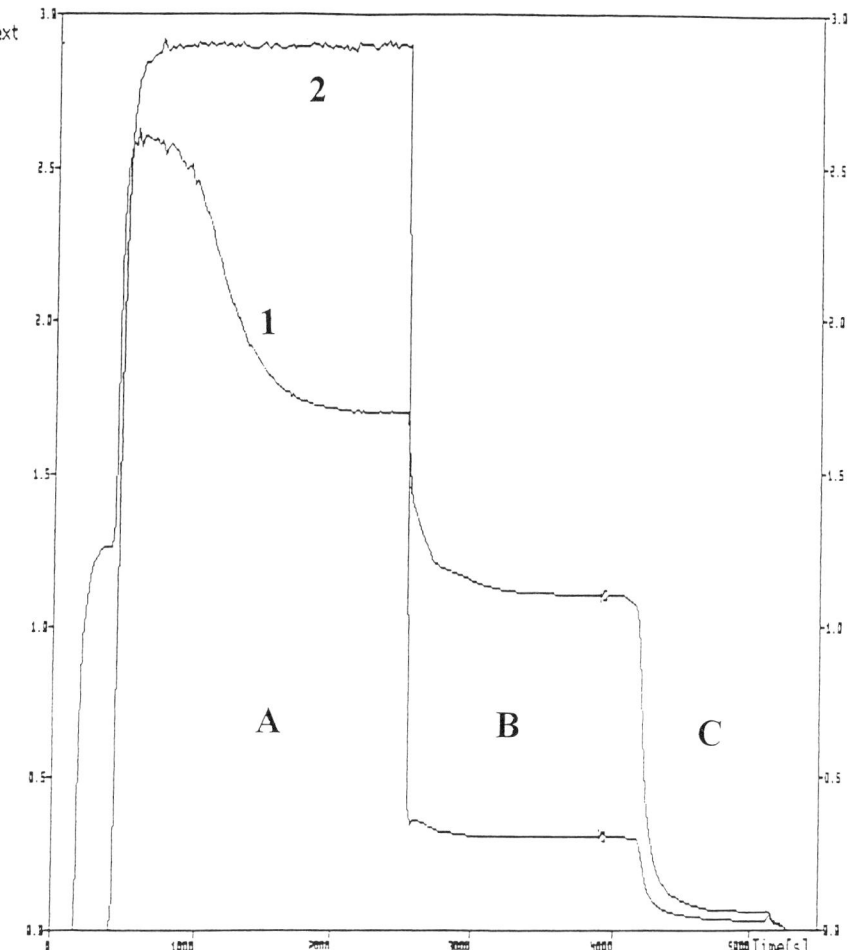

Fig. 1
Time drive UV monitoring: SIS block copolymerization. Styrene polymerization (zone A), Isoprene polymerization (zone B), coupling reaction (zone C).
Curves 1 and 2 registered at 290 nm and 330 nm respectively.

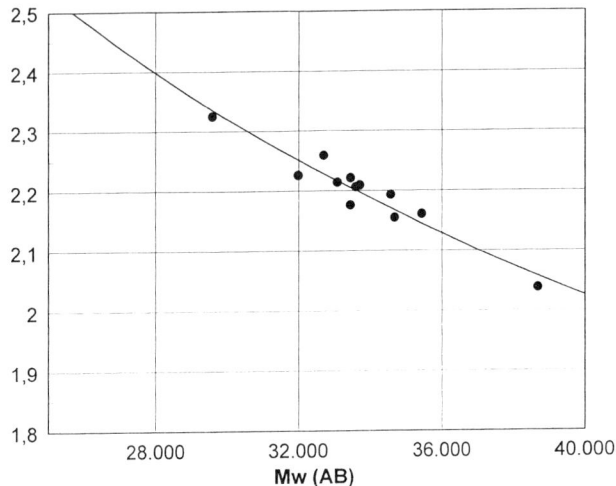

Fig. 2. SBS radial: Mw of (AB) arm vs. UV absorbance.

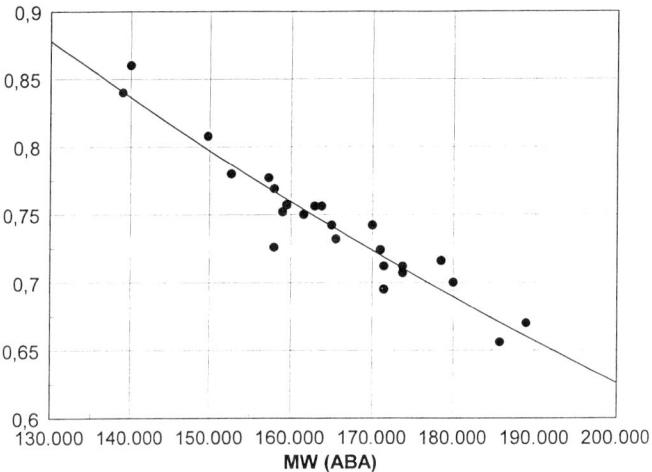

Fig.3. Linear SBS synthesis: Mw of ABA vs. Absorbance.

in Figure 4. For this kind of copolymerization, the absorption band at 310 nm gives a more precise indication than the 290 nm one. This can be explained by the residual styrene absorbance at 290 nm which alters the absorption due to the shoulder of the polybutadienyllithium band. Even though the extinction is lower, decreasing the signal/noise ratio, it is convenient to measure absorption of dienyllithium at 310 nm for polymerization in which residual concentration of styrene could lead to overestimation of concentration at 290 nm.

It is also possible to monitor any kind of deactivation reaction that causes a loss of living centers. Figure 5 shows the relationship between the coupling efficiency as measured by GPC and the change in absorbance (as measured before and after coupling reaction). The good relationship obtained holds when less than stoichiometric amounts of electrophilic reagents are used.

For styrene copolymers it is also possible to follow styrene conversion by monitoring the absorbance variation at 290 nm. It is clear that, due to the very high extinction coefficient of styrene, it is not possible to obtain a linear relationship between absorbance and styrene concentration. However, as shown in Figure 6, the first order derivative of the signal variation gives a good agreement with residual styrene concentration as measured by GC.

Moreover, it is possible to gain from UV spectroscopy applied to pilot plant interesting information which give relevant insights to the process. Some examples pertaining to sequential triblock copolymerization and to random copolymerization of styrene and butadiene under different condition will be briefly discussed here.

With the aim to achieve as high a pure triblock (SBS) copolymer, three different reaction strategies may be followed:
-classical synthesis by sequential monomer addition.
-end linking reaction of SB diblocks with difunctional reactive groups
-starting the polymerization from the middle point of the macromolecule by a difunctional initiator.

For different reasons, depending on dissimilar plant set up some producers prefer either the first, the second or the last strategy. We show in this example that in some case, like the one in which a modification of diene microstructure is needed and obtained by the use of ethers, the first strategy decreases its efficiency. Due to the small differences in MW between SB and SBS produced from sequential polymerization (i.e. 77500 Dalton for SB and 90000 for SBS for a typical SBS copolymer with 30% total styrene content), GPC does not easily distinguish between diblock and triblock copolymer. In Figure 7 it appears clear that the signal of styryllithium in the third step (2° styrene polymerization) of a sequential polymerization is considerably lower than in the first step (1° styrene polymerization). After correction for dilution of the solution due to the addition of butadiene and styrene, it is possible to calculate that, in this particular case, the triblock content is only 80% and not 100% as inferred by GPC. With this information it is then possible to change the process condition to maximize the formation of triblock if needed.

Other useful examples of on-line process monitoring are shown in Figure 8 and Figure 9 in which two different randomizing systems are used for the batch copolymerization of SBR(S). In Figure 8 it can be seen that approaching total conversion styryllithium cromophore concentration (330 nm) increases. This means that possibly styrene blockiness may be formed in the polymerization. This is confirmed by O_3/HPLC analysis[5] which estimates that 10% (related to total styrene) of the styrene was not statistically distributed in the copolymer. In Figure 9 it is shown that copolymerization proceeds in a totally random fashion, taking advantage of a better randomizing system. The concentration of the two cromophores (butadienyllithium and styryllithium) are constant along conversion leading to perfect random copolymer (O_3/HPLC gave block styrene <1%).

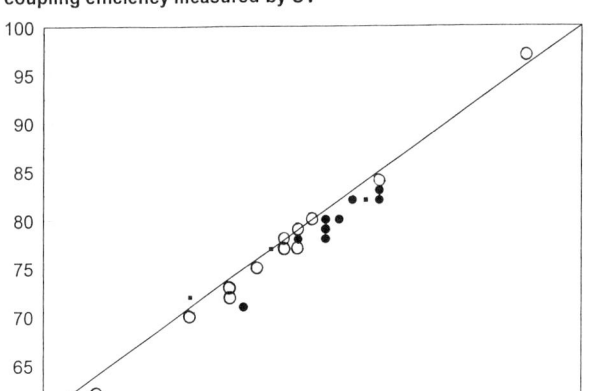

Fig. 5. Evaluation of coupling efficiency (290 nm absorbance before SiCl4 addition).

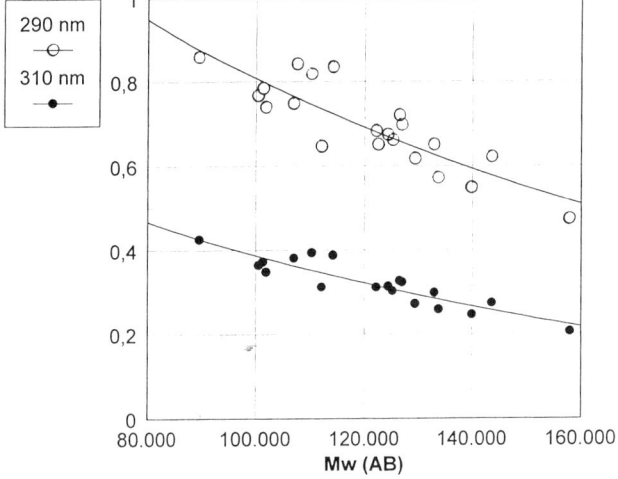

Fig. 4. Diblock copolymerization: comparison between UV signals at 290 and 310 nm.

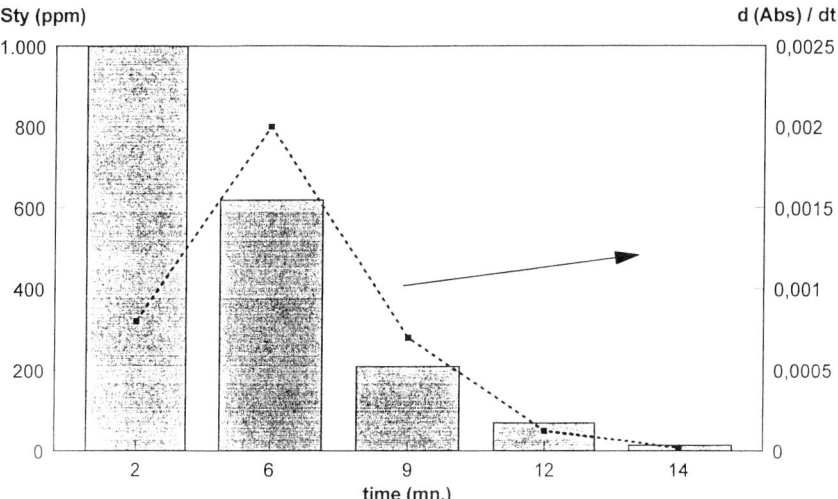

Fig. 6. Kinetics for styrene polymerization

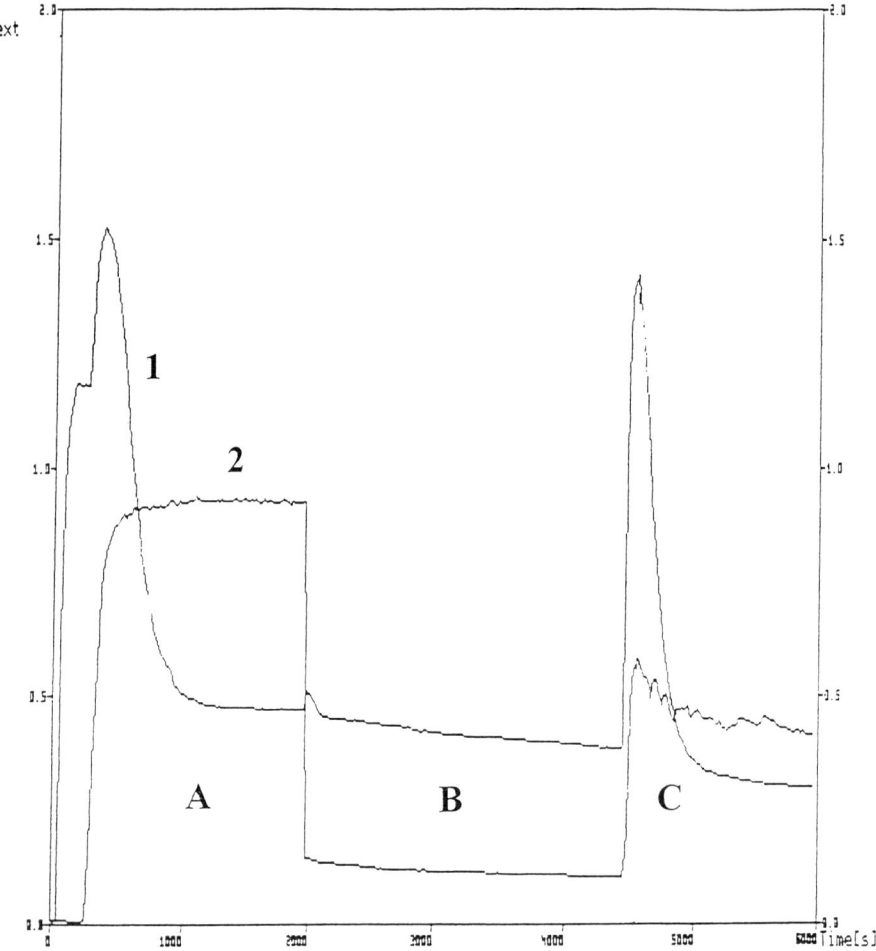

Fig. 7
Time drive UV monitoring: SBS sequential copolymerization. Styrene polymerization (zone A), butadiene polymerization (zone B), styrene polymerization (zone C). Curves 1 and 2 recorded at 290 nm and 330 nm respectively.

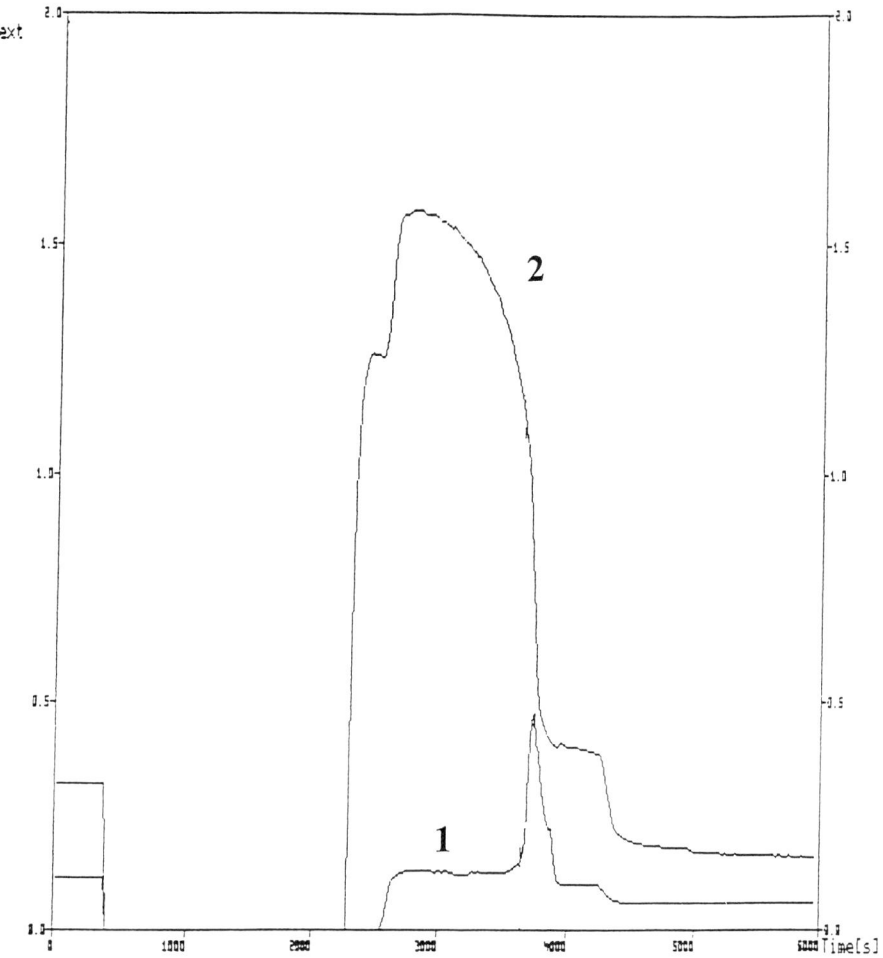

Fig. 8
Time drive UV monitoring: SBR copolymerization (10% Styrene as a block).
Curves 1 and 2 recorded at 290 nm and 330 nm respectively.

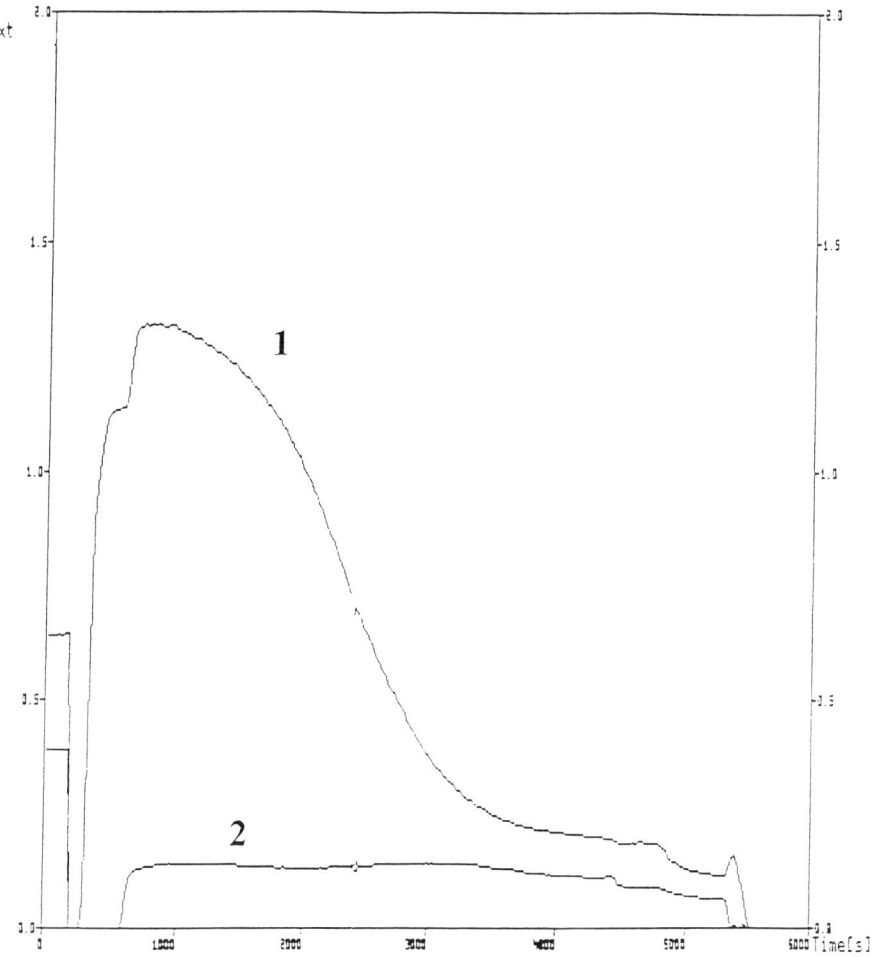

Fig. 9
Time drive UV monitoring: SBR copolymerization (< 1% Styrene as a block).
Curves 1 and 2 recorded at 290 nm and 330 nm respectively.

Also interesting is the monitoring of active chain ends at total conversion. When an efficient randomizing modifier is used, it appears that dienyllithium prevails, as compared to the preference of styryllithium in the unmodified case (Figures 8 and 9). Indeed after some time also in this case the terminal chain end is converted to dienyllithium[8]. This is probably due to the capping action of residual butadiene in the gas phase of reactor.

Some encountered problems are the decrease of precision in the monitoring of chain end concentration at high temperature, where, in absence of monomers, deactivation reaction takes place leading to the formation of new cromophores[6] (i.e. trienyllithium). The superimposition of these cromophores impedes measurement of active chain lithium concentration.

Also the monitoring of coupling efficiency is precluded when these cromophores are present, and the appearing in GPC of multimodal molecular weight distribution can not be anticipated from UV monitoring.

In respect of ruggedness of the system no problems of polymer build-up in the optical cell were encountered and monitoring in the pilot plant was successfully operated for more than 12 months: UV spectrophotometry is now being scaled-up on the industrial plant.

LITERATURE CITED

1) (a) Szwarc M. in *Carbanions, Living Polymers and electron transfer Processes*, Interscience, N.Y., 1968;   (b) Bywater S. *Prog. Poly. Sci.*, **1994**, *19*, 287 and references cited herein.
2) Bean,A.R.; Park, B.; Fraga, D.W.;Beach, R.; Mann, R.H. U.S. Patent 3,553,295 (1971).
3) Hsieh, H.L. *Rubber Chem. Technol* **1976**, *45*, 1305.
4) G. Holden in *Thermoplastic Elastomers, A Comprehensive Review*, N.R. Legge, G. Holden, and H.E. Schroder, Eds., Hanser Publisher, Munich/Vienna/NewYork, 1987.
5) Montalti, A.; Pattuelli, E.; Zazzetta, A.: and Viola, G.T. *Journ. of Chromatog. A* **1994**, *117*,665.
r6) Viola, G.T.;Bortolotti, M.: and Zazzetta A. *Journ. of Polym. Sci.Chem Ed.***1996**, *1*,17.

Chapter 5

Polymerization of Butadiene with Novel *tert*-Amino Group Initiator

Toshihiro Tadaki, Iwakazu Hattori, and Fumio Tsutsumi

Elastomers Laboratory, Yokkaichi Research Laboratories, Japan Synthetic Rubber Company, Ltd., 100 Kawajiri-Cho, Yokkaichi, Mie, 510, Japan

A study was made on a new initiator N,N-dimethyl-aminobenzyllithium (TD-Li) with the purpose of improving functionality of the initial chain end in anionic polymerization. After reacting N,N-dimethyltoluidine (TD) with n-butyllithium (BuLi), polymerization of 1,3-butadiene was conducted in several solvents.
The "Functionality Introduction Ratio" of polybutadiene was influenced by the polarity of the solvents. That is, the higher the polarity, the higher the "Functionality Introduction Ratio." On the basis of on results of a model reaction, in high polarity solvents the reaction between TD and BuLi advanced efficiently, and the polymerization of butadiene advanced by the benzyllithium type initiator (TD-Li). On the other hand, in a solvent having low polarity, because TD-Li formed partially, the polymerization was initiated by the mixed system of TD-Li and BuLi. Furthermore, in the case where TMEDA was added to the reaction system, the "Functionality Introduction Ratio" was improved.

In recent years, lower rolling resistance has become one of the items most talked about in the tire industry. The reason is that "rolling resistance" is a characteristic which is closely related to the fuel economy of cars. Not only original equipment tires which demand good fuel economy, but also replacement tires which boast high grip performance demand better fuel economy these days.
Rolling resistance is primarily related to the hysteresis in the tire tread compound. Thus, the carbon black filled vulcanizates having low hysteresis are in demand as materials for the tire tread, in order to lower the rolling resistance.
As rubber for low hysteresis compositions, various solution polymerized rubbers (S-SBR) of which the chain ends have been chemically modified, have been developed.
They were obtained by the reaction between the living reactive chain ends and some modifiers such as tin compound (*1-2*), isocyanate compound (*3*) or 4,4'-bis-(diethylamino)-benzophenone (*4*).
It is believed that hysteresis characteristics are improved in carbon black vulcanizates using the above SBR because of the following action. That is, the interaction between

the modified molecular ends and the carbon black heightens, and as the numbers of free ends decrease, the dispersion of carbon black is improved.

However, the modifying points of such SBR are only the ends which have completed polymerization. In the case of anionic polymerization, the polymer obtained always has an initiator fragment (residue) at the initiating point. If a functional group can be introduced to the polymerization initiator itself, it will be possible to synthesize "the difunctional polymer" which is expected to more strongly interact with carbon black.

Even up to now, several studies have been made on initiators having functional groups. For instance, alkali metal amide initiators (5-7) or a tributylstannyllithium initiator (8) correspond to this.

In this report, a study is made on a new initiator "N,N-dimethylaminobenzyllithium (TD-Li)" that contains aromatic tertiary amino group, and discussion is made on the experimental results of 1,3-butadiene polymerization based on such an initiator.

The reason we introduced aromatic tertiary amino group as substitution group in the initiator is that we thought the said functional group had strong interaction with carbonblack.

Actually, it is reported that the carbonblack vulcanizates using polymers modified with the aforementioned 4,4'-bis-(diethylamino)-benzophenone bas shown improved hysteresis characteristics.

Therefore, if the initiator containing aromatic tertiary amino group (which is used in this study) effectively initiates polymerization, it is anticipated that the hysteresis loss of the polymers obtained will be lowered significantly.

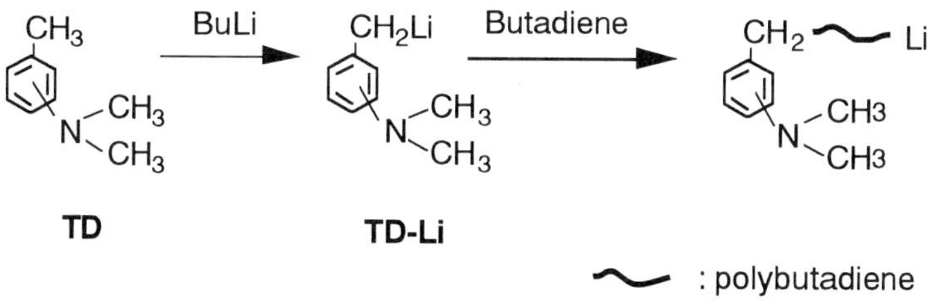

Scheme 1

EXPERIMENTAL

Polymerization of 1,3-Butadiene Using a N,N-Dimethyltoluidine/n-Butyllithium as Initiator. All polymerization reactions were done under a nitrogen atmosphere inside of pressure resistant bottles capped with crown-caps. To various test solutions in which N,N dimethyltoluidine (TD; 39.1 mmol) was dissolved, cyclohexane solution of n-butyllithium (BuLi; 39.1 mmol) was added, and reaction was conducted for specified time in a hot water bath set at 60°C. Twenty-five grams (462 mmol) of 1,3-butadiene was added to the reaction solvent, and reaction was conducted for 60 minutes at 60°C. Subsequently, 2-ethylhexanol (78.2 mmol) was added and the polymerization was stopped. After separating the polymer by making it precipitate in methanol, it was dried under reduced pressure in an oven set at 50°C.

Characterization of the Polymers. The microstructure of the polymer was obtained by the following two methods. That is the infrared spectroscopy (calculated by Morero's Method, (9)) and the calculation of peak area ratio of ^{1}H-NMR.

The molecular weight of the polymer was obtained by gel permeation chromatography (GPC) based on the standard polybutadiene calibration curve. The measurement of GPC was carried out with RI and UV (254 nm) dual detectors.
As it is well known, polybutadiene has no aromatic protons.
Thus, no peaks are observed in the vicinity of d = 6.7~7.1 ppm, and no UV absorption should exist at 254 nm.

Model Experiments for Verifying the Formation Rate of the Initiator Consisting of N,N-Dimethyltoluidine/n-Butyllithium. Similar to the polymerization experiments described above, the reaction was conducted under nitrogen atmosphere inside of pressure resistant bottles capped with crown-caps. To TD (9.37 mmol) dissolved in various test solvents, cyclohexane solution of BuLi (9.37 mmol) was added, and reaction was conducted for a specified time in a hot water bath set to 60°C. By adding a cyclohexane solution of benzylchloride (9.37 mmol), the reaction was quenched. When the reaction solution was left as it was for a while, white precipitation of lithium chloride formed. After washing the reaction solution sufficiently, the white precipitate was dissolved in water and removed. The water layer used for washing was extracted three times with diethyl ether. Then the whole organic layer was washed with saturated brine, and dried on anhydrous magnesium sulfate. The reaction solvent and diethyl ether were removed by reduced pressure distillation, and the reaction product was obtained.
The characterization of the reaction product obtained was done by ^{1}H-NMR measurement and high performance liquid chromatography using THF as the carrier.

RESULTS AND DISCUSSION

Polymerizations of 1,3-Butadiene in Several Solvents. First of all, the preparation of initiators was done (reaction of N,N-dimethyl-o-toluidine and n-butyllithium) and polymerization of butadiene was conducted in various solvents having different polarity.
The solvents used in the study were cyclohexane (CHX: Dielectric Constant = 2.02), diethyl ether (Et$_2$O: ε = 4.34), and tetrahydrofuran (THF: ε = 7.58).
All reactions were carried out at 60°C as described above. The results are shown in Table I.

The all polymerization reaction advanced in a uniform system where there were no precipitation or deposition.
In all cases, high conversions of 75% or above were obtained for the polybutadiene.
In Figure 1, the ^{1}H-NMR spectra of polybutadiene obtained in THF by using "TD/BuLi Initiator," and polybutadiene obtained in the same solution by using only "BuLi Initiator" are shown. The peak close to 7.0 ppm is attributable to aromatic proton, and the sharp peak close to 2.6 ppm conforms to the aromatic dimethylamino group in TD.

The IR spectra of polybutadiene obtained in THF by using only "BuLi Initiator," and polybutadiene obtained in the same solvent by using "TD/BuLi Initiator," are shown in Figures 2 and 3. Polybutadiene using "TD/BuLi Initiator" showed an absorption based on tertiary amino group close to 1300 cm^{-1} and an absorption based on aromatic group close to 750 cm^{-1}. (Figure 3)
In Figure 4, GPC elution curves for polybutadiene obtained in THF by using "TD/BuLi Initiator" and polybutadiene obtained in the same solvent by using only

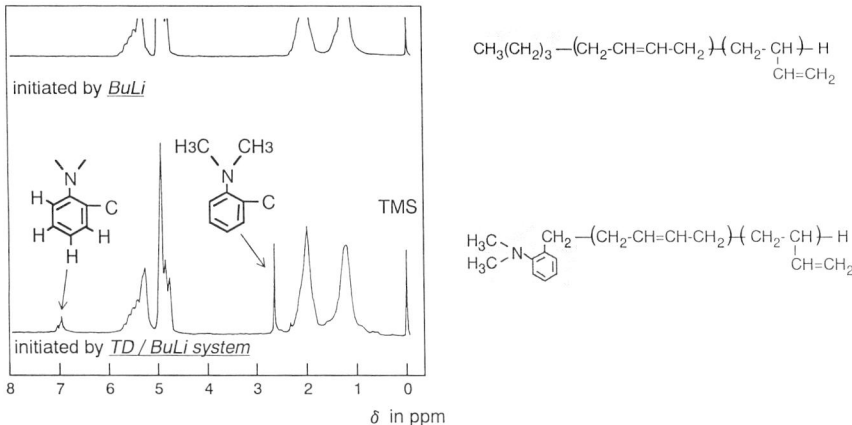

Figure 1. ^{1}H-NMR spectra of polybutadienes.

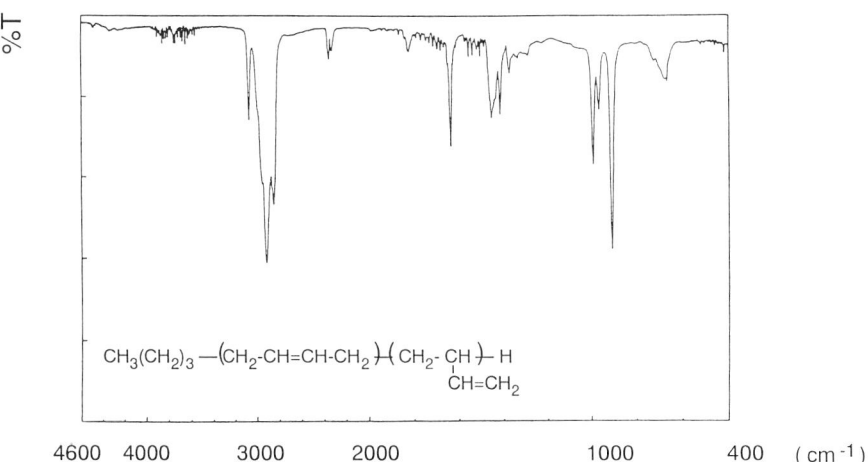

Figure 2. Infrared absorption spectrum of polybutadiene initiated by BuLi Initiator.

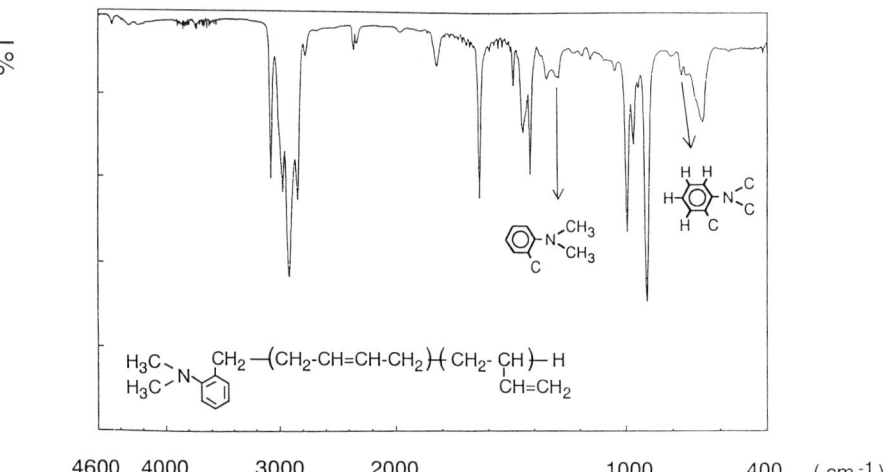

Figure 3. Infrared absorption spectrum of polybutadiene initiated by TD / BuLi Initiator.

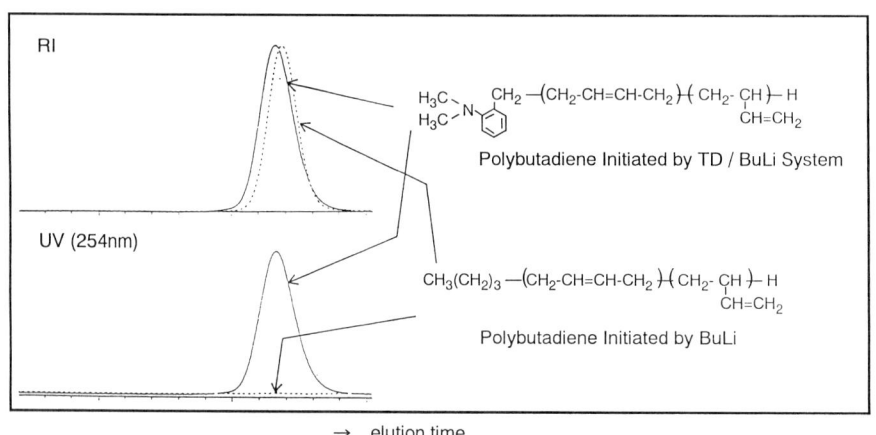

Figure 4. GPC charts of polybutadienes.

"BuLi Initiator" are shown. In the case of the former, GPC molecular weight can be obtained from UV absorption at 254 nm. On the other hand, in the case of the latter, no UV absorption existed at 254 nm.

Table I. Polymerization of 1,3-butadiene using TD/BuLi initiator in several solvents

Initiator Solvent	(a) Electrical Property ε (b)	Conv'n (%)	(c) Microstructures			M_w / M_n	(d) Functionality Introduction Rate (%)
			cis (%)	trans (%)	vinyl (%)		
(1) TD / BuLi							
CHX	2.02	100	30	52	19	1.2	0
Et$_2$O	4.34	100	19	29	52	1.1	49
THF	7.58	79	8	9	83	1.1	91
(2) BuLi							
CHX	2.02	100	31	52	18	1.3	—
THF	7.58	100	5	6	88	1.2	—

(a) ; CHX = cyclohexane, ET$_2$O = diethyleter, THF = tetrahydrofuran
     THF = tetrahydrofuran
(b) ; the dielectric constant
(c) ; measured by Infrared spectroscopy calculated by Morero's Method (6)
(d) ; values obtained by ^{1}H-NMR of polybutadiene

NMR spectra of Figure 1 and IR spectra of Figures 2 and 3 suggest that in the case of polybutadiene obtained by using TD/BuLi initiator in THF, aromatic tertiary amino groups are introduced into the chain. Furthermore, in the GPC elution curves of Figure 4, the fact that UV absorption existing at 254 nm in case of the polybutadiene obtained by using "TD/BuLi Initiator," also supports this result. Based on the results of the above spectra of polybutadiene obtained in THF, we conclude as follows: As shown in Scheme 1, benzyllithium type reaction product is formed, and this acts as the initiator, and the polymerization advances.

As a result of using several solvents, we noted that the reaction of TD and BuLi was affected by the polarity of the solvents. In other word, in THF solvent which has high polarity, aromatic tertiary amino groups were introduced into more than 90% of the polybutadiene obtained.

However, in the case of diethyl ether, the amount of aromatic tertiary amino group introduced into the polymer was about 50%, and in the case cyclohexane was used as the solvent, that was nearly zero.

Moreover, the microstructure of polybutadiene obtained by using TD/BuLi as the initiator has changed according to the polarity of the solvents. The molecular weight distributions were narrow with an Mw/Mn value of 1.1~1.2. They were approximately the same as the polybutadiene obtained by using only BuLi initiator.

Polymerization of 1,3-Butadiene in a Mixed Solvent of Cyclohexane and Tetrahydrofuran. Next, in order to clarify the effect of solvent polarity on the "Functionality Introduction Ratio," polymerization of polybutadiene was conducted in a mixed solvent of CHX and THF using TD/BuLi as the initiator.
Two cases were adopted for the reaction time of TD and BuLi, namely, 60 minutes and 10 minutes. Polymerization time for polybutadiene which followed was 60 minutes. All reactions were done at 60°C.
The results are shown in Table II.

In Figure 5, the relation between the CHX/THF ratio of the mixed solvent and the "Functionality Introduction Ratio" (values obtained by ^{1}H-NMR) are shown. As the THF ratio increases (that is, as the polarity becomes higher) the "Functionality Introduction Ratio" of polymer increases. Furthermore, even when the same solvents are used, if the reaction time of TD and BuLi is shortened to 10 minutes, the "Functionality Introduction Ratio" decreases.

Furthermore, when aromatic tertiary amino groups are introduced into the polybutadiene, UV absorption appears. Thus, by measuring the UV absorption intensity, it should be possible to judge the functionality. In Figure 6, the relation between "CHX/THF ratio of the mixed solvent" and the "Ratio of UV Intensity and RI Intensity of GPC (compensated by the molecular weight)" are shown. This figure shows a similar trend to the preceding Figure 5.

Table II. Polymerization of 1,3-butadiene using TD/BuLi initiator in mixed solvents of cyclohexane and tetrahydrofuran

Solvent CHX / THF (wt / wt)	Initiator Aging Time (min.)	Microstructures (a)		M_w / M_n	Functionality Introduction Ratio (b) (%)
		1,2- (%)	1,4- (%)		
100 / 0	60	19	81	1.2	0
99 / 1	60	47	53	1.1	11
95 / 5	60	69	31	1.2	15
88 / 12	60	69	31	1.1	35
75 / 25	60	81	19	1.1	93
69 / 31	60	82	18	1.1	90
39 / 61	60	83	17	1.1	92
100 / 0	10	22	78	1.2	0
75 / 25	10	76	24	1.1	42
64 / 36	10	83	17	1.1	43
53 / 47	10	86	14	1.1	89
41 / 59	10	87	13	1.1	100

(a) ; measured by ^{1}H-NMR
(b) ; values obtained by ^{1}H-NMR of polybutadiene

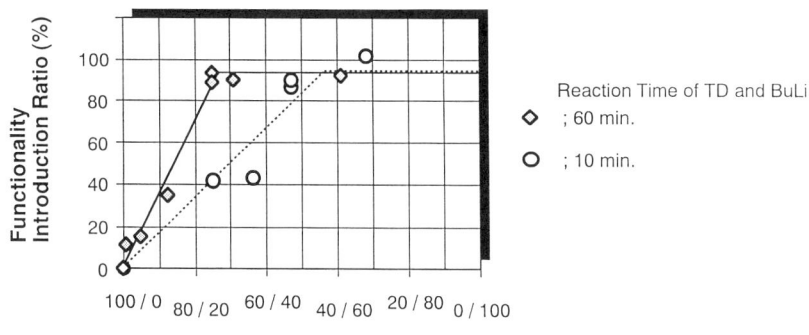

Figure 5. Relationship between "Functionality Introduction Ratio" and polarity of solvent.

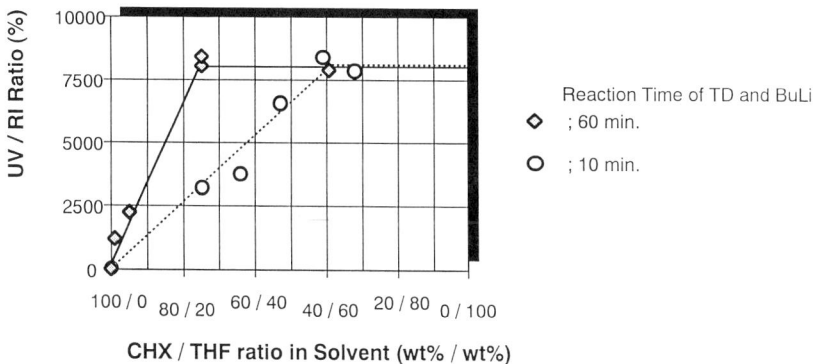

Figure 6. Relationship between UV Intensity of polymer and polarity of solvent.

Analysis of Reaction Product of TD and BuLi. The first reaction shown in Scheme 1 (where TD-Li formed from TD and BuLi) is influenced by the solvent polarity. Thus, it may be hypothesized that the "Functionality Introduction Ratio" of a polymer varies with the polarity of the solvent.

Thus, a slight excess of benzylchloride was added to the solvent in which TD and BuLi were made to react. Then the reaction product was analyzed by ^{1}H-NMR. If benzyllithium type compounds (TD-Li) are formed, upon reaction with benzylchloride N,N-dimethylamino-1,2-diphenylethane (I) should be obtained. On the other hand, if a hydrogen of the methyl group bound to the nitrogen atom is replaced by lithium, compound (II) should be formed by the reaction of benzylchloride. In addition, from the reaction of BuLi and benzylchloride, pentylbenzene (III) is obtained. (Scheme II)

In Figure 7, the ^{1}H-NMR spectrum of TD (N,N-Dimethyl-o-toluidine) which is the raw material of the initiator is shown. The peak at 2.6 ppm is attributable to the dimethylamino group, and the peak at 2.3 ppm is attributable to a methyl group bound to the aromatic ring. The peak area ratios are 2:1.

The ^{1}H-NMR spectrum of the reaction products obtained by using the solvent system of CHX/THF = 100/0 is shown in Figures 8. The peak at 2.6 ppm consists of two kinds of protons. One of them is equivalent to the proton of dimethylamino group, and another is believed to be the proton of benzyl position of (III). The peak area ratio of dimethylamino group protons (at 2.6 ppm) and the peak of 2.3 ppm equivalent to the protons of the methyl group bound to the aromatics was 2:1, similar to TD. Furthermore, peaks believed to be attributable to an n-butyl group were observed near 0.5~2.0 ppm.

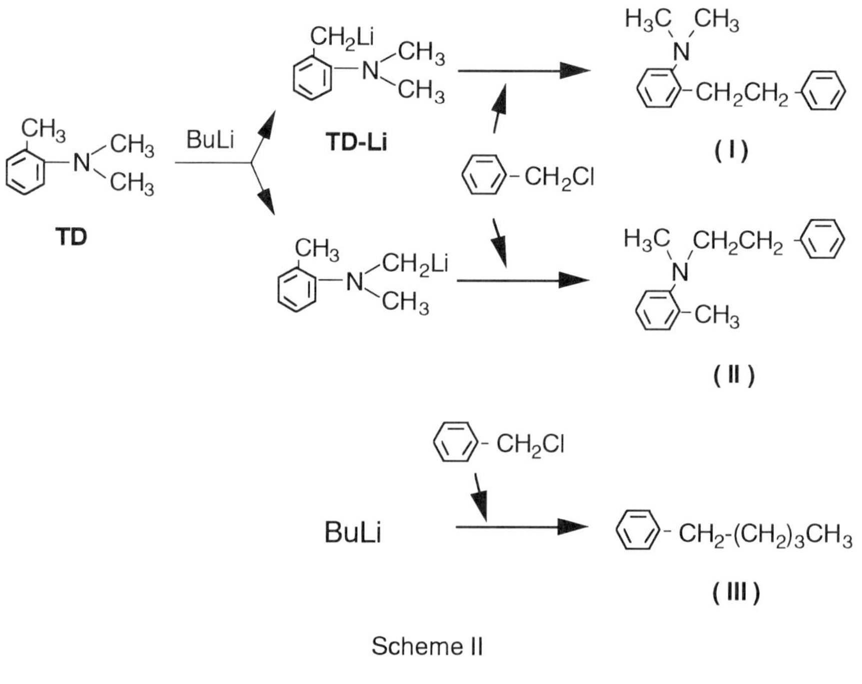

Scheme II

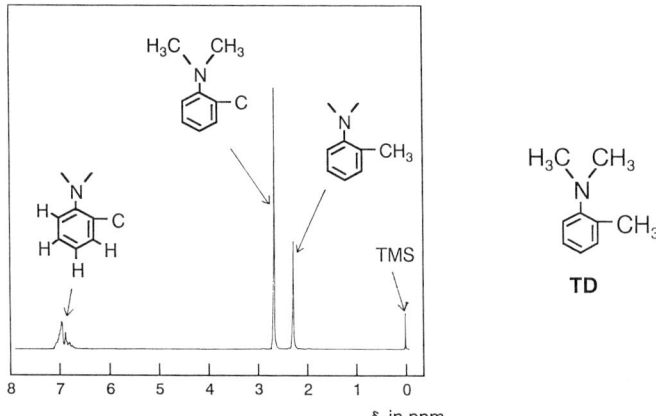

Figure 7. ^{1}H-NMR spectrum of N,N-Dimethyl-o-toluidine.

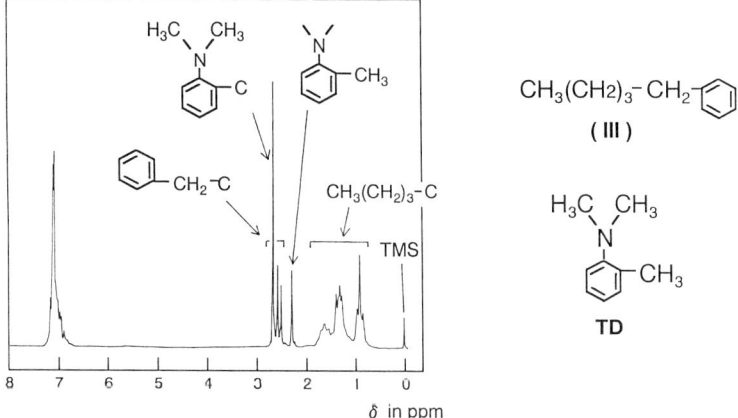

Figure 8. ^{1}H-NMR spectrum of the reaction products obtained in CHX/THF = 100/0.

The ^{1}H-NMR spectrum of the reaction products obtained by using the mixed solvent system of CHX/THF = 75/25 is shown in Figure 9. Contrary to the above, the peak corresponding to the methyl group bound to the aromatics had decreased. (The peak area ratio for the peak at 2.6 ppm and 2.3 ppm was 5.15:1.) A new peak which is believed to be attributable to protons of the benzyl position appeared at 2.9 ~3.0 ppm. In addition, peaks near 0.5 ~ 2.0 ppm became very small. Furthermore, the peak area ratio of the peak of dimethylamino group (at 2.6 ppm) and the peak of aromatic protons was almost 2:3 (6:9). This peak area ratio is in good agreement with the corresponding protons in the structure of (I).

On the basis of the results of the above spectra, the authors conclude as follows:
In the solvent system of CHX/THF =100/0 the reaction of TD and BuLi hardly advanced (i.e., TD remained as it is). The only reaction product with benzylchloride was (III). On the other hand, since in the case of mixed solvent system of CHX/THF = 75/25, the methyl group bound to the aromatics of TD decreased, and protons of benzyl position increased, it was believed that the reaction of TD and BuLi advanced, and TD-Li was formed. In addition, the lithiation reaction of a methyl group bound to nitrogen did not occur in the reaction of TD and BuLi.

Next, measurements were made with a high performance liquid chromatography (HPLC) on the reaction products obtained by using mixed solvent system of CHX/THF at varying ratios.

In Fig. 10 the HPLC curve for the reaction product of TD and BuLi in a solvent system of CHX/THF = 88/12 is shown.

Beside the solvents (CHX) and the peak of TD which is unreacted raw material, two additional peaks were confirmed in Figure. From the respective molecular weight, the peaks are believed to be (I) and (III). The formation ratios were calculated from the peak areas, and the results are shown in Table III.

From Table III, it was found that the formation ratio of (I) and the "Functionality Introduction Ratio" obtained by polymerization of butadiene, matched well.

From the above experiments, the following was made clear. In the case of a mixed solvent of CHX/THF having low concentration of THF, the TD-Li forms partially, so the polymerization of butadiene begins with a mixed system of TD-Li and BuLi. As the concentration of THF rises, the reaction product of TD-Li increases, and the "Functionality Introduction Ratio" of polymer improves.

Addition effect of N,N,N,'N'-Tetramethylethylenediamine. Based on the above results, in order to introduce aromatic tertiary amino groups into the polybutadiene efficiently, the authors found that it was effective if reaction was conducted in a solvent having high polarity. However, the microstructure of polybutadiene thus obtained is also influenced by the polarity of the solvent. The higher the polarity of the solvent, the more the 1,2 bonds of the polybutadiene.

Polybutadiene having high 1,2 content has high glass transition temperature (Tg), and it is believed that this polymer can be used effectively in tires demanding high grip performance. However, if we are to place importance on high abrasion resistance, polybutadiene having low Tg and low 1,2 bond contents is demanded. In particular, products with low 1,2 bond content can be effectively polymerized in solvents having low polarity.

In order to improve the "Functionality Introduction Ratio" of polymer, the efficient promotion of the first reaction of the mechanism shown in Scheme I (Benzyl hydrogen of TD being replaced by lithium) becomes the key.

With the aim of improving the reaction efficiency of TD and BuLi in a low THF concentration, a study was made on the addition of N,N,N,'N'-Tetramethylethylenediamine (TMEDA).

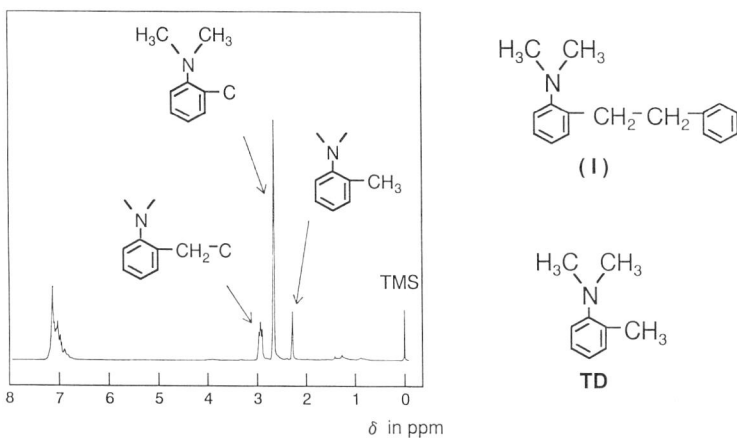

Figure 9. ^{1}H-NMR spectrum of the reaction products obtained in CHX/THF =75/25.

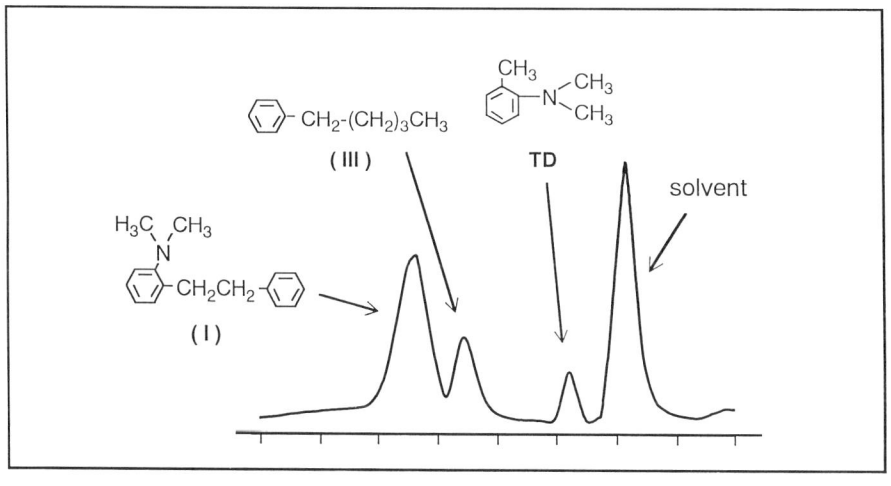

Figure 10. HPLC curve of the reaction products obtained in CHX/THF = 88/12.

Table III. Formation ratio of TD-Li and BuLi in mixed solvents of cyclohexane and tetrahydrofuran

Solvent CHX / THF (wt / wt)	(a) TD-Li / (TD-Li + BuLi) (molar ratio, %)	(b) Functionality Introduction Ratio (%)
100 / 0	0	0
95 / 5	21	15
88 / 12	37	35
75 / 25	84	93
69 / 31	88	90
39 / 61	89	92
18 / 82	88	91

(a) ; calculated from the peak areas of high performance liquid chromatography on the reaction products of TD/BuLi and benzylchloride

(b) ; values obtained by ^{1}H-NMR of polybutadiene

In Figure 11, the relation between the CHX/THF ratio of a solvent and "Functionality Introduction Ratio (value obtained by ^{1}H-NMR)" when the addition amount of TMEDA is varied, is shown.
As we anticipated, by adding TMEDA, the functionality improved significantly. Furthermore, it was found that the more TMEDA is added, the greater the improvement of functionality.
In Fig. 12 the relation between the 1,2 bond content in the polymer and the "Functionality Introduction Ratio" (value obtained by ^{1}H-NMR) is shown.
In a system where TMEDA is added, we found that polybutadiene having aromatic tertiary amino groups efficiently introduced can be obtained at comparatively low 1,2 bond content.

Conclusion

A study was made on a new initiator N,N-dimethylaminobenzyllithium (TD-Li) with the purpose of improving functionality of the initial chain end in anionic polymerization. Experiments on polymerization of 1,3-butadiene using this initiator were made, and the following results were obtained.

1). After reacting TD with BuLi in THF, when polymerization of 1,3-butadiene was conducted, polymers were obtained at a high conversion rate. From ^{1}H-NMR, Infrared Absorption and GPC analyses, it was found that the polymer contained an aromatic tertiary amino group which is the residue of the TD-Li initiator.
2). The "Functionality Introduction Ratio" of the polybutadiene obtained by using the initiator TD/BuLi, was influenced by the polarity of the solvents used in the polymerization. The higher the polarity, the higher the "Functionality Introduction Ratio."
3). Solvent effect would be dominant in the step of the initiator synthesis.
Namely, based on the results of the model reactions, in high polarity solvents the

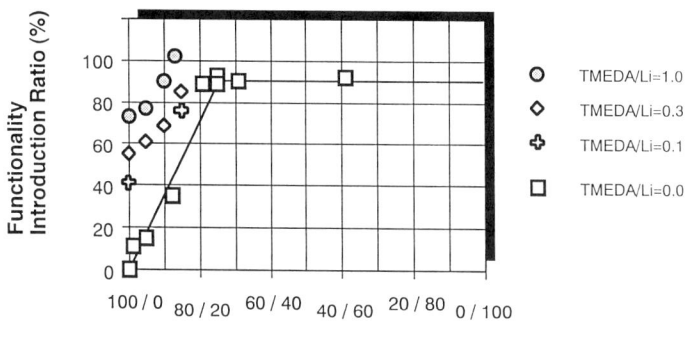

Figure 11. Relationship between "Functionality Introduction Ratio" and polarity of solvent.

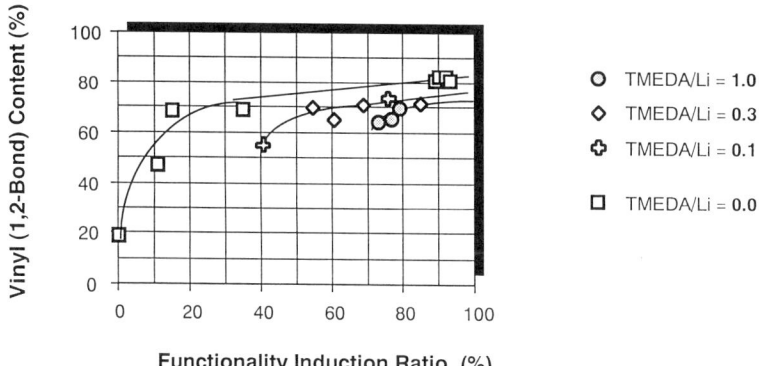

Figure 12. Relation between "Functionality Introduction Ratio" and vinyl content of polymer.

reaction between TD and BuLi succeeded efficiently, and a benzyllithium type compound (TD-Li) was formed. The said compound acted as the initiator, and the polymerization of butadiene proceeded.

On the other hand, in a solvent having low polarity where TD-Li formation was incomplete, the polymerization of butadiene was initiated by the mixed system of TD-Li and BuLi.

4). In the case where TMEDA is added to the reaction system, the "Functionality Introduction Ratio" is improved. Consequently, we found that polybutadiene having aromatic tertiary amino groups efficiently introduced can be obtained at comparatively low 1,2 bond content

REFERENCES

1. N. Ohshima, F. Tsutsumi and M. Sakakibara, *IRC Kyoto, Oct.* **1985**, 16A04.
2. F. Tsutsumi, M. Sakakibara and N. Ohshima, *Rubber Chem. Technol.* **1990**, *63*, 8.
3. T. Tadaki, F. Tsutsumi, M. Sakakibara and I. Hattori, *146th. Meeting of Rubber Div. ACS, Pittsburgh* **1994**, paper No. 64.
4. N. Nagata, T. Kobatake, H. Watanabe, A. Ueda and A. Yoshioka, *Rubber Chem. Technol.* **1987**, *60*, 837.
4. N. I. Nikolayev, N. M. Geller, B. A. Dolgoplosk, V. N. Zgonnik and V. A. Kropachev, *Polym. Sci. USSR* **1963**, *4*,1529.
5. P. A. Vinogradov and N. N. Basseva, *Polym. Sci. USSR* **1963**, *4*,1568.
6. A. C. Angood, S. A. Hurley and P. J. T. Tait, *J. Polym. Sci., Polym. Chem. Ed.* **1973**, *11*, 2777.
7. T. W. Bethea, W. L. Hergenrother, F. J. Clark and S. B. Sarkar, "Novel Tin-containing Elastomers Having Reduced Hysteresis Properties" IISRP, **1994**.
8. D. Morero, A. Santambrogio, L. Porri and F. Ciampelli, *Chim. Ind. (Milan)*, **1959**,*41 (8)*, 758.

Chapter 6

Anionic Polymerization of Dienes Using Homogeneous Lithium Amide (N-Li) Initiators, and Determination of Polymer-Bound Amines

D. F. Lawson[1], D. R. Brumbaugh[1], M. L. Stayer[1], J. R. Schreffler[1],
T. A. Antkowiak[1], D. Saffles[1], K. Morita[2], Y. Ozawa[2], and S. Nakayama[2]

[1]Bridgestone/Firestone Research, Inc., 1200 Firestone Parkway, Akron, OH 44317
[2]Bridgestone Corporation Technical Center, 3-1-1 Ogawahigashi-cho, Kodaira-shi, Tokyo 187, Japan

Polymerizations of dienes have been achieved using N-Li dialkylamide initiators under homogeneous conditions in hexanes. The copolymerization of butadiene and styrene with the reagent N-Li pyrrolidinide·2THF is an example: the products showed narrower polydispersities than those previously described, the polymerizations were reproducible and they proceeded with reasonable concentrations of live ends for further reaction. A number of the products also showed some potential for use in low hysteresis rubber compositions. The approach can be used to synthesize polymers with both head and tail functionality for enhanced hysteresis reduction. Using ^{13}C NMR experiments with lower molecular weight polymers, it was determined that the amine is tethered to the chain through nitrogen. A UV-visible method was successfully used to estimate the concentration of amine ends in high molecular weight polymers. Evidence was found for base-catalyzed retroaddition of the amino group from the living polymer chain at elevated temperatures. The loss is stopped by neutralization, *via* quenching with ROH or end-linking with $SnCl_4$.

Lithium dihydrocarbon amides have been known for some time as alternative initiators for anionic (co)polymerization of dienes. *(1-6)* However, the limited product characterizations reported from polymerizations in hydrocarbon solvents are typified by low conversions and broad polydispersities. The preponderant theme in these reports is that N-Li dialkyl or cyclic amides are insoluble in aliphatic hydrocarbons, and that initiations using these reagents under these conditions are heterogeneous.

The challenge was then establishing conditions that ensure the homogeneity of a N-Li dialkyl or cyclic amide initiator in aliphatic hydrocarbon solvents, in order to overcome these problems. The purpose of our work was to synthesize polymers having reproducible narrow polydispersities, and active lithium terminals capable of further reaction, through the use of anionic polymerization from homogeneous N-Li hydrocarbon amide initiators in aliphatic hydrocarbon solvents. With the functionality incorporated into the initiator, a polymer may be made in which all chains are functionalized at the head. After functional termination, polymers with both head and tail functionalities are possible.

This is of interest to a tire company because in tire compounds the carbon black has a tendency to agglomerate in networks, which are a source of energy losses in the system. Functional polymers, through their end groups, can attach to reactive sites on the carbon black particles, providing a steric stabilization of the dispersion, and preventing re-agglomeration of the dispersed particles. Energy losses via the carbon black network are prevented, giving a measurable reduction in hysteresis, which translates into improved fuel economy. A functional elastomer can thus be a useful raw material for products such as energy-efficient power belts and mechanical goods, as well as fuel-efficient tires.

Experimental

Methods for polymerizations in bottles and small reactors have been described. *(7)* Organometallic reagents were prepared and handled using Schlenk-type procedures.*(8)* Polymeric products were characterized by 300 MHz NMR (Varian Gemini 300), DSC (TA 920), and SEC (standard Waters equipment, using a linear ultrastyragel column). Compounding and methods for evaluating compounded properties have been described elsewhere. *(7-10)* Chain-bound amine levels in polymers of Mn <60000 were estimated by ^{13}C NMR (≥20,000 transients). The amine content of these and higher MW polymers were also determined semiquantitatively at ppm concentrations by UV-visible analysis.

The general procedure described by Palit *(12)* was followed to determine polymer-bound amine. Several changes were made to improve the accuracy of the amine determinations. *(11)* Patent Blue VF (Aldrich Chemicals) was used in place of Disuphine Blue VN150 as the dye. The test solution was agitated vigorously at room temperature for 2.5 hr in order to attain equilibrium. For best separations, centrifugation was necessary, followed by careful pipetting to remove the aqueous phase (aqueous droplets affect the results). The optical density of the organic phase was measured at the same wavelength used in the calibration, about 628 nm. (Since the band is broad in this region, the precise λ_{max} may vary by 1-3 nm without significantly affecting the results.) The test solution should be made to a concentration that gives an absorbance in the range of 0.2-1.2 Absorbance units at this wavelength. The concentration of the amine is estimated from a calibration curve, obtained separately as the linear least squares fit through at least five experimental points in the same absorbance range, correcting for the absorbance of a chloroform blank made with the reagent in the absence of an amino compound. For calibration purposes the N-(1-dodecyl) or other suitable derivative of the identical or very similar amine was used as a reference compound. For example, the reference compound for pyrrolidino head groups was N-(n-dodecyl) pyrrolidine. Amine concentrations are determined as μmol/g. Reliable MW measurements are needed to estimate the polymer functionality.

If the structure of the reference compound is very different from that of the unknown, there may be a significant error in the concentration estimates, owing to differences in the equilibria of different amines in the formation of the dye complexes. UV-visible response is strongly influenced by the nature of the amine, including basicity and steric requirements. Tertiary aliphatic and cycloaliphatic amines gave the most intense responses, obeyed Beer's Law, and could be measured at the lowest concentrations. Primary amines did not follow Beer's Law. The method is appropriate for single amines whose structures are known, or mixtures of amines whose structures and ratios are known. The functionality is estimated by using a calibration curve developed for a reference compound having the same amine structure with a long-chain alkyl group substituent on the nitrogen atom.

Results and Discussion

Polymerizations of 1,3-butadiene and styrene run using N-Li initiators generated from secondary amines as described in the literature *(1-6)* gave erratic results in hexanes.*(8)* We had previously observed that adducts of alkyllithiums with aromatic aldimines formed N-Li type initiators giving consistent results in diene/styrene (co)polymerizations in hexanes. Other N-Li reagents based on simpler amines were sought that formed homogeneous solutions in aliphatic hydrocarbons. *(7)* Three approaches were taken to generate homogeneous initiator reagents: a.) selection of N-Li reagents inherently soluble in the polymerization solvent; b.) the formation of a soluble reagent, either as a complex with near-equivalent amounts of a donor reagent, or by chain-extension of the N-Li reagent immediately upon generation, by oligomerizing with a diene or olefin, resulting in a N-containing C-Li reagent; c.) generation of the initiator in-situ in the presence of monomer. The last method is described in a companion paper. *(13)*

Table I lists some of the amines of this study and the effect of solubilization with two equivalents of tetrahydrofuran (THF). These results showed that most of the N-Li reagents were insoluble at greater than 0.2M in acyclic hexanes. In a few cases, solubility improved in cyclohexane. However, all of the N-Li compounds were solubilized in hexanes when complexed with two equivalents of THF. The abilities of the THF-solubilized reagents to initiate diene (co)polymerizations varied. Initiation was apparently hindered by steric effects: those reagents having branching at the carbon adjacent to N-Li either produced polymers having broad polydispersities, or failed to initiate polymerization within 16 hrs at 50°C (e.g., LDA). LDIBA (lithium diisobutylamide) was soluble in hexane and behaved like a good initiator, but the products did not exhibit reduced hysteresis. Only the cyclic reagents, with the exception of LPIP, gave elastomeric products with reduced hysteresis in carbon black-filled compounds.

The behavior of these reagents is illustrated in the remainder of this paper by the example of the solubilized N-lithiopyrrolidinide·2THF (LPY·2THF) system. LPY is sparingly soluble, at best, when generated in hexanes or cyclohexane. However, when generated in hexanes in the presence of two equivalents of THF, reagent solutions of ~1M are produced that are capable of initiating polymerizations, and are stable for weeks at room temperature (see Scheme I). LPY·2THF was used to initiate polymerization at 45°C, with TMEDA/Li (molar) = 0.5/1,

Table I. Summary of solubilities and polymerization results of selected N-Li dialkylamides.

Structure	Abbr.	Solubility			Initiation ability Mw/Mn	Hysteresis reduction effect
		Hexanes	c-Hex	THF(2eq.)/Hex		
n-BuLi		+	+	+	<1.1	-
(structure)	LDA	+	+	+	no pzn	-
(structure)	LNPA	-	-	+	1.2	-
(structure)	LDIBA	+	+	+	1.1	-
(structure)	LBPA	-	-	+	1.8	-
(structure)	LPY	-	-	+	1.1	+
(structure)	LPIP	-	-	+	<1.1	-
(structure)	LHMI	-	+	+	1.1	+
(structure)	LTHMI	-	-	+	1.1	+
(structure)	LHPMI	-	+	+	1.1	+
(structure)	LDDMI	-	-	+	1.2	+
(structure)	LTABO	+	+	+	1.5	+

following a procedure described elsewhere, *(8)* to produce a medium-vinyl SBR in 99% conversion (see Table II). The product had M_n = 130000, with a relatively narrow polydispersity, M_w/M_n = 1.14. The composition contained 20% styrene, as charged, having 53% of the butadiene content in the 1,2-addition mode, to yield an elastomer with T_g = -38°C (onset temperature). The polymer cement was still substantially living when the polymerization was complete, evidenced by the fact that upon end-linking with $SnCl_4$ the products obtained had 50-65% of their mass at

Scheme I. Generation of LPY·2THF

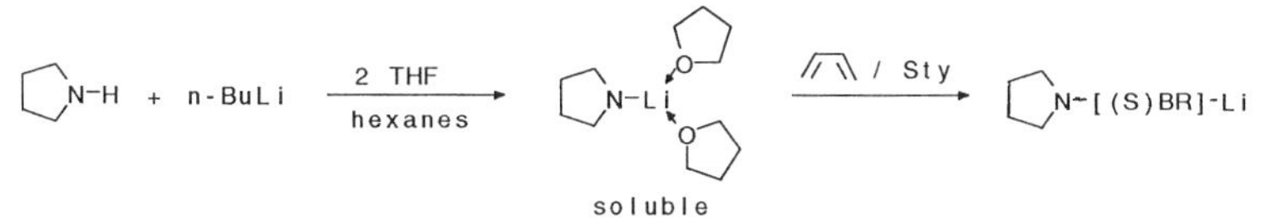

Table II. Summary of characterizations of LPY·2THF-initiated SBR, 45°C, in mixed hexanes (0.98 meq N-Li phgm, 0.35TMEDA/Li, 18.5% solids).

$M_n \times 10^{-3}$	130	242
Polydispersity	1.14	1.79
% vinyl	53	--
%styrene	20	--
T_g (DSC)	-38°C	--
ML/1+4/100, raw / compound	29 / 82	78 / 103
Terminator	ROH	$SnCl_4$

higher molecular weight than the initial polymer. Figure 1 shows a model SEC trace of a polymer end-linked with $SnCl_4$, compared with the corresponding base polymer quenched before end-linking. The amount of end-linking is estimated as the relative area under the high-molecular weight portion of the curve.

The polymer was mixed with carbon black in a typical formulation (see Table III) and cured. The physical properties of the cured compound are listed in Table IV. They showed a 40-50% reduction in hysteresis, as measured by tan δ at 50°C of 0.080, compared to a BuLi-initiated SBR of the same molecular weight. The uncured compound had a bound rubber content of 31% (measured as the weight percentage of rubber content in a carbon black-filled compound remaining insoluble in toluene after two days at room temperature under quiescent conditions), about 50% higher than that of the BuLi-initiated SBR. The hysteresis was further reduced, and the amount of bound rubber markedly increased to 42%, by end-linking with $SnCl_4$. The results of several additional LPY·2THF-initiated polymerizations and functional terminations by $SnCl_4$ coupling or R_3SnCl end-capping are summarized in Table V. The polymerizations continued to show relatively narrow polydispersities (e.g., base polymer examples A, C, and E) and viability of the living C-Li (e.g., end-capped examples B and D). The tan δ results provided additional evidence for reduced hysteresis physical properties in the cured, carbon black-filled compounds prepared according to the recipe in Table III. Figure 2 compares the hysteresis results for polymers initiated with LPY·2THF against those of a series of SBRs which contained

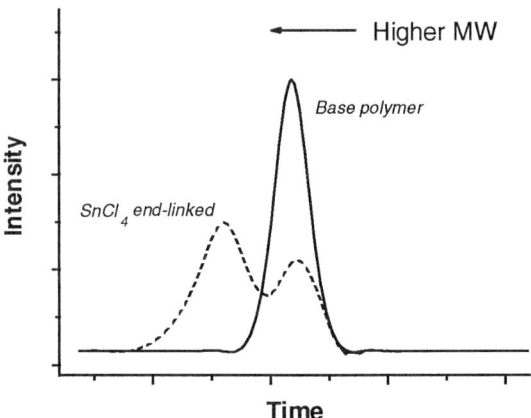

Figure 1. Model SEC traces of polymers made with and without end-linking by SnCl$_4$.

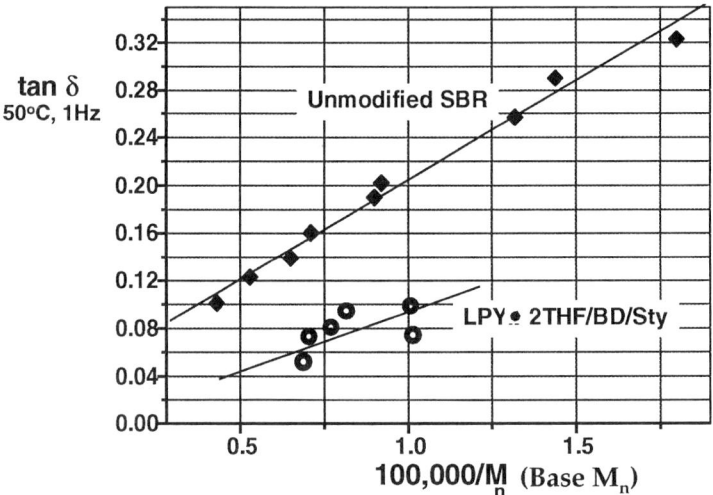

Figure 2. Comparison of hysteresis of cured, compounded LPY·2THF-initiated SBR with unmodified SBR, as a function of molecular weight of the uncoupled base polymers.

Table III. Experimental Test Formulation.

Ingredient	Mix Order	PHR	
Polymer	1	100	
Carbon black, N351 added together	2	55	Masterbatch 60 RPM, 150-170°C
Naphthenic oil	3	10	
ZnO	4	3	
Antioxidant	5	1	
Wax	6	2	
Total masterbatch:		171	
Stearic acid		2	Final, 40 RPM, 77-95°C
Sulfur		1.5	
Accelerator (N-t-butyl-MBT sulfenamide)		1	
Total final:		175.5	

Table IV. Compounded results for LPY·2THF-initiated SBR of Table II.

Termination:	ROH	SnCl$_4$
Tensile strength, psi	2982	2956
E_b, %	431	403
Bound rubber, wt%	31.3	42.0
Tangent δ, (Dynastat, 50°C, 1 Hz)	0.080	0.076

Table V. Characterizations and cured, compounded hysteresis results of LPY·2THF-initiated SBRs, including examples of dicapped polymers.

	A	B	C	D	E
M_n x 10^{-3}	100	251	145	144	127
Polydispersity	1.10	1.48	1.15	1.17	1.27
% styrene / %vinyl	20 / 43	◄	19 / 43	◄	21 / 37
Termination	ROH	SnCl$_4$	ROH	R$_3$SnCl	ROH
Tangent δ, (Dynastat, 50°C, 1 Hz)	0.100	0.079	0.079	0.056	0.006

no functional modification. An inverse relationship between MW and tan δ was apparent, and the entire group of N-Li-initiated SBRs fell well below the curve for theunmodified SBRs. A slight, additional level of crosslinking has sometimes been observed in compounds of the N-Li-initiated polymers. Concerns that hysteresis reductions were obtained as a result of an accelerator effect from the amino moiety are felt to be minimal, since it was found that compounds adjusted to the same cure level showed the same comparisons as those of Figure 2.

The progress of polymerization from LPY·2THF was monitored in a separate experiment by following the ^{13}C NMR of the pyrrolidine head-groups during the polymerization of SBR (see Figure 3). The polymerization conditions used were: 80/20 butadiene/styrene, 1.8 meq N-Li phgm, 16% solids, 50°C. The assignments were confirmed using several decoupling techniques. The amino group was shown to be tethered to the polymer chain at the nitrogen atom. Samples were taken at increasing degrees of conversion, and the ratio of the N-C carbons (53-55 ppm) to the remaining aliphatic carbons was used to estimate molecular weights. Reasonable agreement was obtained between SEC molecular weights and NMR molecular weights up to M_n of about 4-5 x 10^4. Beyond that, the smaller of the two α-C peaks tended to be eliminated by the NMR instrument's scanning algorithm. The results are summarized in Figure 4. It was concluded that the amine was incorporated at the head of the polymer chain, and that N-Li was indeed the site of initiation.

For polymers of molecular weight beyond 50000, e.g., $M_n \geq 10^5$, semiquantitative UV-visible analysis of amine-dye complexes was found to be a more reliable method. *(11)* Polymer-bound pyrrolidine functionalities determined by UV-visible analysis, using calibration curve "PY" in Figure 5, generally confirmed those found by the ^{13}C NMR method (Table VI).

Table VI. Analyses of head-group pyrrolidine functionality in LPY·2THF-initiated SBR by ^{13}C NMR and UV-visible dye extraction methods.

M_n (SEC)	Estimated Functionality from SEC/UV-vis	Estimated Functionality from SEC/NMR
14000	0.93	0.93
27000	0.86	1.38
37000	0.89	0.88
62000	0.75	0.74

In Table VI there is some attrition of head-group apparent in the highest molecular weight polymer, which was sampled two hours after initiation. This was found to result from an apparent elimination or retroaddition process occurring in the living polymer after complete conversion, under monomer-starved conditions. The elimination was markedly accelerated at 80°C by charging additional n-BuLi in a two-fold excess, and is believed to be a base-catalyzed retroaddition of the type described by Tsuruta, et al. (see Scheme II). *(14)* The elimination process was stopped upon end-linking of the living chains with $SnCl_4$ or quenching with a proton source.

Scheme II. Base-catalyzed elimination of amino head-groups.

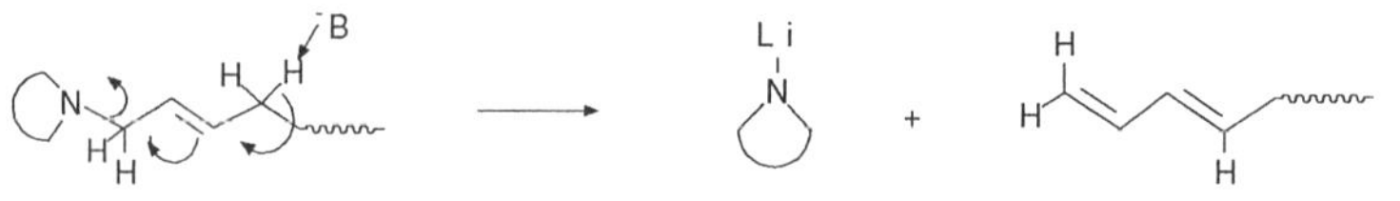

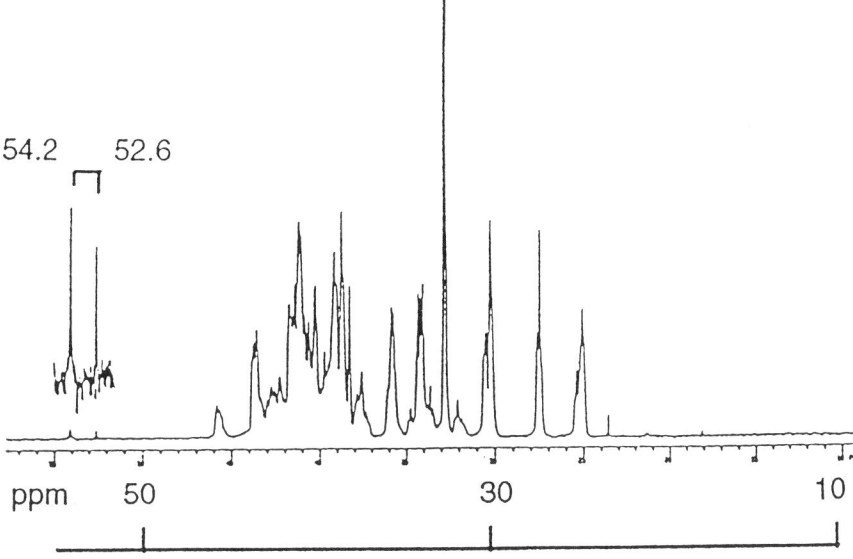

Figure 3. ^{13}C NMR of aliphatic region of PY-SBR-H, M_n = 37000, showing carbons adjacent to nitrogen (inset).

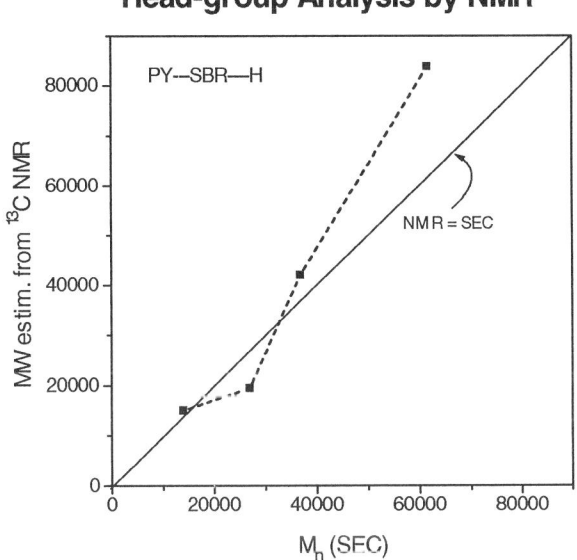

Figure 4. Comparison of ^{13}C NMR and GPC estimates of molecular weight in PY-SBR-H.

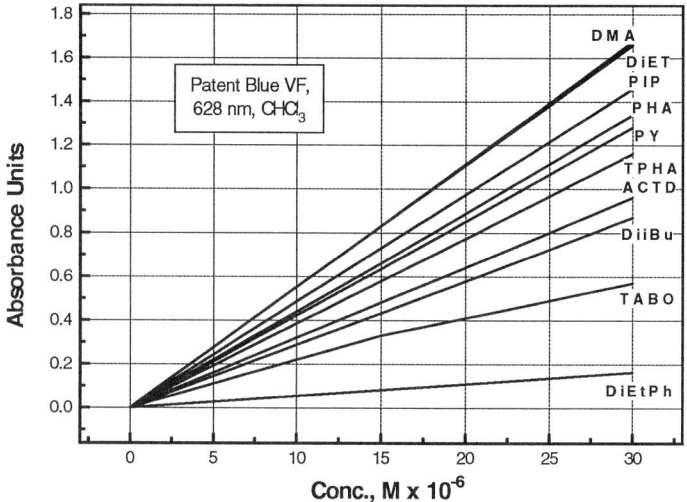

Figure 5. Calibration curves for UV-visible determination of tertiary amines (key to abbreviations is found in Table I).

Acknowledgments

The authors appreciate the contributions of many others to this work, including Mrs. M. J. Hackathorn, Mr. J. E. Hall, Dr. M. W. Hayes, Dr. A. S. Hilton, Mrs. A. Cieszielczyk, Mr. D. Tondra, Dr. R. Fujio, Prof. H. J. Harwood, Mr. T. Fujimaki, and Mr. L. L. Wolf. They are grateful to Bridgestone/Firestone, Inc., for permission to publish this work. Note: no permission is granted or implied for the practice of any invention without appropriate authorization.

Literature Cited

(1.) Foster, F. (to Firestone Tire & Rubber Co.), U.S. 3,317,918 (May 2, 1967); Kibler, R.W. (to Firestone Tire & Rubber Co.), U.S. 2,849,432 (August 26, 1958).
(2.) Angood, A. C., Hurley, S. A., and Tait, P. J. T., *J. Polym. Sci., Polym. Chem. Ed.*, **1973**, *11*, 2777.
(3.) Vinogradov, P. A., and Basayeva, N. N., *Polym. Sci., U.S.S.R.,* **1963**, *4*, 1568.
(4.) Nikolayev, N. I., Geller, N. M., Dolgoplosk, B. A., Zgonnik, V. N., and Kropachev, V. A., *Polym. Sci., U.S.S.R.,* **1963**, *4*, 1529.
(5.) Cheng, T. C. In *Anionic Polymerization. Kinetics, Mechanisms, and Synthesis;* McGrath, J. E., Ed., ACS Symposium Series No. 166, 1981, p. 513.
(6.) Halasa, A. F., Matrana, B. A., and Robertson-Wilcox, S. E., (to Goodyear Tire & Rubber Co.), U.S. 4,935,471 (June 19, 1990).
(7.) Antkowiak, T. A., Lawson, D. F., Koch, R. W., and Stayer, Jr., M. L., (to Bridgestone/Firestone, Inc.), U.S. 5,153,159 (October 6, 1992); U.S. 5,268,413 (December 7, 1993); U.S. 5,354,822 (October 11, 1994).
(8.) Lawson, D. F., Stayer, Jr., M. L., and Harwood, H. J., (to Bridgestone Corp.), U.S. 5,332,810 (July 26, 1994).
(9.) Lawson, D. F., Stayer, Jr., M. L., Morita, K., Ozawa, Y., and Fujio, R., (to Bridgestone Corp.), U.S. 5,329,005 (July 12, 1994).
(10.) Kitamura, T., Lawson, D. F., Morita, K., and Ozawa, Y., (to Bridgestone Corp.), U.S. 5,393,721 (February 28, 1995).
(11.) Palit, S. R., and Mandal, B. M., *J. Macromol. Sci. -- Revs. Macromol. Chem.*, **1968**, *C2*, 225.
(12.) Lawson, D. F., "Analysis of Long-Chain Amines at Low Concentrations," paper presented at Central Regional Meeting, American Chemical Society, Akron, OH, June, 1995.
(13.) Morita, K., Nakayama, S., Ozawa, Y., Ohshima, N., Fujio, R., Fujimaki, T., Lawson, D. F., Schreffler, J. R., and Antkowiak, T. A., *Polymer Preprints*, **1996**, *37*(2), 700.
(14.) Nagasaki, Y., Yoshino, N., and Tsuruta, T., *Makromol. Chem.*, **1985**, *186*, 1335.

BLOCK COPOLYMERS

Chapter 7

The Influence of Uncoupled Styrene–Butadiene Diblock Copolymer on the Physical Properties of SBS Thermoplastic Elastomers

H. R. Lovisi, L. F. Nicolini, A. A. Ferreira, and M. L. S. Martins

Technical Department, Petroflex Indústria e Comércio S.A., 25221-000—Duque de Caxias, Rio de Janeiro, Brazil

Commercial linear SBS copolymers are obtained by anionic polymerization either by a three-step sequential monomer addition or a two-step sequential monomer addition followed by coupling with a difunctional agent. Since 100% coupling is difficult to achieve, typically, 10-25% residual uncoupled SB diblocks remain in the final product. In the present work, we have studied the influence of styrene-butadiene diblock copolymers on the viscoelastic properties of SBS triblock copolymers and their impact on product performance in footwear applications. We have found through dynamic mechanical thermal analysis that SB diblocks decrease the elastic response of SBS/SB blends which will be translated into a decrease in abrasion resistance and hardness and an increase in melt flow index. These properties are extremely relevant for the shoe sole manufacturer. Thus, anionic polymerization processes which minimize the formation of SB diblocks should be preferred for the production of SBS thermoplastic elastomers.

Styrene-butadiene-styrene triblock copolymers are thermoplastic elastomers widely used in footwear, asphalt and plastic modification and adhesives applications. SBS copolymers are obtained by anionic polymerization which allows strict control over molecular weight and block structure (*1-4*). Although intensive research is being conducted on anionic diinitiators (*4-6*), up to this point there are only two commercial processes to produce linear triblock copolymer, e.g., a three-step sequential monomer addition or a two-step sequential monomer addition followed by coupling of the styrene-butadiene diblocks with a difunctional agent(*5*). Since an efficient coupling reaction and a precise control of the stoichiometry between diblocks and coupling agent are difficult to accomplish (*4*), some uncoupled diblocks usually remain in the final product. Morton *et al* (*1*) and Fetters (*7*) alerted about the importance of these contaminants on the physical properties of SBS triblock copolymer and that the high SB content in some commercial products, around 10-25 wt% could not be neglected

(7). Recently, McKay *et al (8)* showed how residual SB diblocks have a detrimental effect on the performance of SBS in asphalt and adhesive applications. Even though some aspects of this subject have already been discussed by other authors *(1-4,8,9)*, we think we can contribute to this discussion by bringing some new additional data.

The main focus of our work will be on TPE performance in shoe sole application and how residual uncoupled SB diblocks affect it. Our special interest in footwear is due to the fact that Brazil is one of the largest footwear producers/exporters in the world. Most of the TPE consumed in Brazil, 78%, goes for the shoe sole industry. So it is quite important to understand how SB diblocks affect the abrasion resistance, hardness and melt flow rate, which are some of the most essential properties for the shoe soles manufacturer. We have also correlated product performance with SBS/SB blends viscoelastic behaviour using dynamic mechanical thermal analysis (DMTA).

Experimental

A commercial SBS triblock copolymer (Product A) with 35 wt % of bound styrene was produced by a three-step sequential monomer addition process . An average molecular weight (Mw) of 150.000 was determined by GPC analysis (Figure 1A), comparing to polystyrene standards. Product B is a commercial SBS with same styrene content as Product A, which was obtained by coupling process. It has an average molecular weight (Mw) of 160.000 for the triblock copolymer and 80.000 for the uncoupled SB diblock (Figure 1B). Both products are extended in paraffinic oil (31wt%), according to thermal gravimetric analysis (TGA).

The SB diblocks, which were used to make blends with Product A, were synthesized separately in order to reproduce a residual uncoupled diblock copolymer of the corresponding triblock copolymer (Product A), with half of its molecular weight and same styrene content. The triblock copolymer A and the diblocks were dissolved in cyclohexane and blended in defined proportions to give mixtures containing 5, 10, 15, 20, 25 and 30 wt% of SB diblock. Antioxidant (1.1 wt%) and oil were added in order to obtain a material similar to commercial products used for shoe sole manufacturing, such as Product B. The material was coagulated in water at 90°C and dried in an oven at 50°C overnight. Final oil content in the SBS/SB blends was checked by thermogravimetric analysis and confirmed to be 31% by weight. Molecular weight and diblock content were determined by gel permeation chromatography .

DIN abrasion of the resulting materials was obtained in a DIN abrader (DIN 53516) . Hardness (Shore A) was measured according to ASTM D2240. Melt Flow Index was determined according to an adapted methodology based on the ISO 1133 but at a lower temperature (e.g. 180°C). Dynamic Mechanical Thermal Analysis (DMTA) was performed in a DMTA MK III from Rheometric Scientific, with a frequency of 1Hz, temperature range -100 to 200°C in a shear mode. GPC analysis was carried out in a Waters 150-C equipment.

Results and Discussion

Figure 1 presents size exclusion chromatograms of two commercial linear SBS thermoplastic elastomers. Product A was obtained by a three-step sequential monomer addition and Product B by coupling SB diblocks with a difunctional agent. It is apparent to see that Product B presents a high content of residual uncoupled diblocks (21%wt), which is typical of this coupling technology.

Figure 2 shows the abrasion resistance of products A and B. The dramatic decrease in abrasion resistance observed for product B can only be explained in terms of its high SB diblock content. Similar performance was found for the corresponding thermoplastic rubber compounds, which means Product A will generate a better quality shoe sole, with a much longer life time.

We then started to investigate the impact of SB diblocks on the physical properties of SBS thermoplastic rubbers and on their performance in footwear applications. It was our intent to quantitatively describe the relationship between performance and SB diblock content. What we have found was a very drastic reduction in abrasion resistance, hardness and an exponential increase in melt flow index for increasing SB diblock content (Figures 3, 4 and 5).

These results are all interrelated and can be explained on the basis of the imperfections imparted by the SB diblocks on the ideal SBS thermoplastic elastomer network.

The unique behaviour of SBS thermoplastic elastomers, combining thermoplasticity and elastic behaviour, is basically due to the incompatibility between its polybutadiene and polystyrene segments. Since they are chemically bonded, they cannot separate in a macroscale, so they separate in a microscale*(4)*. Polystyrene concentrates in domains which act as physical crosslinkings for the rubbery polybutadiene, forming a tridimensional network with elastomeric properties *(4,8,9)*. The elastomeric nature of SBS triblock copolymer depends on having both of its ends anchored in polystyrene domains. On the other hand, the residual styrene-butadiene diblock has only one polystyrene segment, which stays embedded in styrene domains, while its loose polybutadiene endblock behaves as a dangling chain end, disrupting the ideal tridimensional network. The polybutadiene in the SB diblock has more mobility and is able to dissipate energy by chain reptation inside the rubbery matrix, which increases the viscous character of the material *(8,9)*.

So, the sharp decline in abrasion resistance and hardness of the SBS triblock upon addition of SB diblock (Figures 3 and 4) can be interpreted as a reduction of the elastic response. The viscous character imparted by the SB diblock to the rubbery matrix increases the heat build up which supposedly favours the material loss by abrasion *(10)*. Similarly, hardness is a function of the elastic modulus, which depends by its turn on the degree of organization of the SBS network. As the SB diblock disorganizes this system, hardness will be affected consequently.

The increase in melt flow index can also be explained in terms of the higher chain mobility promoted by SB copolymers. (Figure 5). SB diblocks will weaken the intermolecular interactions in the SBS network, making it easier for the chains to flow. SB diblocks behave, in this case, like plasticizers. The effect of SB diblocks on the Melt Flow Rate was so dramatic, that measurements had to be done at 180°C rather than 200°C. These MFI results parallel in a certain way the reduction of the microphase separation temperature (MST) with the addition of SB diblocks observed

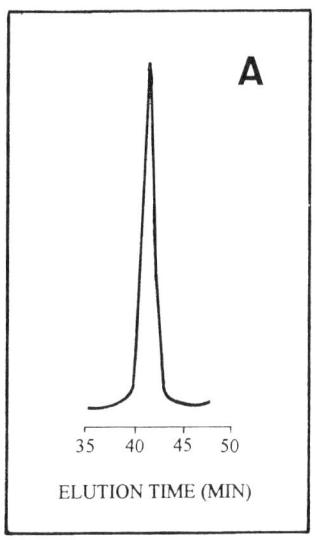 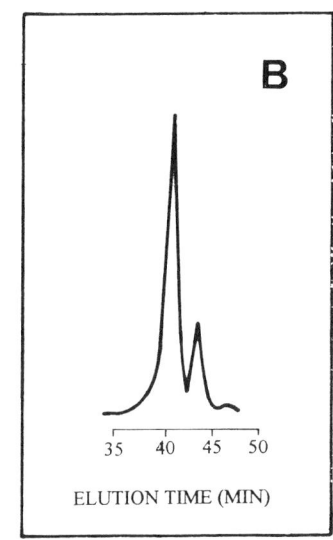

Figure 1. Gel Permeation Chromatogram.
A - Commercial Linear SBS Copolymer (Sequential Addition Process).
B - Commercial Linear SBS Copolymer (Coupling Process).

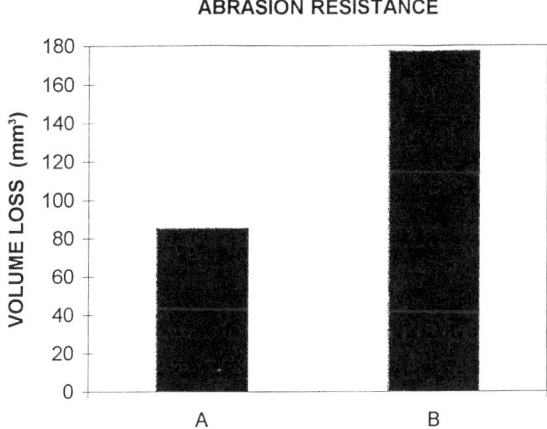

Figure 2. Abrasion Resistance.
A - Commercial Linear SBS Copolymer (Sequential Addition Process).
B - Commercial Linear SBS Copolymer (Coupling Process).

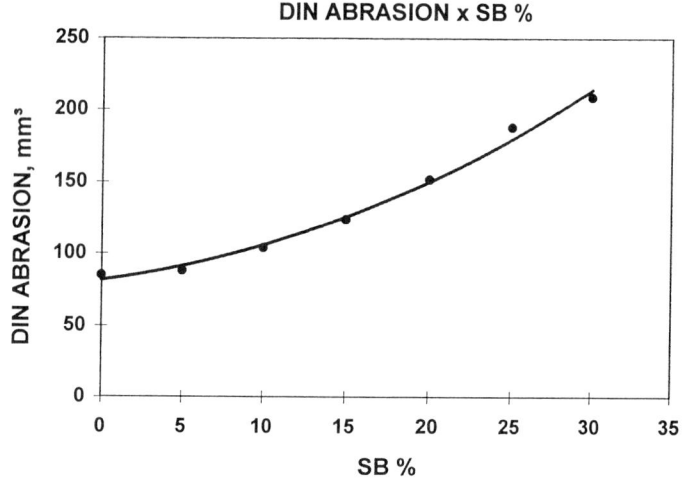

Figure 3. DIN Abrasion vs. SB diblock content (wt %).

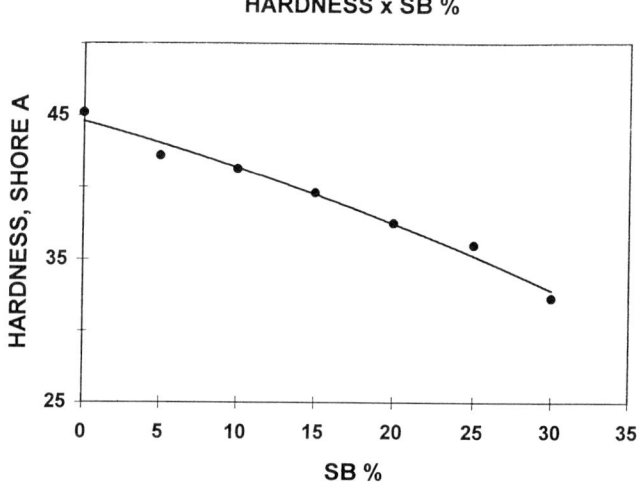

Figure 4. Hardness (Shore A, 15 s) vs. SB diblock content (wt %).

by McKay et al *(8)*. Mc Kay showed that total phase mixing between polystyrene and polybutadiene phases could be achieved at lower temperatures *(8)* for higher SB diblock levels. The SB diblocks would lessen the connectivity between domains in the TPE network, favouring phase mixing.

The viscoelastic behavior of SBS/SB blends was studied with the aid of dynamic mechanical thermal analysis (Figures 6 and 7). If we take a closer look at the rubbery plateau region (Figure 7), we can see that there is a linear relationship between loss tan delta and SB diblock content (Figure 8). So, the higher the SB content, the higher will be the viscous character of the material, with a corresponding decrease in the elastic response . This is in very good agreement with the performance observed in the abrasion resistance test, in the hardness test and melt flow rate measurements.

Our experimental results are also coherent with the work of McKay et al *(8)* and Berglund & McKay *(9)* who have observed a linear rise in the viscous character of SBS or SIS triblock mixtures for increasing SB or SI content and a corresponding decline in product performance for asphalt and adhesives applications.

Conclusions

Styrene-butadiene diblock copolymers (SB) have a deleterious influence on the physical properties of styrene-butadiene-styrene triblock copolymers (SBS). They decrease the elastic response of the final polymer, increasing its viscous character, which will be translated into a decrease in abrasion resistance and hardness and an increase in the melt flow rate. These properties are extremely relevant for the shoe sole manufacturer. Since SBS triblock copolymer is the main constituent of the shoe sole formulation it imparts its properties to the final compound. Thus anionic polymerization processes which minimize the formation of residual SB diblocks are highly desired.

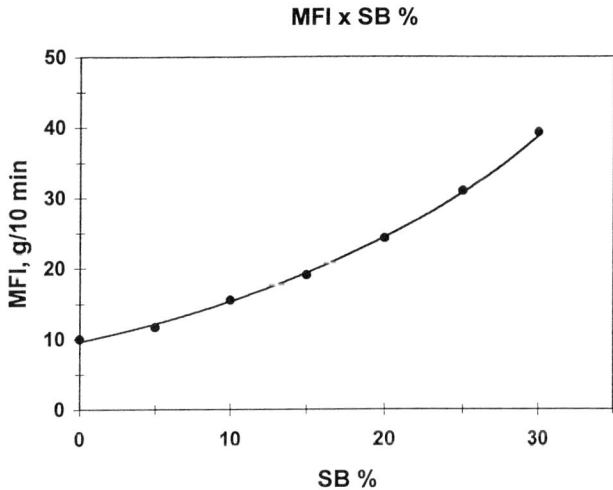

Figure 5 . Melt flow index vs. SB diblock content (wt %) at 180°C/5kg.

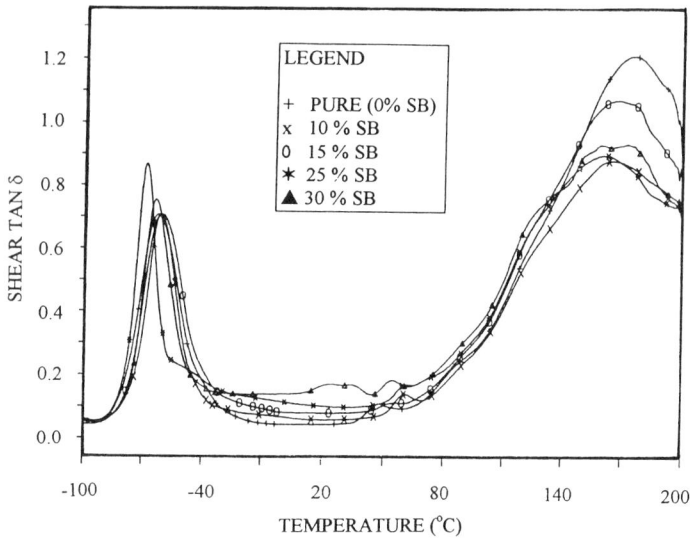

Figure 6. Dynamic mechanical thermal analysis of SBS/SB blends.

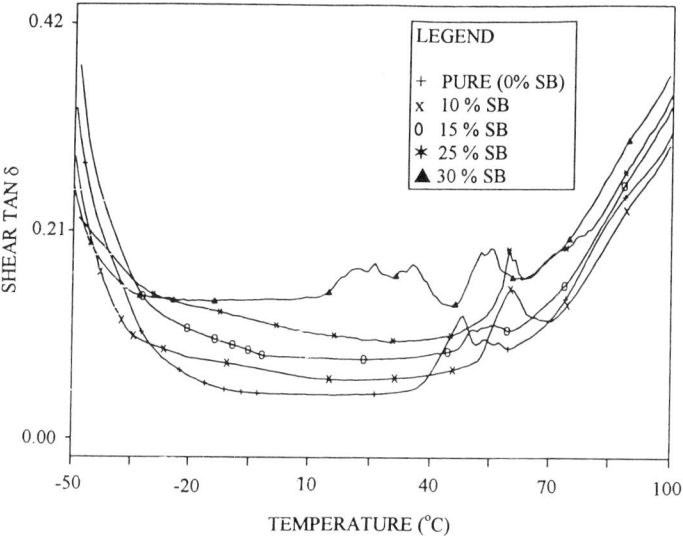

Figure 7. Dynamic mechanical thermal analysis of SBS/SB blends. Rubbery plateau region.

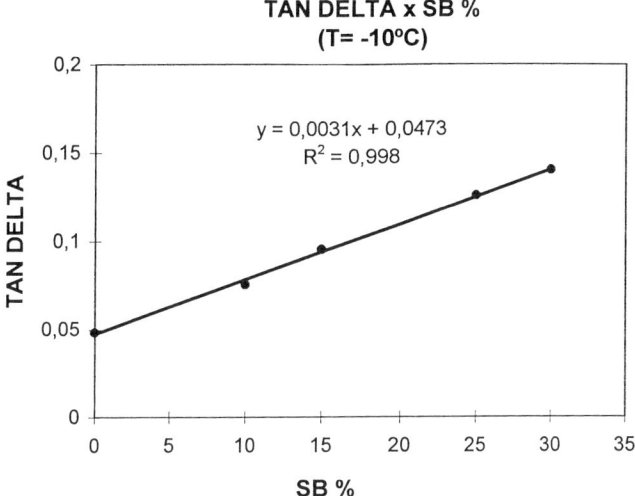

Figure 8. Tan delta vs SB diblock content (wt %).

Acknowledgements

The authors want to thank Prof. Charles Beatty (Univ. of Florida, USA) and Prof. Sebastião Canevarolo (UFSCar, Brazil) for helpful discussions. The authors would also like to acknowledge the work of Édson Romano Marins and Mônica Almeida Sant'anna for tests and analysis performed.

Literature Cited

1- Morton, M.; McGrath, J.E.; Juliano, P.C. *J.Polym.Sci.C,* **1969,** 26, 99.
2- Morton, M.; Quirk, R.P. In *Thermoplastic Elastomers, 2nd Edition-* G.Holden N.R.Legge, R.P. Quirk and H.E.Schroeder,Eds., Hanser: New York, 1996, pp 71-100.
3- Holden, G.; Legge, N.R. In *Thermoplastics Elastomers, 2nd Edition-* G.Holden N.R.Legge, R.P.Quirk and H.E.Schroeder,Eds., Hanser: New York, 1996, pp 47-70.
4- Hsieh, H.L.; Quirk, R.P.; *Anionic Polymerization: Principles and Practical Applications,* Marcel Dekker, Inc: New York, NY, 1996.
5- Van der Steen, F.H. *Polym. Prepr. (Am. Chem. Soc., Div. Polym. Chem.),* **1996,** *37(2),*632-3.
6- Yu, Y.S.; Dubois, Ph.; Jérome,R.; Teyssié,Ph. *Macromolecules,* **1996,** *29(5),* 1753-61.
7- Fetters, L.J.; Meyer, B.H.; McIntyre,D.*J. Appl. Polym. Sci.,* **1972,***16(8),*2079-89.
8- McKay, K.W.; Gros, W.A.; Diehl; C.; *J. Appl. Polym. Sci.,* **1995,** *56(8),*947-58.
9- Berglund, C.A.; McKay, K.W. *Polymer Engineering and Science,* **1993,** *33(18),*1195-1203.
10- Gent, A.N. *Rubber Chem. Technol.,* **1989,** *62,*750-6.

Chapter 8

The Effect of SI Diblock in SIS Block Copolymer on Pressure-Sensitive Adhesive Properties

Tetsuaki Matsubara and Minoru Ishiguro

Research and Development Center, Nippon Zeon Company, Ltd., Kawasaki 210, Japan

We found that Styrene-Isoprene (SI) diblock in Styrene-Isoprene-Styrene (SIS) copolymer improves carton sealability which is one of the important properties of carton sealing tapes. Diblock copolymer reduces storage modulus of adhesive compound. The low modulus gives poor shear adhesion but improves wettability to rough surface which seems to be a predominant factor on carton sealability.

SIS block copolymers produced by living anionic polymerization technique are applied in the largest volume to pressure sensitive adhesive (PSA) tapes, especially to carton sealing tapes. The copolymers produced by coupling method, which is the most commonly used in commercial production processes, contain some uncoupled SI diblock copolymers. There have been some studies of the effect of diblock in SIS copolymers on adhesive properties (1-2). However, the effect on carton sealability has not been discussed while unfolding failure of a carton boxes sealed with a tape is the most serious problem. In this paper, the effect of diblock content in SIS on PSA properties, especially on carton sealing properties, is discussed from rheological viewpoints.

Experimental Section

Three commercial SIS block copolymers were investigated in this study. They are produced by a traditional coupling method which consists of sequential polymerization of styrene and isoprene, and a coupling reaction.

SIS Block Copolymers. The molecular characteristics of the three copolymers are shown in Table I. They are similar in styrene content and Mw which represents overall weight average molecular weight, but not in SI diblock content. It should be noted that the same Mw does not mean the same GPC peak top molecular weight of SIS triblock. Though both of Quintac 3422 and Quintac 3520 have the same Mw, the molecular weight of diblock copolymer or its coupled triblock copolymer in Quintac 3520 is considerably higher than each corresponding molecule of Quintac 3422. Because Quintac 3520 contains much more amount of diblock copolymer whose molecular weight is equal to a half of that for coupled triblock copolymer (Figure 1).

Table I. Molecular Characteristics of the Investigated Polymers

POLYMER		Quintac 3422	Quintac 3433	Quintac 3520
Styrene Content	[wt%]	15	16	15
SI Diblock Content	[wt%]	10	52	78
Mw[a]	[×10³]	200	170	200

[a] overall weight average molecular weight determined with Gel Permeation Chromatography (GPC) which is calibrated with polystyrene standard.

Measurements of PSA Properties. In order to evaluate PSA properties, the SIS copolymers were compounded with an aliphatic hydrocarbon resin with softening point of 96°C (Quintone R100) and a naphthenic process oil (Shellflex 371N). The PSA compounds were coated on a 25 μm polyester film with coating thickness of 25 μm. The PSA properties were measured in accordance with PSTC (Pressure Sensitive Tape Council) standard (3) except carton sealability which was developed by ourselves.

Tack. A PSA tape specimen was arranged adhesive side up in line with the raceway of the PSTC standard incline so that it was free of any wrinkles, creases, or splices. The end of the tape opposite the incline was held to the table with a weight. A steel ball with 14/32 inch diameter was placed on the upper side of the incline. The ball was released and allowed to roll to a stop on the adhesive. This property was measured at 23°C. Tack is expressed in terms of the average of the stopping distance which was determined through five tests for each tape specimen. (unit: mm).

Peel Adhesion (Adhesive Strength). A PSA tape specimen with width of 10mm was stuck on a stainless steel plate. After dwell time of one hour, the tape was peeled off at 23°C in a direction of 180° at a rate of 300mm/min.(unit: g/10mm).

Holding Power (Shear Adhesion). A PSA tape specimen was stuck on a stainless steel panel with contact area of 25mm×10mm. When a paper (Kraft liner) panel was applied as the adherend, the contact area was 1/2inch×1/2inch. The panel was placed in the rack of the tester so that the extending tape hang downward. A load of 1kg was applied to the tape gently without swinging so that no more than the specified weight is acting on the specimen. This property was measured at 50°C with the stainless steel panel, and at 23°C with the paper panel. Holding power is expressed in terms of the time taken for the tape to separate from the panel.(unit: min.).

Carton Sealability. Two pieces of corrugated paper with dimensions of 100mm×50mm and 50mm×50mm were prepared. They were connected lengthwise with glass cloth backing tape which would not stretch at all when tension was applied, then stapled so as to allow free folding. A pair of the connected pieces of corrugated paper were placed side by side lengthwise and connected with a tape specimen with dimension of 50mm long and 25mm wide (Figure 2).

The prepared test assembly was turned over so that the tape specimen faced down, and its ends were fixed tightly to the frame of the tester. A 500 gram weight was placed on the test piece. It would act as a substitute for repelling force of the corrugated paper flaps. A timer with a limit switch being placed below the test assembly was started immediately. When the test assembly failed, the weight fell and stopped the timer automatically (Figure 3).

Six specimens were tested for each PSA compound. The results were reported as the mean time to failure (MTTF) calculated by a statistical method based on the Weibull distribution function instead of a simple arithmetic mean (unit: min.).

Measurements of Rheological Properties. Dynamic viscoelastic properties of SIS copolymers and PSA compounds were measured with frequency of 1Hz by a viscoelastic spectrometer (. Stress relaxation of PSA compounds was measured with 5% strain at 23°C.

Results

Table II shows the test results. The effect of diblock on tack was not so significant. The tackifying resin rich formulation (Formulation 2) showed poorer tack.

The results of peel adhesion and holding power were similar to those in the previous reports *(1-2)*. In peel adhesion of PSA compounds with lower resin content (Formulation 1), a PSA with higher diblock content (Quintac 3433) gave a higher value than one with lower diblock content (Quintac 3422) in the same failure mode, and a PSA with further higher diblock content (Quintac 3520) caused a change of failure mode. In Formulation 2, failure modes were varied by changes in diblock content.

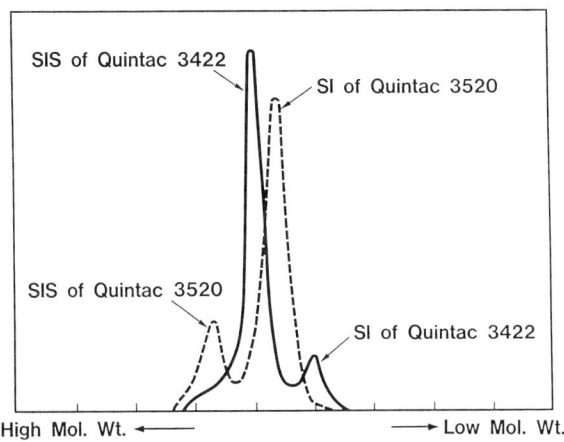

Figure 1. Molecular weight distribution of Quintac 3422 and Quintac 3520
(Reproduced with permission from ref. 4. Copyright 1995 Pressure Sensitive Tape Council.)

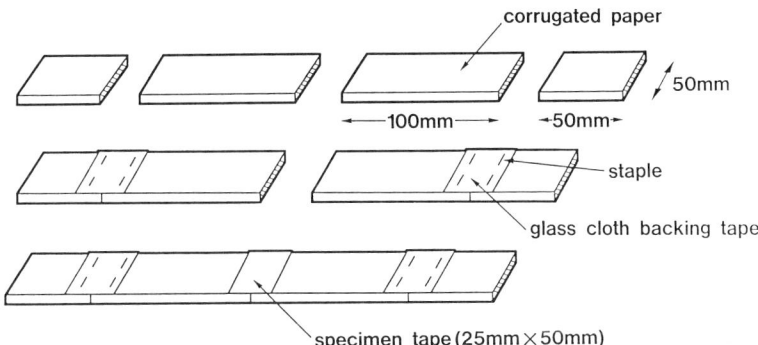

Figure 2. Preparation of a test piece for carton sealability test
(Reproduced with permission from ref. 4. Copyright 1995 Pressure Sensitive Tape Council.)

Both holding powers to steel and to paper decreased with increase of diblock content of SIS used in both formulations. Formulation 2 gave lower holding power than Formulation 1. The difference between the two formulations seemed to be smaller for paper than for steel.

As for carton sealability, higher diblock content apparently gave better carton sealability, and Formulation 2 seemed to improve the property. The difference between the tendency in carton sealability and that in holding power is noteworthy.

Table II. PSA Properties of the Investigated Polymers

Formulation	1	1	1	2	2	2
Ingredients						
Quintac 3422	100	----	----	100	----	----
Quintac 3433	----	100	----	----	100	----
Quintac 3520	----	----	100	----	----	100
Quintone R100	100	100	100	140	140	140
Shellflex 371N	20	20	20	20	20	20
PSA Properties						
Tack (Rolling Ball)	74	51	58	225	147	193
Peel Adhesion	530^a	1350^a	1340^c	1080^a	1540^c	1470^{ss}
Holding Power to Steel	4530	1900	440	1030	410	160
Holding Power to Paper	9460	3940	1370	7860	2670	880
Carton Sealability	230	990	3480	480	1230	3240

^{a}Adhesive Failure, ^{c}Cohesive Failure, ^{ss}Stick Slip

Discussions

Comparing properties of Quintac 3422 and Quintac 3520 in detail, difference between them should be due to their diblock content. Because the two copolymers have completely the same molecular characteristics in styrene content and molecular weight. We noted wettability of PSAs and investigated from rheological aspects.

Wettability to Rough Surface. Curves of the storage modulus of Quintac 3422 and Quintac 3520 are shown in Figure 4. The lower storage modulus should give better wettability to the adherend. In order to prove this idea, we measured peel adhesions with different dwell times. According to the results shown in Figure 5 and Figure 6, longer dwell time gives higher peel adhesion values with Quintac 3422 based PSAs.

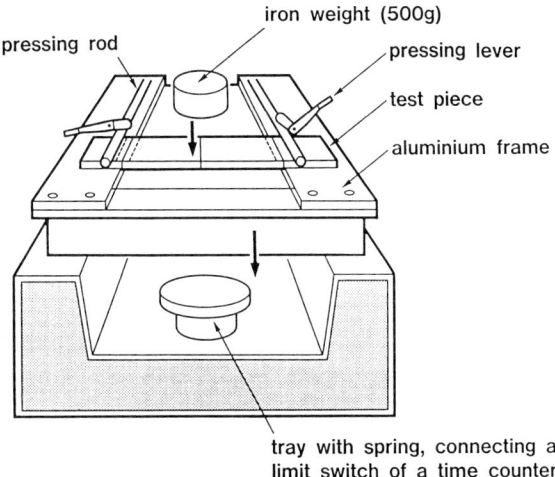

Figure 3. Schematic procedure of carton sealability test
(Reproduced with permission from ref. 4. Copyright 1995 Pressure Sensitive Tape Council.)

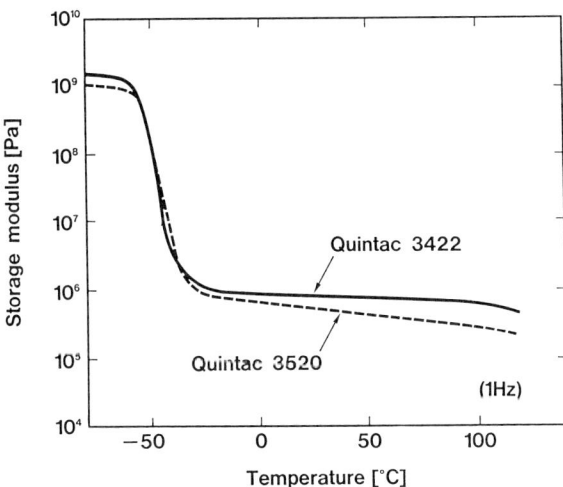

Figure 4. Storage Modulus of Quintac 3422 and Quintac 3520
(Reproduced with permission from ref. 4. Copyright 1995 Pressure Sensitive Tape Council.)

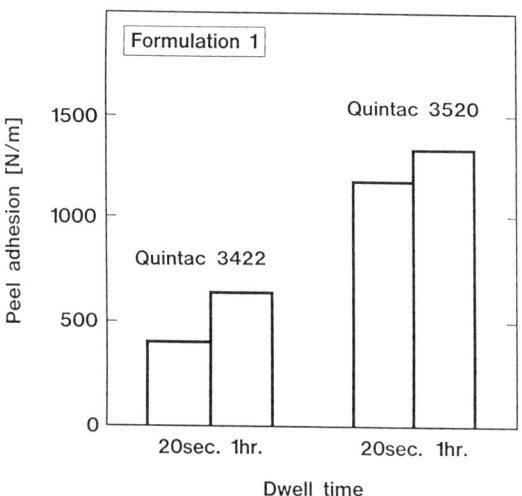

Figure 5. Peel adhesion by dwell time (Formulation 1)
(Reproduced with permission from ref. 4. Copyright 1995 Pressure Sensitive Tape Council.)

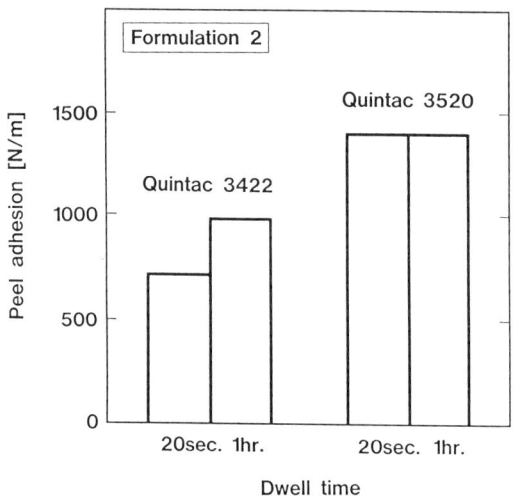

Figure 6. Peel adhesion by dwell time (Formulation 2)
(Reproduced with permission from ref. 4. Copyright 1995 Pressure Sensitive Tape Council.)

On the other hand, very short dwell time can give nearly maximum values with Quintac 3520 based ones. It implies that Quintac 3520 is more wettable than Quintac 3422.

As the differences between two dwell times seem to be smaller in Formulation 2 than in Formulation 1, higher resin content in PSA could give a more wettable PSA. Figure 7 shows curves of storage modulus of Quintac 3422 based PSA compositions. Lower storage modulus of Formulation 2 should relate to better wettability. From all these data, we can conclude that high SI diblock content in SIS or high resin content in PSA gives low storage modulus, which results in good wettability to the adherend.

Peel Adhesion. The reason for the high peel adhesion of Quintac 3520 compared with Quintac 3422 should be explained by high loss tangent (Figure 8). However, it could be interpreted by tensile properties, because they show a physical response to large deformation under high deformation rate which coincides well with peel test conditions. Figure 9 shows tensile properties of Quintac 3422 and Quintac 3520. The stress on Quintac 3422 rapidly increases as strain increases while the stress change on Quintac 3520 is gentle. Elongation of Quintac 3520 is larger than that of Quintac 3422.

We also measured peel adhesion at various peel rate from 0.1mm/min. to 1,000mm/min. in order to investigate peel phenomena in detail. The results are shown in Figure 10 and Figure 11. Quintac 3520 shows higher peel values than Quintac 3422 at any peel rate while its tensile strength is lower.

Generally, adhesive strings are observed along the peel line during peel test. In an adhesive failure, the strings are pulled apart from the surface of the adherend before they are broken. On the other hand, they should be broken in a cohesive failure. The Quintac 3422 based PSA fails adhesively over the entire range of the peel rates with continuous change of the peel values. For the Quintac 3520 based PSA, the failure mode varies as peel rate increases and adhesive failure is unexpectedly observed with discontinuous change of peel adhesion value at very low peel rate. Actually, this result is the opposite of the normal pattern reported before. Some previous studies pointed out that PSAs should fail cohesively at low peel rate from dynamic rheological viewpoints (5-6). However, it should be considered that there are a lot of complicated factors of peel adhesion or of failure mode. At least, neither dynamic rheological properties nor tensile properties could not determine the peel adhesion values.
Meanwhile, wettability should affect on the peel values as well as rheology of the PSA compounds. High peel adhesion values in PSAs based on high diblock content SIS sould be due to their excellent wettability.

Carton Sealability. In the measurement of carton sealability, a weight is placed on a test assembly instead of experiencing the repelling force of carton flaps. When the weight is placed, the test piece sinks slightly and a very small angle is generated

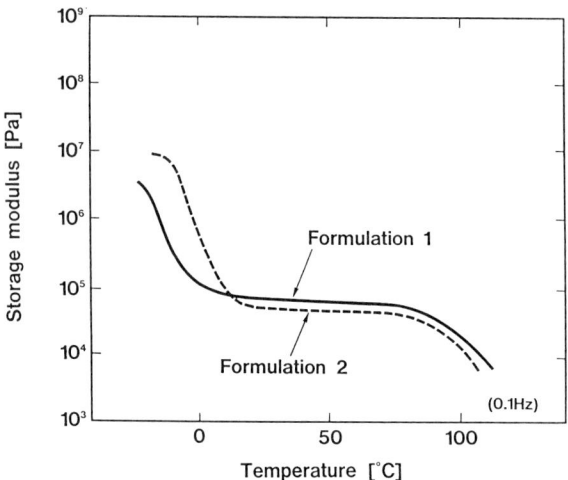

Figure 7. Storage Modulus of Quintac 3422 based adhesive
(Reproduced with permission from ref. 4. Copyright 1995 Pressure Sensitive Tape Council.)

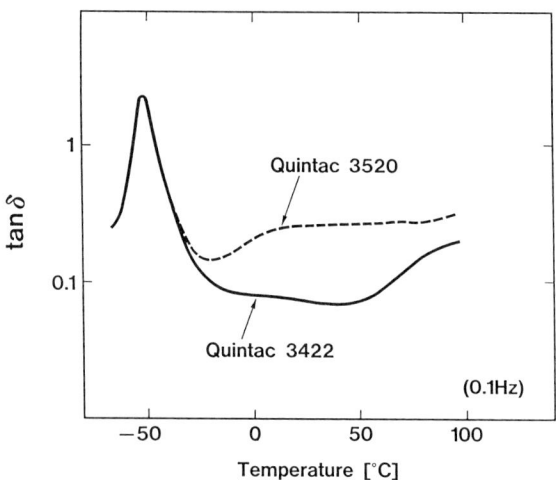

Figure 8. Loss tangent of Quintac 3422 and Quintac 3520
(Reproduced with permission from ref. 4. Copyright 1995 Pressure Sensitive Tape Council.)

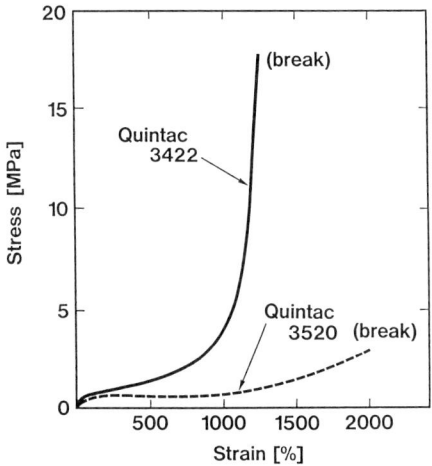

Figure 9. Tensile properties of Quintac 3422 and Quintac 3520
(Reproduced with permission from ref. 4. Copyright 1995 Pressure Sensitive Tape Council.)

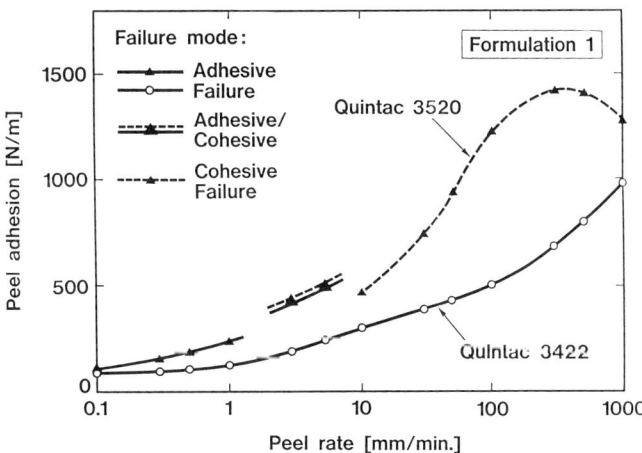

Figure 10. Peel adhesion vs. peel rate (Formulation 1)
(Reproduced with permission from ref. 4. Copyright 1995 Pressure Sensitive Tape Council.)

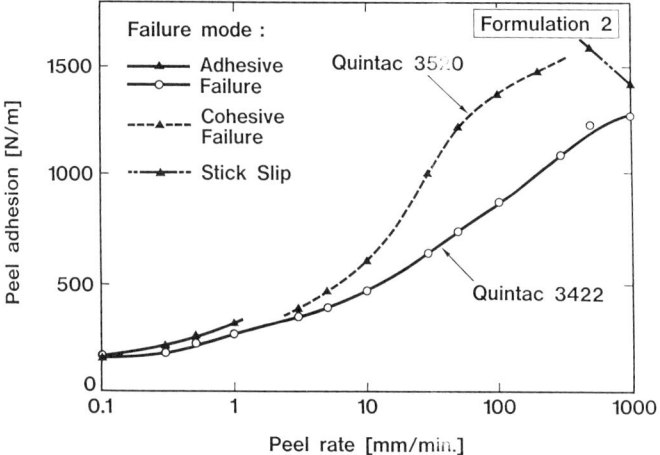

Figure 11. Peel adhesion vs. peel rate (Formulation 2)
(Reproduced with permission from ref. 4. Copyright 1995 Pressure Sensitive Tape Council.)

between the test piece surface and the horizontal plane (Figure 12). Tension T is generated along the specimen tape widthwise and acts as shear stress to the adhesive layer. Load W and tension T should balance according to equation 1.

$$W = 2T\sin\theta \qquad (1)$$

T would be much larger than W as θ is very small. The value of T rapidly decreases as θ increases from zero. For example, T becomes half when θ increases 1 degree to 2 degrees. Therefore, larger θ is more advantageous from the viewpoint of force balance, but adhesive layer is forced a larger deformation at the same time. A PSA based on low diblock SIS like Quintac 3422 can take only very small θ because of its high shear strength, and it tends to be pulled apart easily due to its poor wettability. On the other hand, a PSA based on high diblock SIS such as Quintac 3520 can take larger θ than that based on low diblock SIS because it can deform easily. Furthermore, its excellent wetting to the adherend also contributes to improved carton sealability.

In order to support this consideration, we observed stress relaxation of four PSA compounds (Figure 13). The stress on PSAs with Formulation 2 relaxes very quickly at the beginning. A considerable differences between Quintac 3422 based PSAs and Quintac 3520 based ones are also observed. The significant fall of stress or relaxation modulus on Quintac 3520 based PSA or resin richer formulation should contribute to the large deformation at the beginning of carton sealability test and to wettability.

Holding Power. In the case of carton sealability, the adhesive layer of a tape specimen bears some deformation at the beginning of measurement process. But the situation is different with holding power to paper. When a weight is hanged, no distinct shear deformation is observed at least at the beginning of measurement. A specimen tape coated with low diblock SIS based PSA fails suddenly without any evidence of cohesive failure. On the other hand, A soft PSA based on high diblock SIS fails cohesively to some extent in this test. It seems that the predominant factor affecting this property is not wettability to the paper but essentially the shear strength of the PSA.

Conclusions

High diblock SIS based PSAs tend to deform easily and show excellent wettability to the adherend. This behavior is advantageous for peel adhesion and carton sealability of a tape applied only to the closed flaps of a carton box. On the other hand, for holding power to paper, which is considered as corresponding to sealing property of both ends of a tape applied to both sides of a carton box, low diblock content is advantageous.

It is difficult to answer how much diblock content is optimum for box sealing tape.

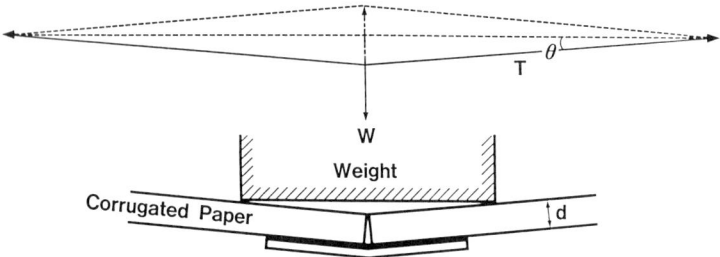

Figure 12. Force balance in Measurement of carton sealability
(Reproduced with permission from ref. 4. Copyright 1995 Pressure Sensitive Tape Council.)

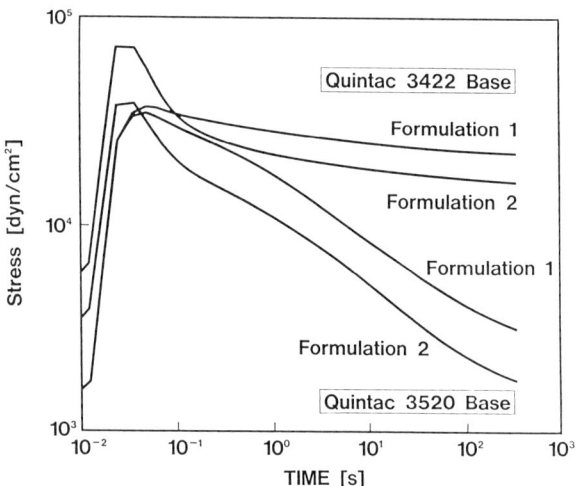

Figure 13. Stress relaxation curve of adhesives

Because, the repelling force of carton boxes is not constant, and because the contributing ratio of carton sealability and holding power to total sealing properties may vary with the dimension of a carton box, width of a tape and length of tape ends applied to both sides of a box. But we can point out that a PSA based on a conventional SIS with low diblock content is generally hard, difficult to deform and poor in wettability.

Acknowledgments

We would like to thank Mr. M. Nakamura of Nippon Zeon Co. Ltd. for useful discussions.

References

1. Johnsen, K. E. *Adhesives Age* **1985**, *29*.
2. *The Effect of Diblock on Pressure Sensitive Adhesive Properties;* Dillman, S. H.; PSTC Technical Seminar Proceedings, **1991**.
3. *Test Methods for Pressure Sensitive Adhesive Tapes,* 10th Edition, Pressure Sensitive Tape Council, **1992**.
4. *The Effect of Diblock on Carton Sealability of Packaging Tape;* Ishiguro, M.; PSTC Technical Seminar Proceedings, **1995**.
5. Parsons, W. F.; Faust M. A.; Brady L. E. *J. Poly. Sci.; Poly. Phys. Ed.* **1978**, *16*, 775.
6. Aubrey D. W. *J. Poly. Sci.; Chem. Ed.* **1980**, *15*, 2597.

Chapter 9

Styroflex: A New Transparent Styrene–Butadiene Copolymer with High Flexibility

Synthesis, Applications, and Synergism with Other Styrene Polymers

Konrad Knoll[1] and Norbert Nießner[2]

[1]Polymers Laboratory ZKT/I–B 1 and [2] Polymers Laboratory ZKT/C–B 1, BASF Aktiengesellschaft, D-67056 Ludwigshafen, Germany

Styroflex, a new styrenic thermoplastic elastomer, is obtained by sequential anionic polymerization of styrene and butadiene having a block sequence S-SB-S. It proved to match the property profile of plasticized PVC in the most important features such as transparency, mechanics, processability, and permeability. Impact strength and elastic recovery are even more pronounced. Styroflex is currently being tested as material for the food packaging industry. Together with Styrolux and general purpose polystyrene it forms a unit construction system e.g. for transparent film materials and injection molded parts allowing finetuning of hardness and toughness.

Styroflex is a new experimental product based mainly on styrene and butadiene. Our goal was to create a resin for extrusion and injection molding with similar or even improved characteristics compared to plasticized PVC. The new experimental product should exhibit the mechanics of a thermoplastic elastomer, e.g. low modulus and yield strength, high elongation and excellent recovery. It should be suited for high speed processing, especially for thin films. The latter aspect requires an intrinsic high thermal stability in order to avoid gel formation during processing.

Structure and Synthesis

To meet these goals we chose the block sequence hard-soft-hard typical for TPEs with a block length ratio of about 15:70:15 (*1*). The hard segments consist of polystyrene, but instead of a commonly used butadiene soft segment we introduced a statistical SB sequence with a glass transition temperature similar to that of plasticized PVC, located between -10 and -30 °C and an S/B ratio of approximately 1. Thus the overall styrene content of Styroflex reaches almost 70 % which is in a range known for transparent, impact modified polystyrene like Styrolux. To obtain an easily processable material having a melt flow rate of about 15 g/10 min. (200°C) the molar mass has been adjusted to about 140 000 g/mol (*2-4*).

The polymer is prepared by butyllithium-initiated sequential anionic polymerization in cyclohexane (Figure1). In order to generate the statistical SB block the presence of a randomizer is required (*5*). Commonly at least 0.2 to 0.5 vol% tetrahydrofuran (THF) is added, based on cyclohexane. Using less THF renders a styrene end block which changes the hard-soft-ratio and results in the loss of elastic recovery. Unfortunately THF changes the butadiene microstructure raising the 1,2-vinyl content from about 8 % in pure cyclohexane up to 30 % in the presence of 0.25 vol% THF. Because 1,2-vinyl units undergo crosslinking more readily than 1,4 units the thermal stability of the SB rubber is reduced. In this respect the THF content has to be kept as low as possible, accepting some gradient from butadiene to styrene. Dividing the SB block into several short blocks supports on the other hand the formation of a homogeneous soft phase. As an alternative, potassium alkoxides can be added as randomizers. Most favorable are sterically hindered tertiary alkoxides like triethyl carboxide (*6*). Best results are obtained with a K/Li molar ratio of 1:30 (*6,7*). The even incorporation of styrene and butadiene could be shown by taking samples during the copolymerization and analyzing the butadiene content.

There are three basic synthesis routes to Styroflex (Figure 1): sequential polymerization, coupling of a living S-SB diblock, or bifunctional initiation (*8*). Bifunctional coupling agents X such as dichlorodimethylsilane and butanediol diglycidyl ether give almost quantitative coupling yields. Carboxylic esters like ethyl acetate work only well when donor solvents are not used. In the presence of THF the coupling yield drops to 30-40%.

Essentially ether-free bifunctional initiators giving highly bifunctional growing polymers have been developed by BASF (*9*). At the time our goal was to find a starting system for the preparation of 1,4-polybutadienediol. Starting materials for the bifunctional initiators are alkylated, particularly methylated, 1,1-diphenylethenes, and β-alkylstyrenes. Reductive coupling with lithium metal in an ethylbenzene/ethyl ether mixture yields the desired 1,4-dilithiobutane derivatives in quantitative yield. The ether serves as charge transfer catalyst. In the case of 1,1-diphenylethylene at least two methyl groups are required to achieve solubility in ethylbenzene. Best suited in terms of accessability and side reactions is 1-(3,4-dimethylphenyl)-1-phenylethene, which can be readily made by addition of ortho-xylene to styrene in the presence of sulphuric acid followed by dehydrogenation. After the reductive coupling the ether and ethylbenzene are stripped under reduced pressure leaving the dilithiobutane as a deep red, viscous liquid, which tends to crystallize after several hours. Hydrolysis and GC analysis did not show any ether remaining. At least one equivalent THF per dilithiobutane remains ligated however, if THF instead of ether is used. The dilithiobutane derivatives themselves proved to be poor bifunctional initiators. About 60% monofunctional growth has been observed. A possible explanation is fast lithium hydride elimination during the initiation step. In order to obtain a useful bifunctional initiator the dilithiobutane is reacted at 0°C with 10 equivalents of butadiene which plays the role of both activator and solvent. The resulting pale yellow oligomeric initiator, which is obtained as a viscous liquid, can be diluted with cyclohexane, and virtually quantitative bifunctional initiation of butadiene and/or styrene is observed.

Morphology

Figure 2 shows the TEM micrograph of a compression-molded Styroflex sample. Due to the phase/volume ratio of about 30 % polystyrene in the block polymer, the

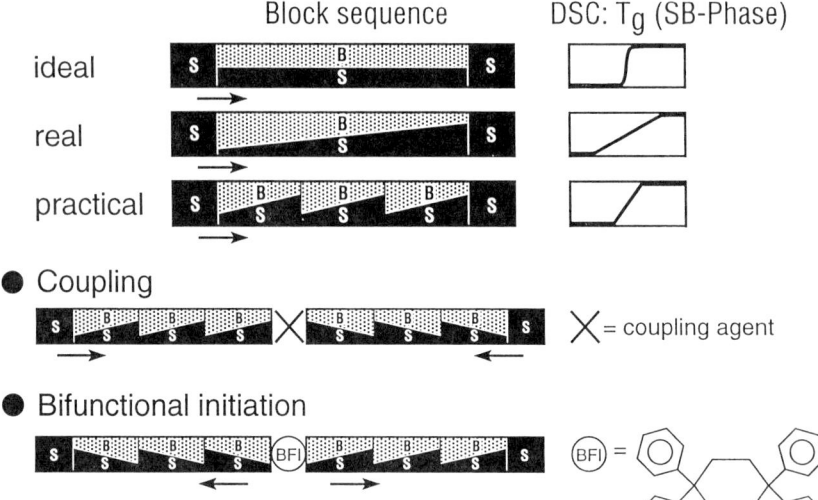

Figure 1. Synthesis routes to Styroflex. The bars indicate the chemical composition along the polymer chain. The light gray areas within the bars symbolize butadiene, the dark areas styrene.

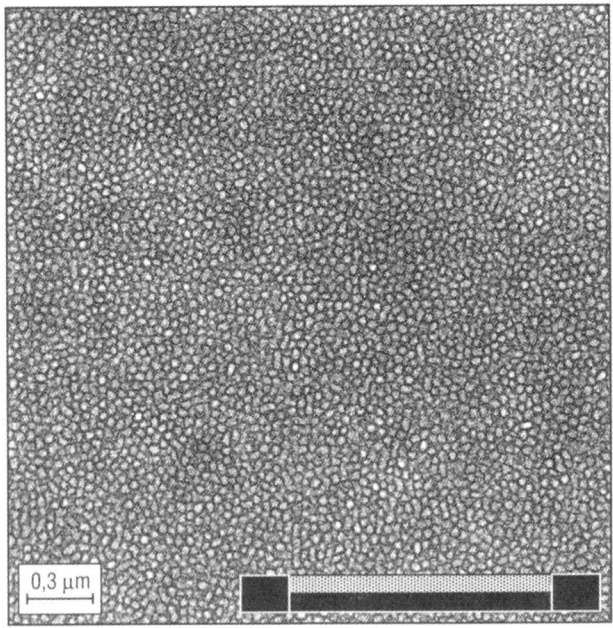

Figure 2. Transmission elecron micrograph of Styroflex.

transmission electron micrograph of Styroflex depicts spherical morphology with the SB rubber as matrix and polystyrene as spheres. The fuzzy borders of the spheres indicate an extended interphase typical of a system close to the order-disorder transition (ODT). The repulsive interaction between the polystyrene and the SB phase is greatly diminished compared to a polystyrene/polybutadiene system. Styroflex has however a higher molecular weight than commercial SBS TPEs, which in turn favors phase separation. When cooling from melt within a few minutes no long range order could develop, but the arrangement of the styrene domains is quite regular yet. Styrene domains are predominantly hexagonally surrounded by their neighbors. The domain identity period is about 39 nm.

Thermal Properties

The DSC analysis (Figure 3) shows the glass transition temperature of the soft phase at around -17 °C. The long flat slope up to 80 °C indicates an extended interphase ending in a barely separated hard phase. We proved that this is not an effect of the heterogeneity of the SB rubber phase by synthesizing and analyzing the pure rubber block. In this case the glass transition was limited to a temperature range between -30 and 0°C.

DMTA measurements (Figure 3) on Styroflex show two softening points around
-20°C and +70°C, which are in good agreement with the DSC measurement.

Rheology

Figure 4 depicts the relationship between the melt viscosities at 190 °C and the shear rate for three different triblock copolymers. Styroflex is compared with Styrolux KR 2691 (BASF), which is a symmetrical SBS triblock with a molecular weight (MW) of approx. 70 000 g/mol and a butadiene content of about 26 %, and with Kraton D 1102 (Shell), an SBS-type thermoplastic elastomer with a MW of approx. 70 000 g/mol and a butadiene content of about 70 %. It can be seen that the three triblock copolymers behave quite differently. At low shear rates the melt viscosities of both Styroflex and Kraton D show little rate dependence in contrast to Styrolux KR 2691, where a decreasing rate is accompanied by a pronounced increase in viscosity, which is characteristic of thixotropic behavior. At high shear rates both Styrolux KR 2691 and Styroflex show significantly more shear thinning compared to Kraton D 1102, indicating good processability.

Determination of the Order-Disorder Transition Temperature (ODT). Rheological measurements have been performed between 110 and 220°C in order to determine the ODT. Figure 5 depicts the the reduced storage and loss moduli at 190°C reference temperature. Between 150 and 220°C the time-temperature shift works perfectly well indicating a homogeneous melt. This notion is also supported by the slope ratio of 2. Below 150°C a deviation to higher viscosities occurs.

Plotting the reduced storage modulus at a frequence of 3 rad/s against the temperature allows fairly accurately the determination of the ODT, which is found at 145°C (Figure 6). This temperature is well below the favored Styroflex processing temperature (170 - 210°C). Thus, the production of thin films with diminished residual melt history is facilitated. The low ODT is a key advantage over conventional SBS block copolymers.

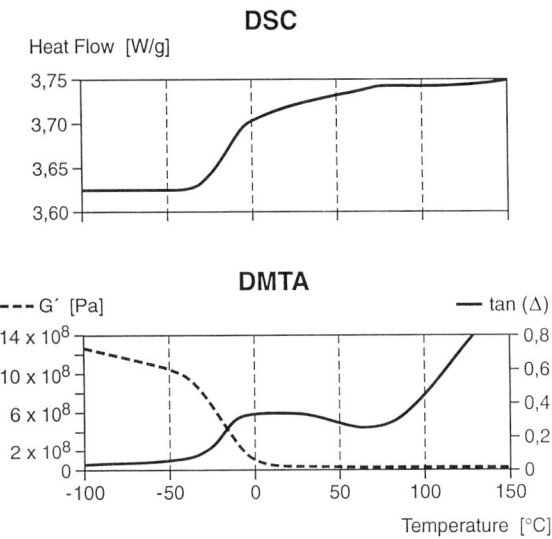

Figure 3. Top: DSC analysis of Styroflex. Sample quenched from melt; heating rate 20 K/min. Bottom: Dynamical mechanical thermoanalysis (DMTA) of Styroflex.

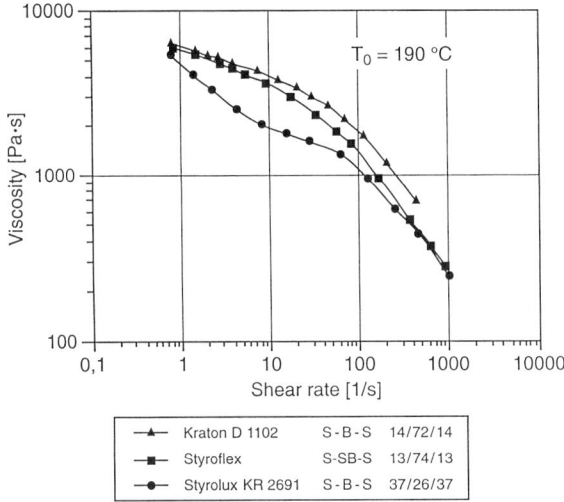

Figure 4. Rheological behavior of different symmetrical triblock copolymers at 190°C.

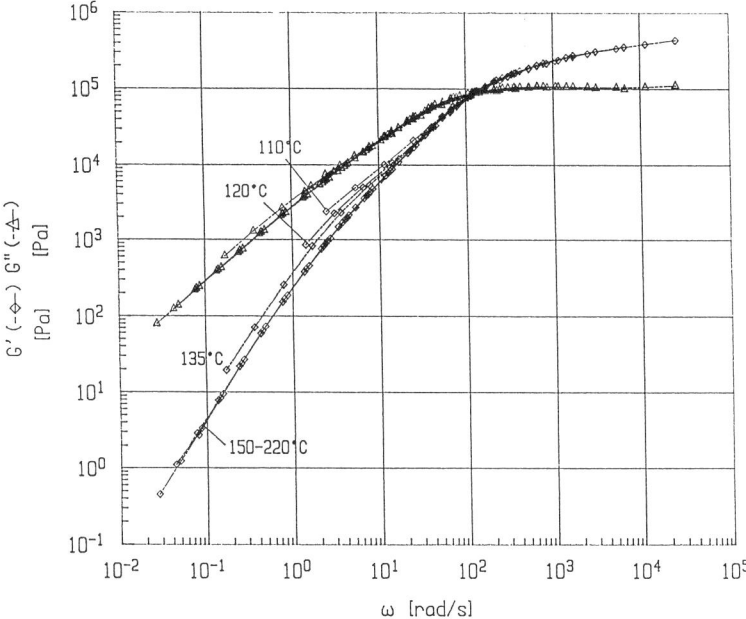

Figure 5. Rheology of Styroflex: reduced storage (G_r') and loss (G_r'') moduli shifted to reference temperature $T_0 = 190°C$.

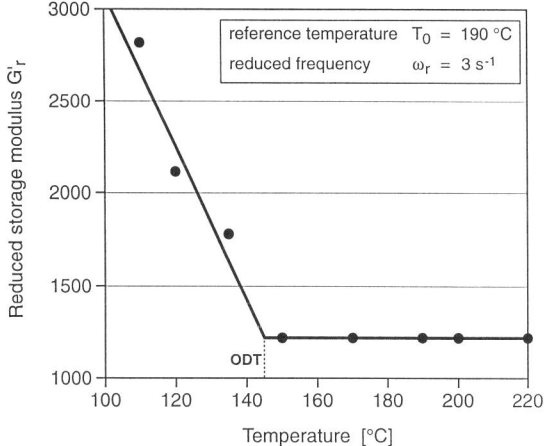

Figure 6. Determination of the order-disorder transition temperature (ODT): temperature dependence of the reduced storage modulus G_r' at frequence $\omega_r = 3\ s^{-1}$ and reference temperature $T_0 = 190°C$.

Mechanical Properties

The mechanical behavior of Styroflex is that of a typical thermoplastic elastomer (Figure 7). It appears to be somewhat harder than a classical SBS TPE. With an annealed specimen, an ultimate elongation at break of approx. 900 % can be achieved, whereas typical SBS polymers fail at elongations of about 1000 %. Similar behavior is observed with metallocene polyethylene.

At elongations well below failure, Styroflex follows Hooke's law. In contrast Styrolux KR 2691, a highly transparent and tough polystyrene with symmetrical SBS triblock structure and a butadiene content comparable to that of Styroflex, has a pronounced yield point followed by plastic deformation. The tensile strength of PVC depends on the annealing time - indicated by the dashed line. In films Styroflex achieves a similar hardness to plasticized PVC due to orientation.

Films designed particularly for food wrapping should maintain their smooth, optically attractive surface over an extended period of time even after touching, stapling and other manipulations, which might cause indentations. Therefore a preferentially complete recovery of the stretched film is desirable.

Adapted hysteresis experiments have been performed in order to compare the recovery of Styroflex with other film materials (Figure 8). The sample is elongated to 200% and released to stress zero at a constant rate of 100% per min. The further recovery is then monitored for five minutes. The experiment is repeated up to 300 and finally 400%. The measurements prove that the recovery of Styroflex is in the range typical of SBS-TPEs and far better than plasticized PVC. In fact after 30 min. the residual deformation is reduced to 3%. In comparison even the least crystalline metallocene linear low-density polyethylene (density 0.903) does not show a good recovery. The ratio of plastic to elastic deformation increases rapidly with density. Conventional SBS TPEs behave comparable to Styroflex, however their major drawbacks are poor thermoplastic processability and extensive thermal crosslinking. Thus, they are not destinated for sophisticated extruded and molded parts.

One of the predominant features of Styroflex is its great toughness. We compared 14 micron Styroflex film with 19 micron plasticized PVC film and 15 micron metallocene PE film. The dart drop test proves its superior performance over plasticized PVC (note: oriented, extruded blown film was used for mechanical tests, see Table I). Due to the orientation the Shore D hardness is roughly the same as with PVC. The great elongation at break makes Styroflex the material of choice for stretch and cling film applications, particularly food packaging.

Processability

Due to the fact that Styroflex has the same phase/volume ratio as that of commercial SBS TPEs, but with a high (70 %) styrenic content, it proved to be readily processable by all common methods. The blown and cast film extrusion is especially of commercial interest, and places highest demands on thermal stability. Experiments on a typical 60 mm ∅/25 D single screw extruder at output rates of 80 kg/h showed that Styroflex has superior temperature resistance compared to other SBS elastomers (Table II). Under typical blown film conditions (approx. 200 °C mass temperature, 220 °C die temperature) SBS elastomers tend to build up crosslinked gel particles which prevent thin film gauges. Under the same conditions, Styroflex is stable, and can be blown down to 10 - 15 µm thickness.

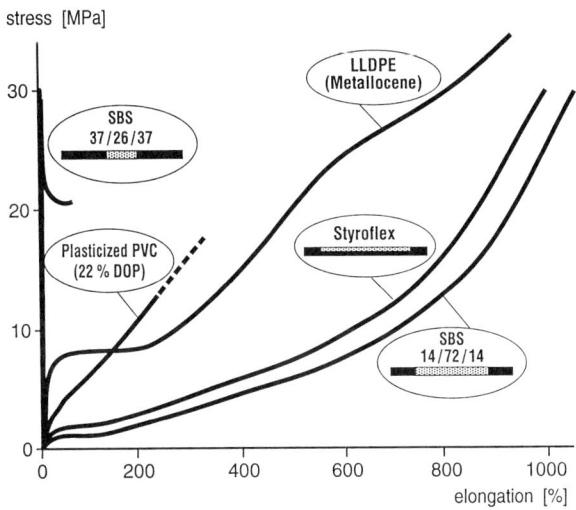

Figure 7. Stress-strain diagram of compression-molded flexible materials.

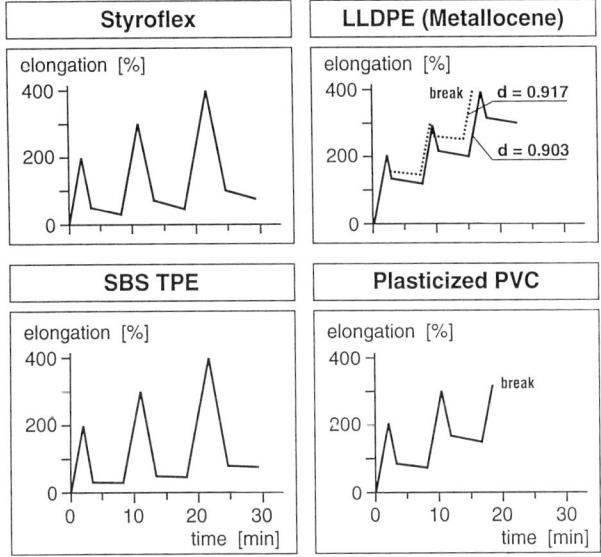

Figure 8. Hysteresis of film materials.

Table I. Properties of Styroflex vs. Plasticized PVC and Metallocene PE: Extruded Blown Film

	Styroflex	plasticized PVC (22 % dioctyl adipate)	metallocene polyethylene
Film gauge (μm)	14 ± 5	19 ± 3	14 ± 4
Dart drop test (g) [a]	600	290	500
Shore D hardness [b]	39	36	42
Vicat A [b] (°C)	40	-	87
Elastic recovery	++	++	+/-
Elongation at break (%) parallel / transverse to extrusion direction	250 / 350	180 / 210	

[a] Weight (g) of a falling dart until failure of film sample. [b] Determined on compression-molded specimen (T = 200°C / 5 min).

Table II. Extrusion Conditions of Materials for Packaging

	Typical processing temperatures [°C]	Max. processing temperatures [°C]
metallocene PE	200	240
plasticized PVC	190-200	220
Styroflex	170-210	230

Permeability

It is well known that polymers with a continuous rubbery phase are very permeable to gases and water. Styroflex contains approx. 70 vol % of a butadiene-co-styrene rubber phase and is therefore a highly permeable material (Figure 9). In food packaging, permeability is a crucial parameter besides mechanical strengh. For some applications like fresh meat packaging, high permeability to oxygen is necessary to maintain the red color of myoglobin over a storage time of several days. This important application is dominated by plasticized PVC films combining high permeability with impact resistance and recovery. Since Styroflex matches this combination of properties very well, one of its target applications will be fresh food packaging.

Competitive Analysis of Commercial Fresh Meat Packaging Films

Although plasticized PVC film dominates the fresh food packaging sector e.g. beef, pork, poultry, other films based on polyolefins are emerging. Typically, these films consist of coextruded structures, based on low density polyethylene as the major part of the core layer, and commonly ethylene-vinyl acetate top layers (Table III). Styroflex as core layer enhances the impact strength (dart drop test) as well as the elastic recovery of such coextruded structures. The comparision clearly shows that only Styroflex fulfills the high requirements of today's high performance packaging films.

Table III. Comparison of Styroflex-Containing Fresh Food Packaging Films with State-of-the-Art Technical Films

	EVA^a/Styroflex/EVA	Plasticized PVC	EVA/PE-LLD/EVA	Metallocene PE
Structure (μm)	2 / 10 / 2	14	2 / 10 / 2	14
Dart drop test (g)	600	210	250	500
Recovery^b (%)	95	80	40	50
Oxygen permeability^c	30 000	12 000	20 - 25 000	20 000

^{a}EVA = copolymer of 8 % vinyl acetate with ethylene (density: 0.928 g/ml, MVR (190/2,16) = 4,0). ^{b}Recovery 5 min after 200 % elongation of a tight film by a sharp tip like a nail. 100% means complete recovery, no visible deformation. ^{c}(ml/m^2·d·bar).

Synergistic Blends with other Styrene Polymers

80 % of Styrolux, the transparent and tough polystyrene of BASF with styrene butadiene block structure, is blended with general-purpose polystyrene (GPPS) in applications like packaging film, beakers and injection molded parts (*10*). This type of polymer was introduced by Phillips Petroleum in the late fifties and is now a steadily growing speciality in a.m. applications (*11*). Adding Styroflex as a third blend component opens up new possibilities (Figure 10).

In order to understand the role of transparent and tough polystyrene like Styrolux as blend component, its molecular design is discussed briefly. All Styrolux grades exept the already mentioned KR 2691, a linear, symmetrical S-B-S triblock, consist of unsymmetrical star block copolymers. When subsequently referred to Styrolux star polymer grades are meant. The Styrolux grades cover a butadiene content from about 20 to 30%, differing in their toughness/stiffness ratio. Styrolux combines a pronounced yield point (25-35 MPa) with plastic deformation up to 300%.

Synthesis, Structure and Morphology of Styrolux. Transparent and tough polystyrene is prepared by sequential anionic polymerization, where butyllithium may be added in more than one charge. A possible synthesis consists of the formation of a long styrene block in the first step, followed by further addition of butyllithium and styrene, yielding short styrene blocks. The molar ratio of short to long chains is significantly larger than 1. In Styrolux, a mixture of styrene and butadiene is added finally, resulting in a butadiene/styrene block with a tapered block transition. This mixture consisting of short, butadiene-rich and long, styrene-rich triblocks is coupled with an oligofunctional coupling agent giving on average an unsymmetical star polymer with about 4 arms (Figure 11) (*12*). The coupling reaction is of course a statistical process and all possible combinations of long and short arms are observed in an amount predicted by Pascal's triangle, taking the molar ratio into consideration.

The morphology of Styrolux does not fit into the common morphology scheme of styrene/butadiene block copolymers with spheres, cylinders and lamellae, as basic morphologies. Figure 12 shows TEM micrographs of a compression-molded Styrolux

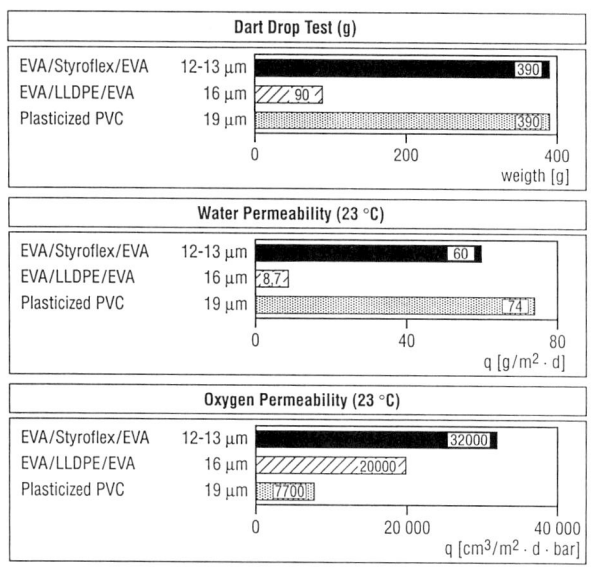

Figure 9. Water and oxygen permeability of packaging films, 16 microns, 23°C, 0% rel. humidity.

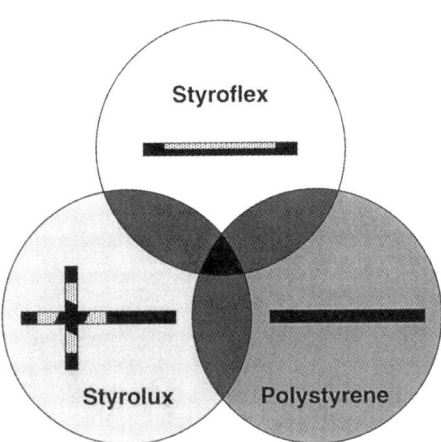

Figure 10. Styrene polymers for blends with tailor-made properties.

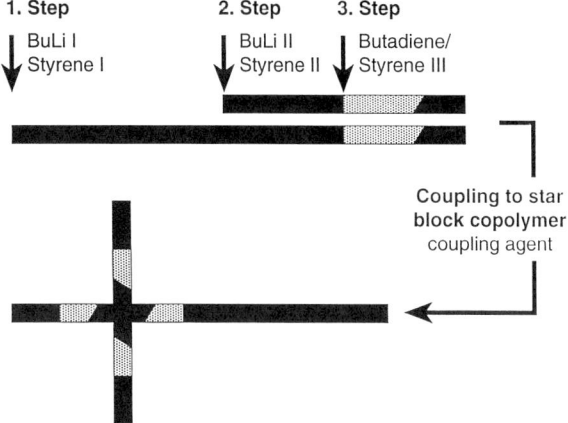

Figure 11. Synthesis of Styrolux.

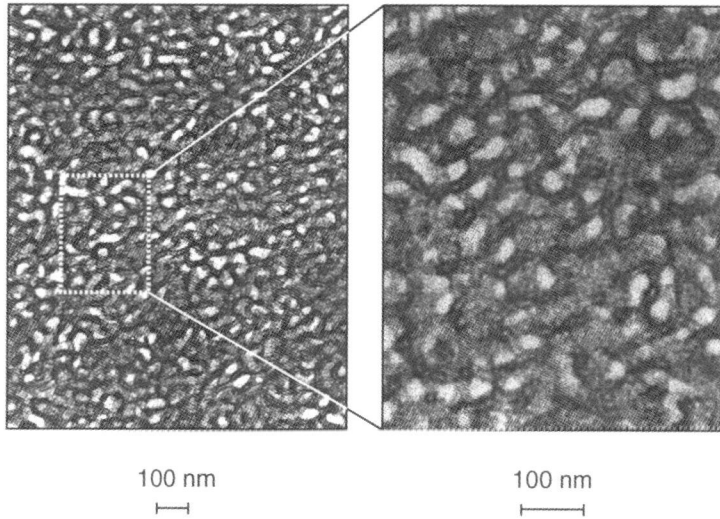

Figure 12. TEM micrograph of Styrolux.

sample. At first glance, the impression is of an irregular, wormlike morphology. Styrene and butadiene form an interpenetrating network with styrene as the predominant phase (Figure 12, left). A closer look however reveals that the butadiene "lamellae" are often split and appear to contain small styrene cylinders and spheres. Remarkable is the difference in scale (diameter): large styrene "worms": 25 nm, small styrene inclusions in butadiene: 6 nm, butadiene „lamellae": 8 nm. One explanation might be that long and short styrene blocks partly demix. Particularly the short styrene block in the core of the star polymer might form a separate phase. [This notion is also supported by DSC measurements discussed later. The glass transition temperature of the hard phase is unusually broad and stretches from 105 down to 30°C suggesting an at least inhomogeneous polystyrene phase.] In blends with polystyrene the "double lamellae" continue to exist unchanged while only the large styrene domains are widened.

Styroflex / Styrolux Blend System. Extruder blends of Styroflex and Styrolux have been prepared for a number of mixture ratios. The transparency was virtually unimpaired by the blend composition and remained at the high level of the pure materials (Figure 13), indicating homogeneity of the blends well below the wavelength of visible light. The mechanical properties such as modulus and elongation at break vary almost linearly with composition. This enables finetuning of hardness, toughness and flexibility of films and injection-molded parts.

In Figure 14 the morphologies of pure Styrolux and Styroflex (top and bottom TEM) are compared with two blends. Styrolux (bottom) exhibits the already discussed complex wormlike morphology containing two different types of styrene domains with the long PS arms concentrated in the thick „lamellae". When blended with Styroflex two phases become apparent. One phase is butadiene-rich and contains spherical PS domains, the other phase large styrene worms. The phase ratio does not reflect the mixture ratio of Styroflex and Styrolux, but the "Styroflex phase" seems to have a far larger volume ratio. The domain identity period is at roughly 15 nm less than half that of Styroflex. In view of the fact that the small styrene cylinders embedded in the butadiene lamellae disappear with increasing Styroflex content it appears most likely that Styroflex extracts the butadiene-rich stars containing only short arms and forms a new phase. Since this separation process occurs on a scale of about 80 nm transparency does not decrease.

The DSC analysis supports the conclusions from the TEM interpretations (Figure 15). Only one soft phase is visible in both blends containing 20 and 60 % Styroflex respectively, with average glass transition temperatures.

Styroflex / Styrolux / Polystyrene Blend System. The already discussed Styroflex /Styrolux blend system enables the property range between transparent polystyrene and soft thermoplastic elastomers to be covered which has recently been named "plastomer" by metallocene LLDPE producers due to the characteristic large plastic deformation. The stress-strain diagrams for a series of blends is depicted in Figure 16 center, showing how the balance of elastic to plastic deformation shifts with increasing Styrolux content. The bottom diagram elucidates the transition from a tough, relatively soft to a stiff, but fairly brittle thermoplast. Mixing Styroflex with small amounts of general purpose polystyrene (GPPS) increases the hardness but retains the elasticity (Figure 16 top). We also investigated the ternary blend system and found it to be a predictable and calculable unit construction system e.g. for tailormade film materials (Table IV). Styrolux plays mainly the role of a compatibilizer between Styroflex and GPPS. The fact that a certain degree of

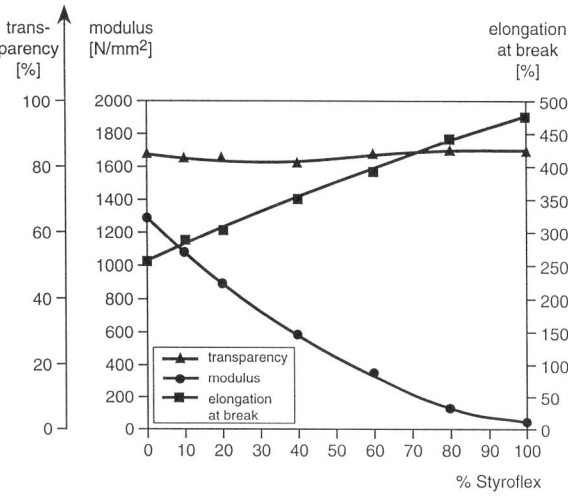

Figure 13. Styroflex / Styrolux blend system: mechanical and optical properties.

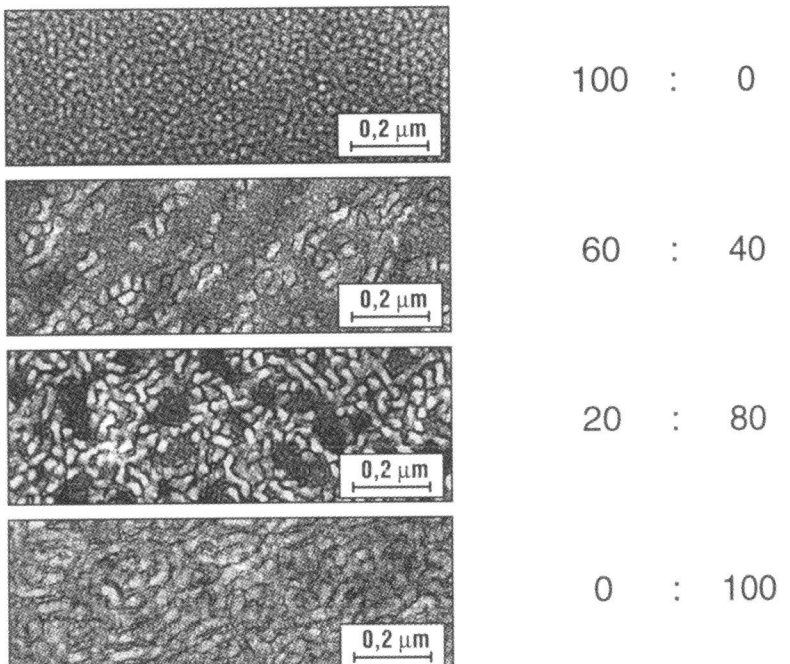

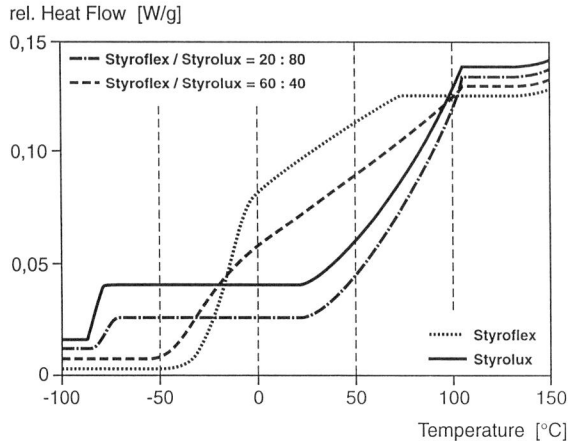

Figure 15. Styroflex / Styrolux blend system: DSC analysis.

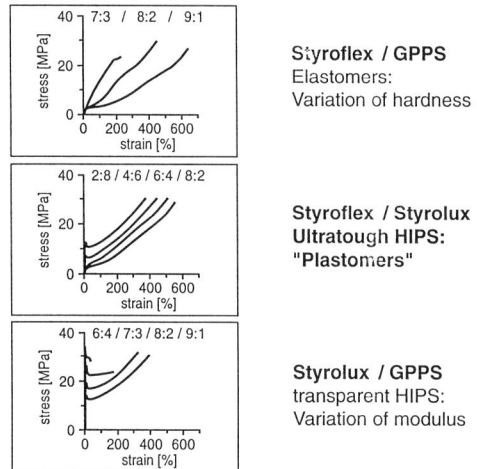

Figure 16. Styroflex / Styrolux / GPPS blend system: stress-strain diagrams.

Table IV. Properties of Styroflex / Styrolux / Polystyrene Blends[a]

Mixing Ratio[b]			MVR[c]	VST[d]		Modulus	YP [e,f]	Break[f]:		Transparency[g] Y
				A	B		Stress	Stress	/ Strain	
flex	lux	PS		(°C)		(MPa)	(MPa)	(MPa)	(%)	(%)
100	0	0	11,6	35,2	-	34,9	2,2	14,1	479	84,4
80	20	0	10,6	39,3	-	125	3,5	16,7	445	85,1
60	40	0	10,4	47,6	-	349	5,7	18,6	395	83,8
40	60	0	10,7	56,3	-	712	9,3	19,8	353	80,9
20	80	0	10,9	66,1	38,7	892	14,7	20,8	305	82,6
10	90	0	11,5	69,7	42,9	1081	18,6	20,9	291	82,4
0	100	0	12,3	75,6	48,7	1288	21,6	20	257	83,8
80	0	20	9,8	48,9	-	270	7,8	17,2	163	78,5
60	0	40	8,8	69,8	39	736	15,8	19,3	118	66,5
40	0	60	7,8	91,8	61,5	1807	29,7	23,8	36	63,4
20	0	80	5,2	103	89,4	2750	50,7	50,7	34	62,6
0	0	100	3	106	101	3200	55	55	3	90,0
0	80	20	9,4	85,1	60,4	1763	29,9	17,9	20,4	83,3
0	60	40	7,2	94,8	73,3	2161	34,8	23,7	20,6	77,9
0	40	60	6	99,7	84,7	2542	40,8	30,7	11,3	71,3
0	20	80	4,7	104	95,9	2988	49,7	45,4	3,9	72,2
75	5	20	9,1	54,2	37,5	316	7,8	17,9	264	78,8
55	5	40	8,4	70	43,8	832	16,2	19,6	112	68,2
60	20	20	8,6	56,8	37,8	550	8,8	18,6	235	78
40	40	20	8,6	64,9	38,5	812	11,2	20,6	224	73,9
20	60	20	9	75,7	48,5	1354	20,4	21,1	195	73,6
10	70	20	9,2	80,9	54,4	1579	26,1	17,7	43	77,4
40	20	40	7,6	77,4	48	1159	18,6	20,2	105	58
20	40	40	7,5	88,3	60,3	1679	25,9	20,5	55,9	55,7
10	50	40	7,5	91,5	66,8	1945	32,7	20,1	24,5	65,7
20	20	60	7	97,4	76,1	2223	39,2	24	19,6	50,3
10	30	60	6,2	99,2	80,4	2415	43,4	27,8	8,6	57,9

[a]Blending: ZSK 25 twin screw extruder, 150 r.p.m., mass temperature: 200°C, output: 10 kg/h; sample preparation: injection molding, 220°C mass temperature, 45°C mold temperature, in the case of „soft" specimens 25°C. The pure components have been extruded and molded in the same fashion in order to obtain comparable results. In the case of the pure block copolymers stress and strain at break are about 2/3 compared to the values found for compression molded samples; the the difference decreases with increasing PS content. [b]flex = Styroflex; lux = Styrolux 693 D (star shaped asymmetrical SBS block copolymer according to Fig. 11, with 25% butadiene content); PS = general purpose polystyrene Polystyrol 158 K (BASF). [c]Melt volume rate @ 200°C, 5 kp, [ml·10 min^{-1}] (ISO 1133). [d]Vicat softening temperature @ 50 K/h (ISO 306). [e]Yield Point. For Styroflex rich blends only a shoulder instead of a maximum is observed. In this case the height of the shoulder has been taken. [f]Determination of tensile properties (ISO 527). [g]2 mm plate thickness.

toughness can be achieved with a higher GPPS content as with binary Styrolux/GPPS blends makes the ternary system particularly attractive in respect to thermal stability and cost reduction.

Conclusion

Styroflex, the new BASF experimental styrenic polymer, combines the advantages of SBS elastomers (high toughness, recovery) with the properties of transparent, impact-modified SBS polymers like Styrolux (good processability, thermal stability, high transparency).

It is obtained by sequential anionic polymerization of styrene and butadiene. It proved to match the property profile of plasticized PVC in the most important features such as mechanics, processability and permeability. Impact strength and memory effect are even more pronounced. Styroflex is currently being tested as material for the food packaging industry and as impact modifier for styrene polymers.

Together with Styrolux and general purpose polystyrene it forms a unit construction system e.g. for transparent film materials and injection molded parts allowing finetuning of hardness and toughness.

References

1. Hashimoto, T. in Holden, G.; Legge, N. R.; Quirk, R. P.; Schroeder, H. E. *Thermoplastic Elastomers*, 2nd ed.; Carl Hanser: Munich, 1996; Chapter 15A, pp 429.
2. Knoll, K.; Gausepohl, H.; Nießner, N.; Bender, D.; Naegele P. *DE-A 4420952*.
3. Knoll, K.; Nießner, N. *ACS Polym. Prep.* **1996**, *37(2)*, 688.
4. Nießner, N.; Knoll, K.; Skupin, G.; Naegele, P.; Beumelburg, C. *Kunststoffe* **1996**, *87*, 66.
5. Hsieh, H. L.; Quirk, R. P. *Anionic Polymerization, Principles and Practical Applications*; Marcel Dekker: New York, 1996, pp 237.
6. Smith, S.D.; Ashraf, A. *ACS Polym. Prep.* **1994**, *35(2)*, 466.
   Smith, S.D.; Ashraf, A.; Clarson, S.J. *ACS Polym. Prep.* **1993**, *34(2)*, 672.
7. Wofford, C.F.; Hsieh, H.L. *J. Polym. Sci., Part A-1* **1969**, *7*, 461.
8. Lit.cit.5., pp. 307
9. Bronstert, K.; Knoll, K.; Hädicke, E. *EP 0477679*
10. Fahrbach, G.; Gerberding, K.; Mittnacht, H.; Seiler, E.; Stein, D. *DE 2610068*.
11. Phillips Petroleum *US 3078254*
12. Fahrbach, G.; Gerberding, K.; Seiler, E.; Stein, D. *DE 2550227*.

Chapter 10

Tapered Block Copolymers of Styrene and Butadiene: Synthesis, Structure, and Properties

Sergio A. Moctezuma, Enrico N. Martínez, Rodolfo Flores, and Enrique Fernández-Fassnacht

Industrias Negromex, S.A. de C.V., P.O. Box 257-C, Tampico, Tams. 89000, México

> In this work we describe the synthesis, structure and properties of four tapered block copolymers of styrene and butadiene (TBC's) that were tested as asphalt and high impact polystyrene (HIPS) modifiers. In both cases, the polymers improved significantly the properties such as the softening point temperature in asphalt-TBC mixtures and the Izod impact in HIPS-TBC blends prepared by mechanical mixing. Rheological measurements on pure asphalt as well as 4% by weight polymer modified asphalts were carried out and a noticeable rise in storage modulus G' with a lowering of Tan Delta was achieved by TBC modification of the asphalt. No difference attributable to structure differences between the several polymers was observed.

Among a variety of elastomers synthesized by anionic polymerization, Industrias Negromex produces tapered block copolymers of styrene and butadiene (TBC's), which are used for different applications including asphalt modification, plastic modification, pressure sensitive adhesives and shoe soles. For this reason, we are interested in knowledge of the performance of those polymers in their applications, as a means of making increasingly better products for both specific and general uses. In this work we describe the evaluation of four different TBC's in asphalt-polymer mixtures and HIPS-polymer blends.

Anionic polymerization is used to synthesize of polymers with very narrow molecular weight distribution and to control molecular weights, microstructures and morphologies (*1-4*). Tapered block copolymers of styrene and butadiene (TBC's) are synthesized by anionic polymerization and therefore have a well-defined structure. They are copolymers of the SB type and are extensively used in asphalt modification for paving and in plastic modification for HIPS and ABS production.

The synthesis, structure and properties of these block copolymers have been reviewed and it has been reported that because of their structure the TBC's exhibit characteristics such as high modulus and hardness, low shrinkage, good extrusion, high resistance to abrasion, low brittle point, transparent and glossy appearance and thermoplastic behavior (5,6).

For road pavements it has been mentioned that the use of asphalt with different materials such as sulfur, carbon black, asbestos, glass fibers, adhesion promoters and particularly rubbers improves toughness and crack resistance. Recent studies in the asphalt paving industry have pointed out the need for polymers in order to obtain softer mixtures at low temperatures, which increase flexibility and decrease cracking. On the other hand, polymers allow also stiffer mixtures at high temperatures which reduces flow and therefore permanent deformation (rutting). For paving application, the use of block copolymers and other thermoplastics, synthetic and natural rubbers and others (ground tire, fibers, etc.) has been reported. Among other polymers the following have been mentioned: SB, SBS, SIS, SEBS, SEPS, LDPE, EVA, EPDM and APP (7-10). For SB copolymers, it has been mentioned that there is a need to add sulfur to the mixture with asphalt, in order to improve the resulting mechanical properties (8).

Rheological studies of asphalt-SBS mixtures for roads have recently been reported, which emphasize one of the most important failures of roads submitted to high temperatures and/or high traffic loadings: rutting or permanent deformation. For roofing membranes and other water-proofing systems, the use of radial thermoplastic block copolymers, i.e. Solprene 411, is well known to modify the asphalt into a rubber-like substance with 10-14 wt % of rubber content. This block copolymer may prove to be too expensive for road surfacing, depending on the concentration necessary to meet the specifications for road surfacing (11-12).

Block copolymers of styrene and butadiene have been reported as impact modifiers for PS, both in polymerization "in situ" and in mixing by extrusion. In addition, these copolymers have been used to improve the properties of high impact polystyrene (HIPS) by extrusion mixing. Traditionally, elastomers have been used to increase the toughness of plastics because of their properties at low temperature which diminish the brittleness and improve the impact behavior of the mixtures at low temperature. On the other hand, because of their structure and morphology, block copolymers can lend transparency and gloss to certain grades of HIPS. Annealing of extruder blends of PS and block copolymers to decrease brittleness is said to reduce transparency because the regular arrangement of domains is crucial for the transparency of the blends (13).

In the first part of this work we describe the structure of the TBC's that were studied. The second part deals with the asphalt-TBC mixtures. Particularly, the properties of the asphalt used and the preparation of mixtures, as well as the method for the rheological measurements are presented. In the third part we report the preparation of the HIPS-TBC blends and their properties. The results and discussion are given in the fourth part, where performance properties in both applications are analyzed.

Experimental

Tapered Block Copolymers Characteristics. The TBC's used in this work are listed in Table I. The main differences between these polymers are styrene content and molecular weight.

Table I. Characteristics of the TBC's

TBC	Molecular Weight[a]		T_g^c	Bound Styrene	Polystyrene Block	
	Mn^b	Mw / Mn	[°C]	[wt %]	[wt %]	$Mw^{a,b}$
A	230	1.09	-89.8	15.0	11.0	15
B	110	1.10	-87.2	25.0	18.0	12
C	220	1.06	-86.3	30.0	22.0	32
D	300	1.06	-86.8	40.0	30.0	52

[a] PS Standards.
[b] Thousands.
[c] Polybutadiene part.

Relative molecular weights and molecular weight distributions were measured via GPC, in a Millenium-Waters 410 chromatograph with a refractive index detector; the columns used were Millipore KF-804, KF-805 and KF-807, and polystyrene standards were used for the calibration curve. The polymers were dissolved in THF which was also used as the carrier at flow rate of 1 ml/min. The concentration of the sample was 1.5 mg/ml and the injection volume was 25 µl.

The glass transition temperature were measured according to ASTM E-1356-91 method in a DSC TA Instruments 2910 calorimeter. Bound styrene and block polystyrene were measured according to ASTM D-1416 and ASTM D-3314, respectively.

GPC chromatograms of these TBC's showed a narrow distribution and very similar molecular weights for samples A and C (Figure 1). TBC's B, C and D have increasingly higher molecular weight and styrene content.

Asphalt-Tapered Block Copolymer Mixtures. The composition of the asphalt used in the mixtures with TBC's is shown in Table II. Clay-Gel analysis was employed to determine such composition, according to ASTM D-4124.

Sample Preparation. Mixtures were similarly prepared for all copolymers. A ROSS High Shear Rate Mixer was used at temperatures between 185-195 °C and 2500 rpm. The mixing procedure consisted of heating the asphalt to 160-180 °C, followed by gradual addition of the polymer in a maximum time of 10 minutes with continuous stirring. The temperature was then increased and kept within the control limits during the time necessary to observe complete incorporation of the polymer. By mixing 10 minutes more after this the complete dispersion of the rubber in the asphalt was assured. This procedure was used to prepare mixtures with concentrations of 4, 6, 8, 10 and 12 % by weight for each TBC. The mixing times were 90-100 minutes for the first two concentrations, and 120-130 minutes for the other three.

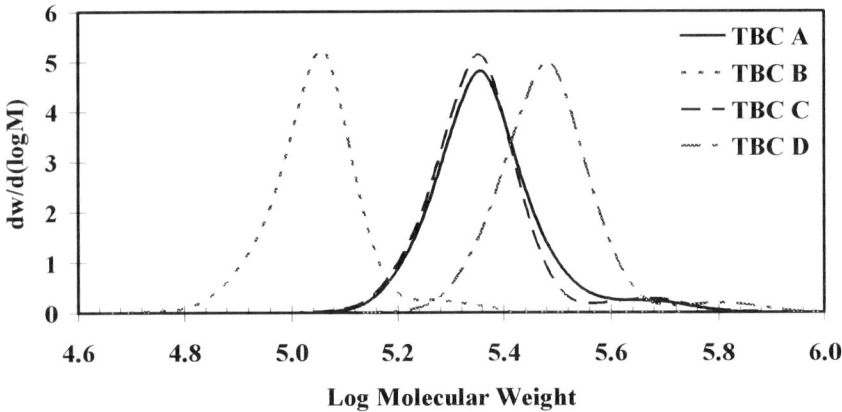

Figure 1. Relative molecular weight distribution of the TBC's.

Table II. Asphalt Composition[a]

Composition	% Weight
Asphaltenes	23.0
Saturates	9.6
Polars	54.0
Aromatics	13.4

[a] by Clay Gel analysis (ASTM D-4124).

Typical Properties of Mixtures. Brookfield viscosity at 160 °C and Ring & Ball softening point temperature of both pure asphalt and mixtures were measured according to ASTM D-4402 and ASTM D-36, respectively. This was done to refer our observations to properties which are very commonly used in paving.

Rheological Measurements. The dynamic mechanical analysis of the pure and modified asphalt samples was performed with a Paar-Physica 200 Universal Dynamic Spectrometer. Small angle oscillatory shear experiments were done at 2% of strain amplitude using a parallel plate geometry with a 12.5 mm radius and a 1 mm gap height at 60 °C; the mixtures were stored at room temperature. At the strain amplitude applied all samples were within the linear viscoelastic range, and therefore G' is independent of the applied strain as can be seen in Figures 2 and 3 which show the strain sweeps for pure and modified asphalt samples at 10 rad/s. Samples were allowed to stabilize for 10 minutes at the preset evaluation temperature before running the isothermal tests. In all cases, the isoconfigurational state of the materials was achieved and checked.

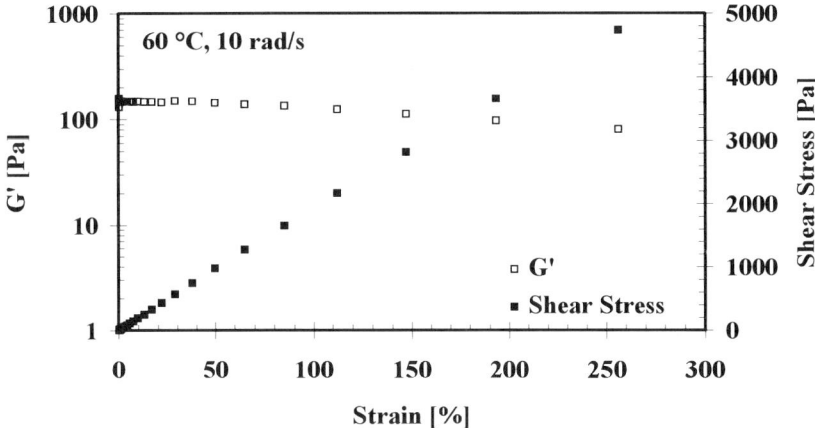

Figure 2. Storage modulus and shear stress of Asphalt as a function of strain amplitude levels.

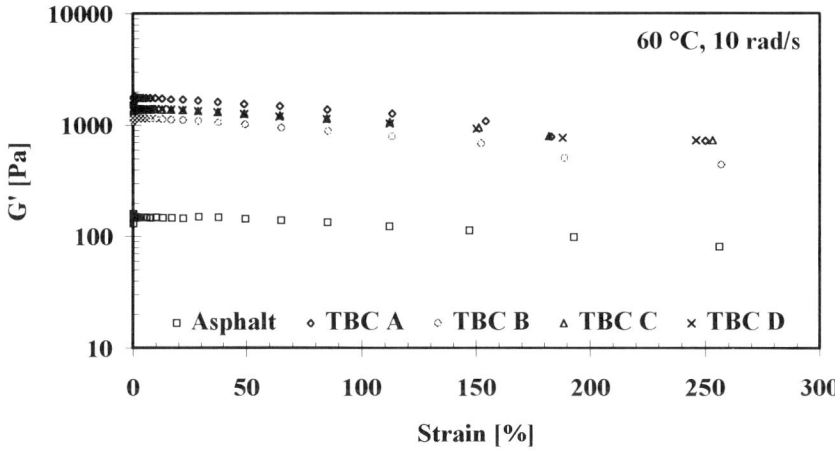

Figure 3. Strain dependency of storage modulus of Asphalt and Asphalt-TBC mixtures (4 %wt).

HIPS-TBC blends. A high impact polystyrene (Table III) was used to test the effectiveness of the several TBC's as impact modifiers by mechanical mixing. A Haake Büchler TW-100 counter rotating twin screw extruder was used to prepare the mixtures, at 75 rpm and with a 190/200/200/200 °C temperature profile. The TBC's were fed continuously to the extruder in powder form, while the HIPS was used as pellets; TBC concentrations were 5, 10 and 13 % by weight. The mixtures were stabilized with 0.2 phr BHT plus 0.2 phr Irganox-1010; Loxamide (1 phr) and magnesium stearate (3 phr) were used as lubricants. The extruded blends were used to prepare injection molded test specimens in a Negri-Bossi 70 injector at 1400 psi, 200 °C, 150 rpm and 70 s cycle time.

In order to determine the Izod Impact values for the HIPS-TBC blends, a Tinius Olsen 66 Impact tester was used according to ASTM D-256 method. Tensile and flexural moduli were measured by means of a Zwick Dynamometer with ASTM D-638 and D-790 methods.

Tabla III. HIPS Characterization

	Molecular Weight [a]		Tg [°C]
	Mn [b]	Mw / Mn	
HIPS	72	2.98	97.9

[a] PS Standards.
[b] Thousands.

Results and Discussion

Figure 4 shows the Brookfield viscosity values for the modified asphalt samples as a function of copolymer concentration. An exponential rise in viscosity is observed with increasing TBC concentration. Also, blend viscosity increases with polymer MW as well as with the block polystyrene content in the TBC. Mixing times increased with MW also. Softening point temperature as a function of polymer concentration is shown in Figure 5; increasing MW and block polystyrene content cause a rise in blend softening points, while higher polymer concentrations also lead to higher softening points. Within the range of concentrations studied it is clear that all TBC samples were capable of modifying the asphalt properties, the correct choice depends upon the particular requirements of the application as well as economics.

The G' isotherms at 60 °C for pure asphalt and 4 % by weight polymer modified asphalts are shown in Figure 6, while the Tan Delta profiles are given in Figure 7. A noticeable rise in storage modulus G' with a lowering of Tan Delta is achieved by TBC modification of the asphalt. Therefore, TBC-asphalt blends present a higher resistance to flow than pure asphalt which is very desirable to reduce permanent deformation in road paving. Moreover, all modified asphalt blends show a very similar behavior with no difference attributable to structure differences between the several polymers used.

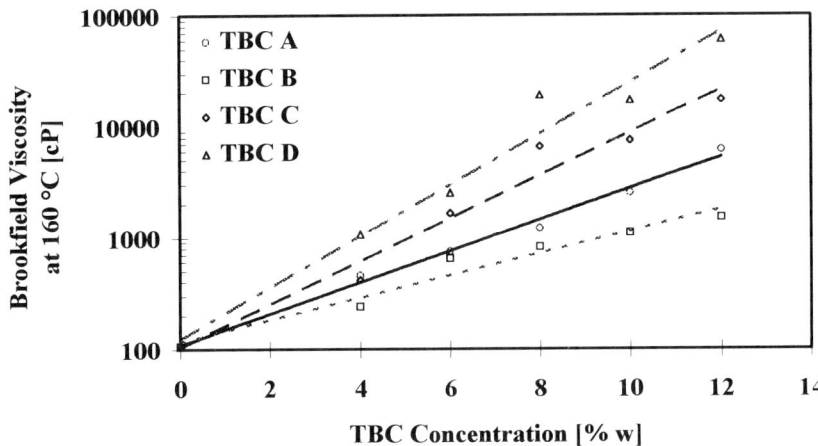

Figure 4. Brookfield viscosities of Asphalt-TBC mixtures as a function of TBC concentration and type of TBC.

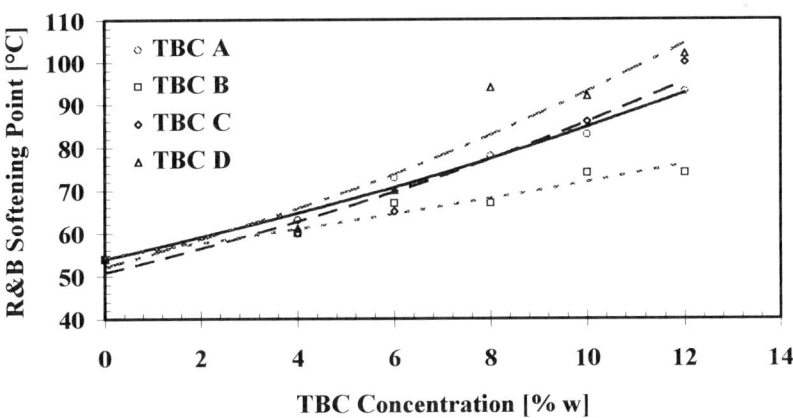

Figure 5. Softening point temperatures of Asphalt-TBC mixtures as a function of TBC concentration and type of TBC.

However, for the cases of the higher MW and block polystyrene content, the value of G' tends to remain constant at frequencies below 1 rad/s which would indicate better properties at high temperatures.

Figure 8 shows the Izod impact values for the HIPS-TBC blends at varying concentrations. Impact increases with increasing TBC content and the best results are achieved with polymer C.

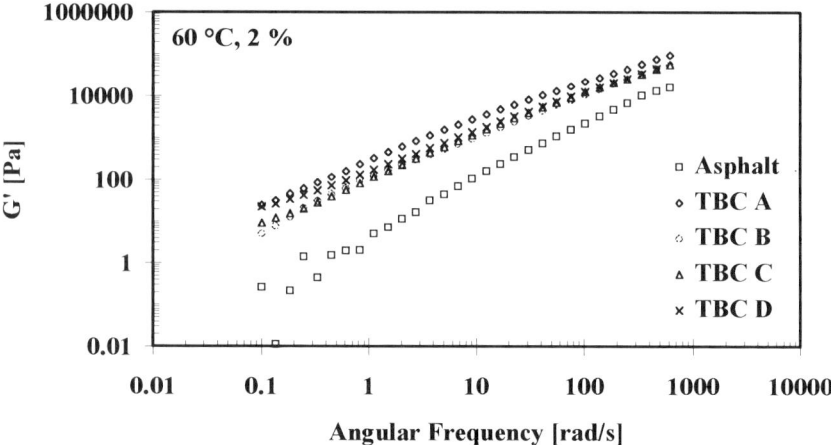

Figure 6. Storage modulus of Asphalt-TBC mixtures as a function of frequency, 4 % wt of TBC.

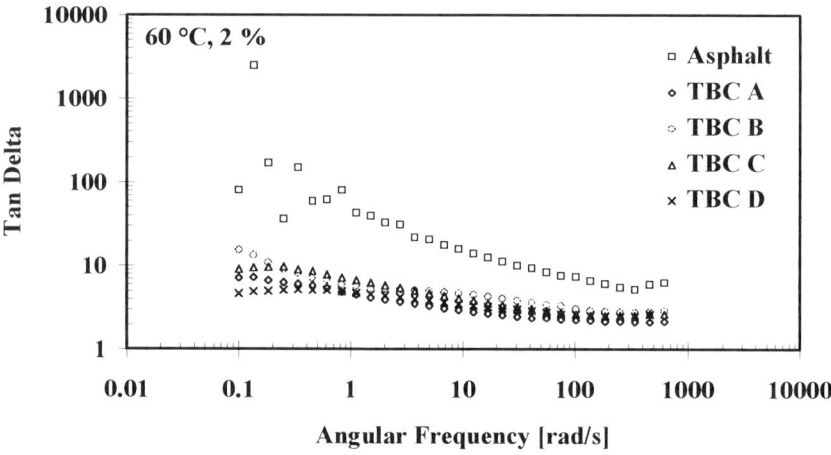

Figure 7. Tan Delta of Asphalt-TBC mixtures as a function of frequency, 4 % wt of TBC.

In Figures 9 and 10, tensile and flexural moduli are given, respectively, for the different blends. In both cases, the modulus goes down with increasing TBC concentration, with less dramatic reductions for the blends with samples C and D (higher MW and block polystyrene content).

Clearly, the best modifying balance is shown by polymer C, with the highest Izod impact results and a smaller effect on tensile and flexural moduli which indicates there must be an optimum MW and block polystyrene content for the TBC to be a good impact modifier by mechanical mixing.

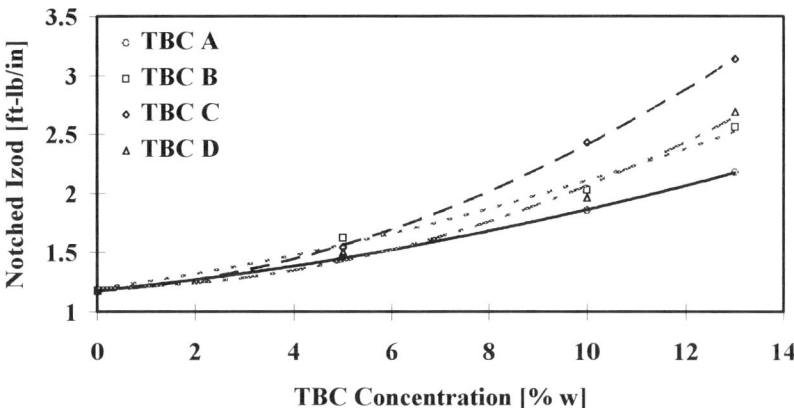

Figure 8. Notched Izod Impact (1/2") of HIPS-TBC blends as a function of TBC concentration and type of TBC.

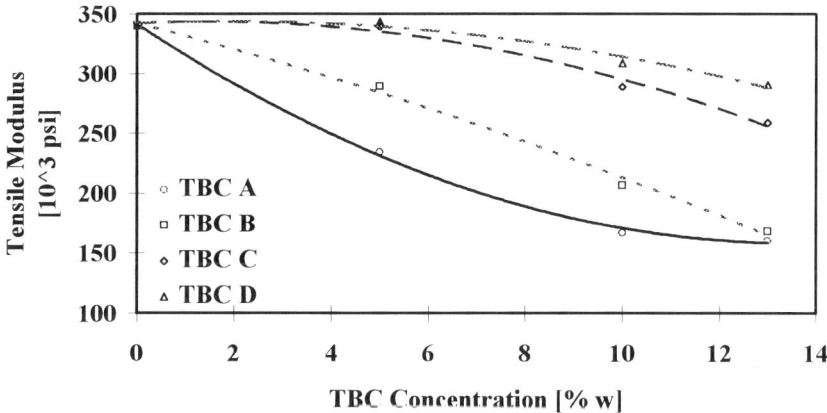

Figure 9. Tensile modulus of HIPS-TBC blends as a function of TBC concentration and type of TBC.

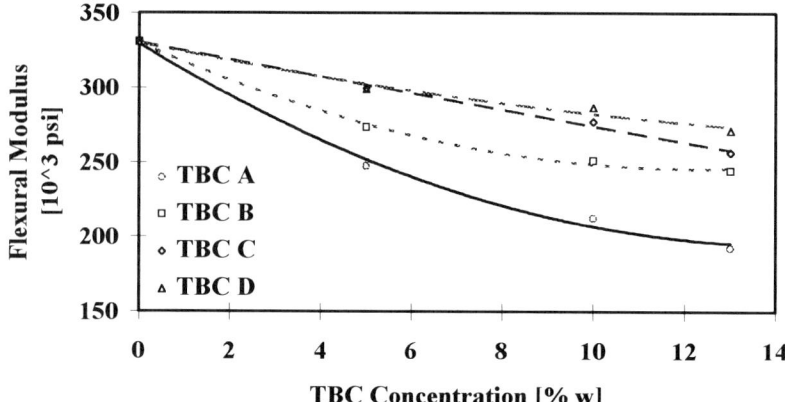

Figure 10. Flexural modulus of HIPS-TBC blends as a function of TBC concentration and type of TBC.

Conclusions

This work shows that relatively simple to produce and inexpensive, linear TBC's can be good asphalt modifiers as well as high impact modifiers for HIPS.

In the case of modified asphalts, there does not seem to be a clear difference in performance between the several TBC's in the range of variables studied and with the particular asphalt sample chosen. Further studies in which the blends will be vulcanized and analyzed by fluorescence microscopy are forthcoming.

All TBC's showed interesting impact modifying capabilities for HIPS extruded mixtures. Polymer C, with an intermediate MW and block polystyrene content gave the best results by reasonably maintaining the tensile and flex moduli while increasing Izod impact by more than 300 % yielding blends with the best balance of toughness, brittleness and stiffness.

Acknowledgments

The authors wish to express their gratitude to Industrias Negromex S. A. for permitting the publication of this work.

References

1. Van Beylen, M.; Bywater, S.; Smetz, G.; Szwarc, M.; Worsfold, D. J. *Adv. Polym. Sci.* **1988,** 86, 87.
2. Odian, G. *Principles of Polimerization;* John Wiley & Sons: New York, NY, 1991.
3. Szwarc, M.; Van Beylen, M. *Ionic Polymerization and Living Polymers;* Chapman & Hall: New York, NY, 1993.
4. Quirk, R. P.; Kim, J. *Rubber Chem. Tech.* **1991,** 64, 450.

5. Hsieh, H. L.; Quirk, R. P. *Anionic Polymerization: Principles and Practical Applications;* Marcel Dekker, Inc.: New York, NY, 1996.
6. Zelinski, R. P. *U.S. Patent 2,975,160 (1961),* assigned to Phillips Petroleum Co.
7. Shim-Ton, J.; Kennedy, K. A.; Piggott, M. R.; Woodhams, R. T. *Rubber Chemistry and Technology* **1980,** 53(1), 88-105.
8. Lewandowski, L. H. *Rubber Chenistry and Technology* **1994,** 67(3), 447-480.
9. Hagenbach, G. V.; Maldonado, P.; Maurice, J. P. *U.S. Patent 4,554,313 (1985),* assigned to Elf France.
10. Gagle, D. W.; Draper, H. L. *U.S. Patent 4,130,516 (1978),* assigned to Phillips Petroleum Co.
11. Bouldin, M. G.; Collins, J. H. *Rubber Chemistry and Technology* **1991,** 64(4), 577-600.
12. Kraus, G. *Rubber Chemistry and Technology* **1982,** 55(5), 1389-1402.
13. *Rubber-Toughened Plastics,* Riew, C. K., Ed.; Advances in Chemistry Series 222; American Chemical Society: Washington, DC, 1989; Chap. 2.

STAR POLYMERS

Chapter 11

High-Temperature Anionic Polymerization Processes: Synthesis of Star-Branched Polymers

F. Schué[1], R. Aznar[1], J. Couve[1], R. Dobreva-Schué[1], and P. Nicol[2]

[1]Laboratoire de Chimie Macromoléculaire, Université Montpellier II, Place E. Bataillon, 34095 Montpellier Cedex 5, France
[2]G.R.L. Groupement de Recherche de Lacq (ELF Atochem) Lacq, 64170 Artix, France

> Basic studies on the stability of polystyryllithium and polybutadienyllithium species at high temperature, have been of great interest to further synthesis of well defined styrene-butadiene copolymers.
> Subsequent works on the coupling of living poly(styrene-b-butadienyl)lithium with diesters have led to the control of star shaped structures. The molecular characteristics of these polymers called CLIPS (clear Impact polystyrene) have been checked with regard to their synthesis. Asymmetrical star-shaped copolymers have been developed in a high temperature, high solid content process. A good concordance is observed between the theoretical and the experimental values.

Much kinetic data are available which elucidate the initiation and propagation reactions of anionic polymerization systems. They are described as "living polymerizations", and the presence of a termination reaction is often neglected. These polymerizations result in the formation of narrow molecular-weight distribution polymers (1-3). The absence of a termination reaction is a good assumption when impurities are absent and low temperatures are involved. However, some polymerization processes frequently operate at much higher temperatures (50° to 100°C) than these at which kinetic data have been reported. Under such process conditions, termination does occur along with other complex reactions which significantly alter the molecular-weight distribution of the polymer (4, 5). There is also a significant amount of kinetic data describing the thermal stability of alkyllithium compounds (6-8).The mechanism of these elimination reactions is conveniently elucidated by manometric methods, and mechanistic interpretations have been proposed. However, there are less published accounts of the thermal stability of polymer lithium systems (2-9). There is a need to have kinetic data describing the thermal termination of "living" polymers at lithium concentrations approaching those of commercial polymerization systems. This paper will discuss the thermal stability and chain transfer to solvent of concentrated solutions in ethylbenzene of polystyryllithium, polybutadienyllithium and poly(styryl-b-butadienyl)lithium. Finally, we examine the practical case for a high concentration of monomers for the polymerization of styrene at 70°C followed by a block copolymerization with butadiene at 90°C with lithium as counterion. The resulting living block copolymers were finally reacted with diesters and epoxidized soyaben oil to produce star-branched polymers.

Results and discussion

Thermolytic behaviour of polybutadienyllithium and polystyryllithium in ethylbenzene at high concentrations

Polystyryllithium. In a first experiment we synthesized low Mw polystyryllithium respectively in pure decalin, decalin + 10 % (by vol.) ethylbenzene and in pure ethylbenzene with polymerization conducted at room temperature (n-BuLi = 1.6 x 10^{-2} mol.L^{-1}, solvent = 87 % by weight). The thermal stability of polystyryllithium as well as the chain transfer to ethylbenzene were estimated (Figure 1). The UV titration data indicated that the decrease in polymer lithium concentration is slightly enhanced when changing from pure decaline to pure ethylbenzene. These results may suggest that the degradation reactions are in fact the predominant ones in ethylbenzene compared to chain transfer. Indeed the occurrence of transfer was confirmed by Sigwalt and co. in the case of the anionic polymerization of styrene in toluene (10). In a second set of experiments, the polystyryllithium was prepared using n-butyllithium as a catalyst. Polymerization of the styrene was initiated at room temperature and maintained at that temperature through out the polymerization. After polymerization was complete, the concentration of the living polymer in ethylbenzene was about 67 % (by weight) and the temperature of the reactor was increased to 80°C desired for thermolysis. The 340 nm peak characteristic of the polystyryllithium end - groups, gradually disappears and a new peak appears in the region of 440 nm together with some tails of absorption between 500 and 600 nm. Indeed polystyryllithium solution changed from yellow to dark red upon heating. The thermolysis of polystyryllithium can involve an elimination reaction with the formation of unsaturation in the polymer chain and lithium hydride.

$$\sim\!\sim\!\sim CH_2-CH^-Li^+(Ph) \longrightarrow \sim\!\sim\!\sim CH=CH(Ph) + LiH$$

As shown by Spach et al (11) the new polymer formed, with the C=C end-group, possess a relatively acidic hydrogen which may be transfered to the carbanion of the non-isomerized polymer.

$$\sim\!\sim\!\sim CH(Ph)-CH=CH + \sim\!\sim\!\sim CH_2-CH^-Li^+(Ph) \longrightarrow$$

$$\sim\!\sim\!\sim \overset{Li^+}{C^-}(Ph)-CH=CH + \sim\!\sim\!\sim CH_2-CH_2(Ph)$$

But this reaction is responsible for the absorption at 535 nm, which does not appear clearly on our spectra (Figure 2). It is also impossible in our experimental conditions to detect a new absorption related to the formation of a metalated ethylbenzene, which probably proceeds more or less at high temperature, and is able to initiate styrene and/or butadiene polymerization. The kinetic analysis (Figure 3) of the titration data measured by UV spectrophotometry shows a slower rate of decrease in polymer-lithium concentration as the concentration of polystyrene increases. If we

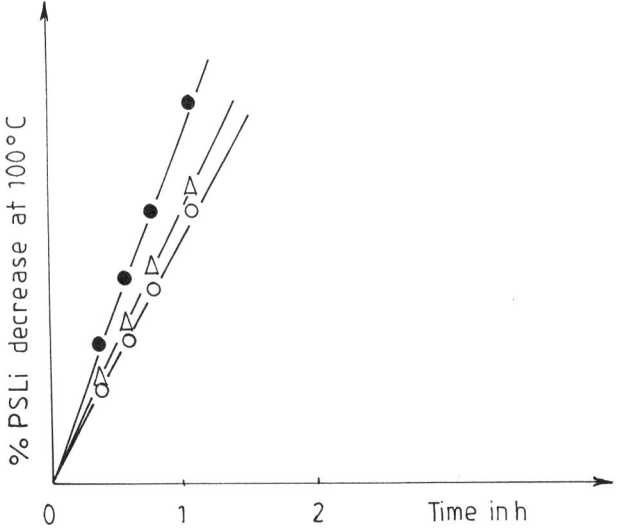

Figure 1. Decrease in polymer-lithium concentration followed at 340 nm as a function of heating time at 100°C for polystyryllithium. (Concentration of living species 1.6×10^{-2} mol.L^{-1}.
Concentration of polymer = 13 % in weight). : pure decalin ; : decalin + 10 % (vol) ethylbenzene ; : pure ethylbenzene.

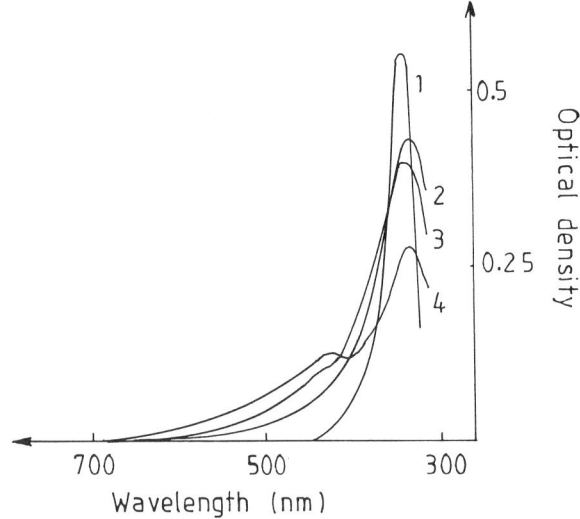

Figure 2. Variation of the UV-visible spectrum of polystyryllithium (67 % in weight of polystyrene in ethylbenzene) at 80°C versus time. (1) End of the polymerization at room temperature.
(2) After 1.5h of heating time. (3) After 2h of heating time.
(4) After 3.5h of heating time. (Concentration of living species 1.1×10^{-2} mol.L^{-1}).

assume that the decrease in polymer-lithium concentration as a function of heating time is mostly due to a complex mechanism involving elimination and metalation reactions, we can suppose that the slower rate mentioned previously can be attributed to an increase of the viscosity of the solution. Moreover, the occurrence of the chain transfer to ethylbenzene decreasing with increasing polystyrene concentration, can also contribute to the observed data.

Polybutadienyllithium. Initially, polybutadienyllithium solutions were colorless to light yellow, but upon heating the color changed through dark yellow to dark amber. The dark color of heated polymer solutions was destroyed by complete termination (4). Ultraviolet spectrophotometric analysis of polybutadienyllithium solution in ethylbenzene at 100°C showed a peak at about 300 nm which gradually disappears together with some tails of absorption appearing between 350 and 500 nm (Figure4). The titration data in figures 5 and 6 show the carbon-bound lithium concentration of polybutadienyllithium and poly(styryl-b-butadienyl)lithium solutions as a function of heating time at 80°C and 100°C and for several concentrations of polymer in ethylbenzene. Antkowiak and Nentwig (12) and Sinn (5) have studied the thermolytically induced reactions for poly(butadienyl-) and (isoprenyllithium) in hydrocarbon solvents. Their combined findings are shown in the following equations :

$$2 \sim CH_2-CH=CH-CH_2Li \xrightarrow{\Delta} \sim CH_2-CH\overset{CH}{\underset{Li}{\diagup\diagdown}}CHLi$$

$$+ \sim CH_2-CH=CH-CH_3$$

$$\sim CH_2-CH=CH-CH_2Li \xrightarrow{\Delta} \sim CH=CH-CH=CH_2 + LiH\downarrow$$

These results show that the elimination of lithium hydride can occur and that dilithiated chain ends can be formed. The formation of such species may account for the dark red to brown coloration these solutions develop upon heating with no monomer present. As for polystyryllithium, we observe a slower rate of decrease in carbon-bound lithium concentration as the concentration of polybutadiene or poly(styrene-b-butadiene) increases. In order to explain such a behaviour, the same conclusions can be drawn as before. The kinetic data for polystyryllithium and polybutadienyllithium can be compared. In the same conditions of concentrations of living species, polymer and of temperature and time, polybutadienyllithium is three times more stable than polystyryllithium. The degrees of association of polystyryllithium and polybutadienyllithium in aromatic solvents have been proposed to be dimeric and tetrameric (13). It is interesting to note that the observed thermal stabilities are consistent with these suggested degrees of association. We believe also that, even if we cannot conclude definitively for the time being, the chain transfer to ethylbenzene is less effective for polybutadienyllithium, compared to polystyryllithium.

Polymer and copolymer synthesis.

Finally, we examine the important practical case in which pure ethylbenzene is used as solvent as well as high concentration of monomers for the polymerization of styrene at 70°C followed by a block copolymerization with butadiene at 90°C with lithium as counterion. The importance of transfer and thermolytic reactions of

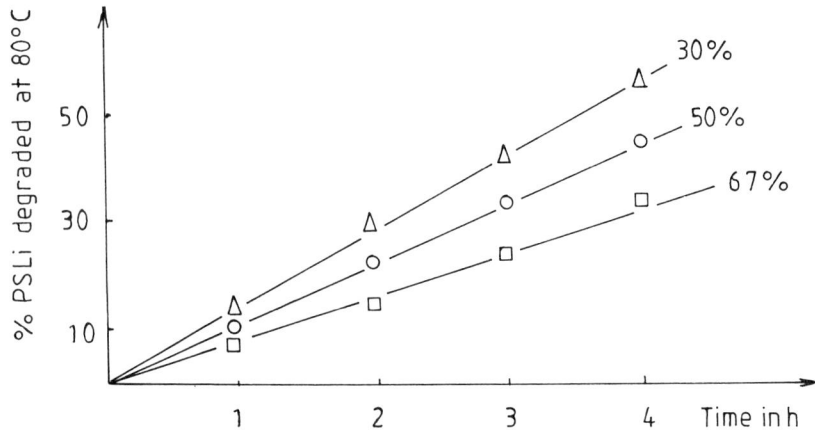

Figure 3. Decrease in polymer-lithium concentration followed at 340 nm as a function of heating time at 80°C for 30 %, 50 % and 67 % by weight of polystyrene in ethylbenzene. (Concentration of living species 1.1×10^{-2} mol.L^{-1}).

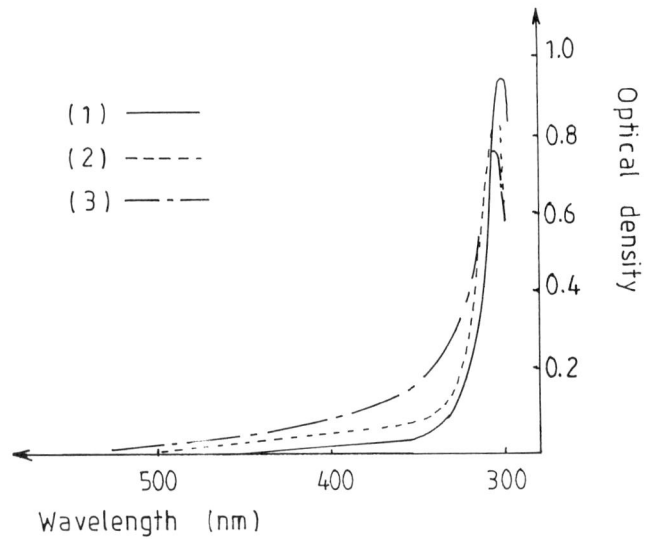

Figure 4. Variation of the UV-visible spectrum of polybutadienyllithium (20 % by weight of polybutadiene in ethylbenzene) at 100°C versus time. (1) End of the polymerization at room temperature. (2) After 1.5h of heating time. (3) After 2.5h of heating time. (Concentration of living species : 1.6×10^{-2} mol.L^{-1}).

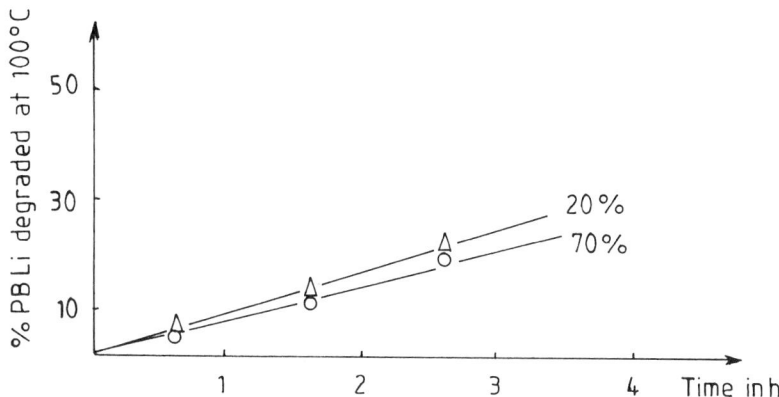

Figure 5. Decrease in polymer-lithium concentration followed at 300 nm as a function of heating time at 100°C for 20 % and 70 % by weight of polybutadiene in ethylbenzene. (Concentration of living species : 1.6×10^{-2} mol.L^{-1}).

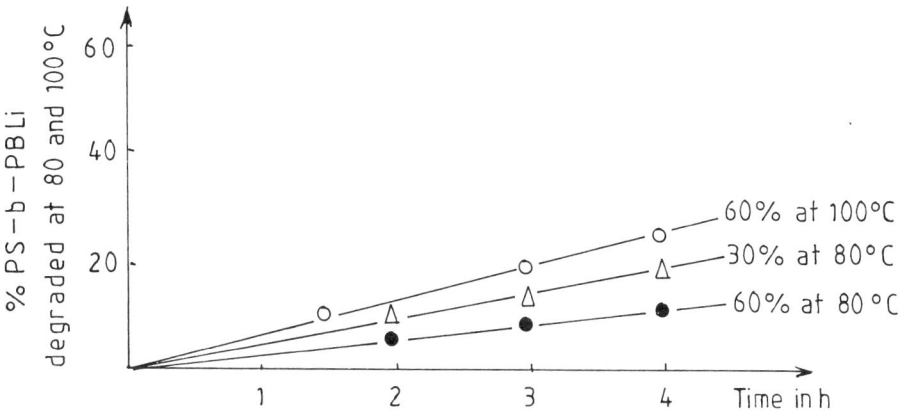

Figure 6. Decrease in polymer-lithium concentration followed at 300 nm as a function of heating time at 80°C for 30 % and 60 % in weight and at 100°C for 60 % by weight of poly(styryl-b-butadiene) in ethylbenzene. (Concentration of living species : 1.2×10^{-2} mol.L^{-1}).

polystyryllithium and polybutadienyllithium was deduced from the experimental molecular weight and the molecular-weight distribution observed. In a first experiment we synthesized a polystyryllithium in pure ethylbenzene at 70°C (initiator : n-BuLi, styrene = 60 % by weight, time of polymerization : two hours). The resulting mixture was then devided into two parts, the first for molecular-weight characterization, the second for adding 16 % by weight of butadiene at 90°C for one hour. GPC analysis of the first sample shows that the polymer had a unimodal molecular-weight distribution (Figure 7a) with $M_{n1theor.}$ = 38,000 ; $M_{n1exp.}$ = 40,000 ; and I = 1.2. The sample obtained after the addition of butadiene shows always a unimodal molecular-weight distribution (Figure 7b) with $M_{n2theor.}$ = 52,400 ; $M_{n2exp.}$ = 55,000 ; and I = 1.2

The increase for the experimental molecular weights compared to the theoretical ones are partly in accordance with the decrease in polymer-lithium concentrations for polystyryllithium (about 15 % at 67 % by weight of polystyrene at 80°C) and for polybutadienyllithium (about 7 % at 70 % by weight of polybutadiene at 100°C). The fairly low polydispersity index of the polymers obtained and the absence of a second GPC signal are indicative respectively of a low decrease of carbon-lithium concentration and apparently nearly no transfer to ethylbenzene. Indeed, if the polymerization is realized during a relatively short time, the extent of transfer is small because the transfer constant is lower than the propagation constant (14, 15). Moreover, the solvent concentration being low, the occurence of transfer can be neglected.

Synthesis of star-branched oligomers by reaction of oligobutadienyllithium with diesters and epoxidized soyabean oil.

Anionic polymerization can produce branches with low molecular weight and narrow molecular weight distributions (16-21). Post polymerization linking involving the still active chain ends can then be done to produce branched oligomers with a predetermined number of arms. These species are then excellent model compounds for testing high molecular weight branches. Here we chose the reaction of oligobutadienyllithium with mono, diester and epoxidized soybean oil (22-24).

Oligobutadienyllithium. In a first experiment, we synthesized in pure toluene (20 ml) a very low molecular-weight oligobutadienyllithium with polymerization made at room temperature (n-BuLi = 0.028 mol. ; butadiene = 0.14 mol.) followed by a deactivation with methanol. The following equations can be written :

$$CH_3\text{-}(CH_2)_3\text{-}Li \ + \ x\ CH_2=CH\text{-}CH=CH_2 \longrightarrow$$

$$CH_3\text{-}(CH_2)_3\text{-}(CH_2\text{-}CH)_{(x-y)}\text{-}(CH_2\text{-}CH=CH\text{-}CH_2)_{(y-1)}CH_2\text{-}CH=CH\text{-}CH_2^{\ominus}Li^{\oplus}$$
$$| \atop CH \atop \| \atop CH_2 \qquad \qquad \Big\downarrow CH_3OH$$

$$CH_3\text{-}(CH_2)_3\text{-}(CH_2\text{-}CH)_{(x-y)}\text{-}(CH_2\text{-}CH=CH\text{-}CH_2)_{(y-1)}\text{-}(CH_2\text{-}CH=CH\text{-}CH_3)$$
$$| \atop CH \atop \| \atop CH_2 \qquad + CH_3O^{\ominus}Li^{\oplus}$$

From the ^{1}H NMR spectrum (Figure 8), the ratio of the vinyl group to the aliphatic methyl groups indicated a $(DP_n)_{exp.}$ close to the theoretical one and a percentage of 1,4 addition equal to 69 % (table I). GPC analysis (Figure 9) shows that the polymer had a unimodal molecular weight distribution and a number average molecular

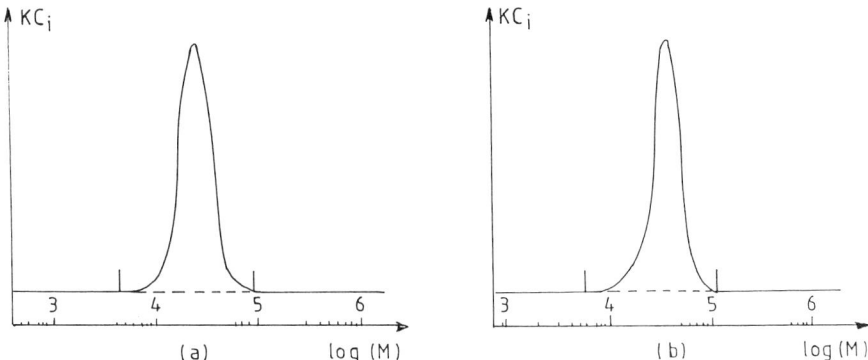

Figure 7. Molecular-weight distribution of polystyrene (a) and poly(styrene-b-butadiene) (b) synthesized in ethylbenzene at high temperature.

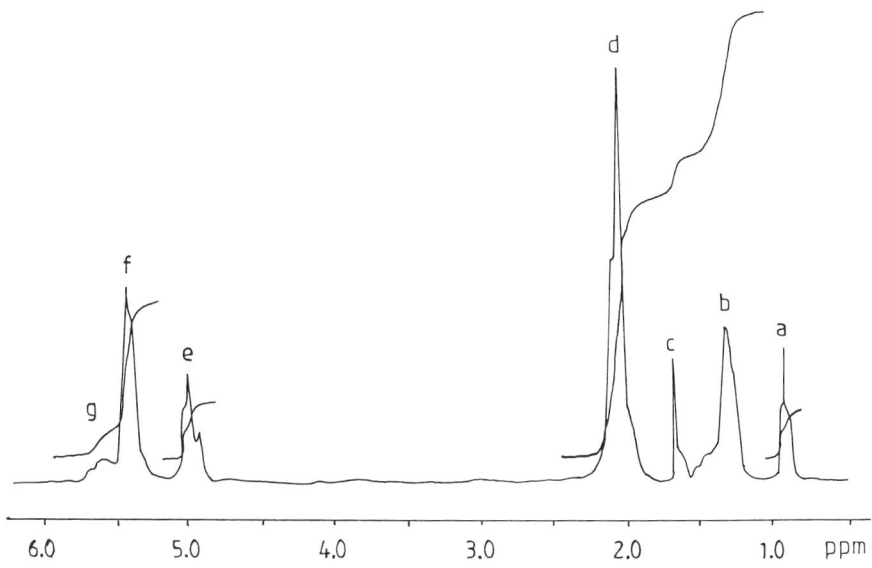

Figure 8. ^{1}H NMR Spectrum in CDCl$_3$ of oligobutadienyllithium.

weight equal to 652 (calibration curve being made with polystyrene) which has to be divided by 1.96 in order to fit the real molecular weight obtained by NMR.

Table I. Characterization by ^{1}H NMR and GPC of an oligobutadienyllithium deactivated by MeOH.

	^{1}H NMR	GPC (polystyrène calibration)
DP$_n$ (theor.)	5	
DP$_n$ (exp.)	5.1	
Mn (exp.)	332	652
Mw (exp.)		780
I		1.2

Coupling with a monoester : ethylacetate (EA). The former living oligobutadienyllithium was added at room temperature to EA in a molar ratio 2/1 and stirred for 48 hours. The ^{1}H NMR spectrum of the coupled oligomer showed the absence of the -OCH$_2$CH$_3$ group at 4.01 ppm indicative of the complete leaving of the alkoxide group (Figure 10). Moreover, GPC analysis (Figure 9) shows that the oligomer has a M$_{n(corrected)}$ = 669 and a M$_{w(corrected)}$ = 723, leading to a degree of coupling equal to 2. In these conditions, the normal product from the reaction of an oligobutadienyllithium compound with an ester remains a tertiary alcohol according to the following equations (BD = 1,4 or 1,2 addition of butadiene :

$$CH_3-\overset{O}{\underset{\|}{C}}-OCH_2CH_3 + CH_3-(CH_2)_3-(BD)_4-BD^{\ominus} Li^{\oplus}$$

$$\downarrow$$

$$CH_3-\overset{O}{\underset{\|}{C}}-(BD)_5-(CH_2)_3-CH_3 + CH_3CH_2O^{\ominus}Li^{\oplus}$$

$$\downarrow CH_3-(CH_2)_3-(BD)_4-BD^{\ominus} Li^{\oplus}$$

$$\begin{array}{c} O^{\ominus}Li^{\oplus} \\ | \\ CH_3-C-(BD)_5-(CH_2)_3-CH_3 \\ | \\ (BD)_5 \\ | \\ (CH_2)_3 \\ | \\ CH_3 \end{array} \xrightarrow{CH_3OH} \begin{array}{c} OH \\ | \\ CH_3-C-(BD)_5-(CH_2)_3-CH_3 \\ | \\ (BD)_5 \\ | \\ (CH_2)_3 \\ | \\ CH_3 \end{array} + CH_3O^{\ominus}Li^{\oplus}$$

Coupling with a diester :

Dimethyladipate (DMA). Several molar ratios "R" : 4/1, 3/1 and 2/1 of oligobutadienyllithium over DMA were tested. The mixture was stirred for 48 hours at room temperature. In all cases, the ^{1}H NMR spectrum showed the absence of the -OCH$_3$ at 3.6 ppm and allows to estimate the degree of coupling. Therefore, the product resulting from a molar ratio 2/1, having a degree of coupling equal to 2, remains a diketone :

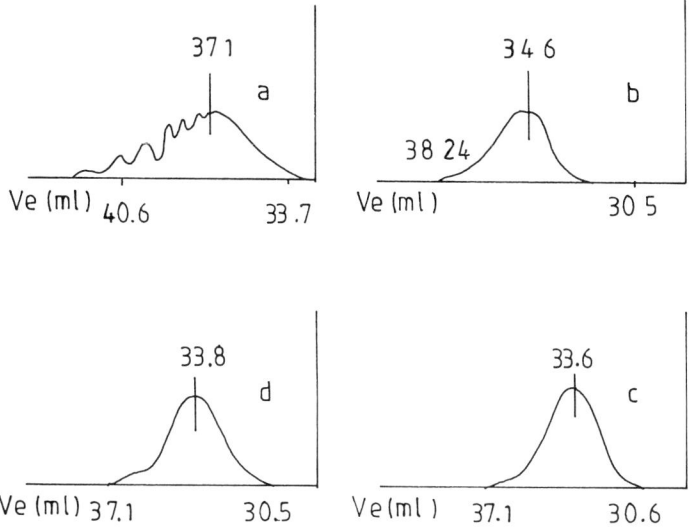

Figure 9. GPC Curves of the precursor oligomer (a) (before coupling) and of products from coupling with ethyl acetate (b), dimethylphthalate (c) and epoxidized soybean oil (d).

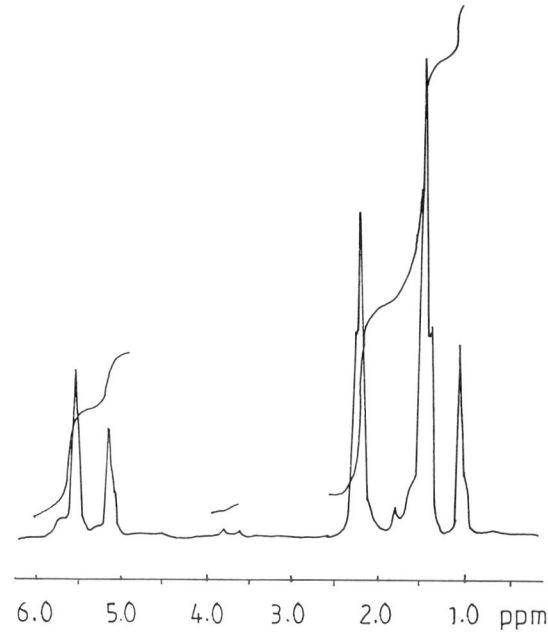

Figure 10. ^{1}H NMR Spectrum in CDCl$_3$ of oligobutadienyllithium coupled with ethyl acetate.

$$CH_3+CH_2\!\!\!\xrightarrow{}_{\overline{3}}\!\!\!+BD)_5\overset{\overset{O}{\|}}{C}+CH_2\!\!\!\xrightarrow{}_{\overline{4}}\!\!\!\overset{\overset{O}{\|}}{C}+BD\!\!\!\xrightarrow{}_{\overline{5}}\!\!\!+CH_2\!\!\!\xrightarrow{}_{\overline{3}}\!\!\!CH_3$$

For R = 3 and 4, the corresponding degrees of coupling are respectively 2.2 and 2.4. It seems that some steric hindrance decreases the rate of addition onto the carbonyl group.

Diethyladipate (DEA). A similar result is obtained for R = 4, and for which the degree of coupling estimated by ^{1}H NMR equals 2.7, close to the one determined by GPC, 2.78.

Dimethylphthalate (DMP). For R = 3, the degree of coupling was estimated by ^{1}H NMR and GPC (Figure 3). In both cases the value obtained was close to 2.5.

Coupling with an epoxidized soyabean oil (ESO). The ESO used in our coupling reaction has a molecular weight of about 930 and a number of epoxy groups per mole equal to 4.3. In Figure 11 we reported the ^{1}H NMR of ESO with all the assignments. The living oligobutadienyllithium was added to ESO in several molar ratio "R" : 4.3/1, 6/1, 8/1 and 10/1.
Table II shows the efficiency of the coupling reaction determined by ^{1}H NMR and GPC as R increases.

Table II. Characterization by ^{1}H NMR and GPC of an oligobutadienyllithium reacted with ESO.

R	Theoretical degree of coupling	Degree of coupling estimated by ^{1}H NMR	Degree of coupling estimated by GPC
4.3	1.43	1.9	1.6 (Fig. 9)
6	2	2.3	2.1
8	2.67	1.8	2.1
10	3.3	2.25	2.5

When R increases from 4.3 to 10, the efficiency of the coupling reaction increases too from 1.6 to 2.5 estimated by GPC. There is also a fairly good fit between these results, those determined by ^{1}H NMR, and theoreticall. We also have to point out that whatever the value of R, we note the absence in the ^{1}H NMR spectrum of the -O-$\underline{CH}_2$- group at around 4.3 ppm. This may also be true for the $-\overset{|}{\underset{}{O}}-\underline{CH}$ group at 5.4 ppm well hidden by the oligobutadiene. Moreover, in all cases, we noted the presence at around 3.1 of some unreacted epoxy group $-\overset{\overset{O}{\diagup\diagdown}}{CH-CH}-$. According to these observations, we can conclude that in the presence of oligobutadienyllithium, the rate of cleavage of the ester group is faster than the addition to the epoxy group by ring opening. For R = 6 we observed the absence of the $-\underline{CH}_2-\overset{\overset{O}{\|}}{C}-O-$ group at 2.4 ppm. At these conditions, the normal product from the reaction of the oligobutadienyllithium with ESO remains a tertiary alcohol with an almost unreacted epoxy group, according to the following equations (BD = 1,4 or 1,2 addition of butadiene) :

$$6 \quad CH_3-(CH_2)_3-(BD)_4-BD^{\ominus} Li^{\oplus}$$

$$+$$

$$\begin{array}{l} CH_2-O-\overset{O}{\underset{\|}{C}}-CH_2-(CH_2)_{(x-1)}-(CH-CH)_{1.43}-(CH_2)_y\ CH_3 \\ | \\ CH-O-\overset{O}{\underset{\|}{C}}-CH_2-(CH_2)_{(x-1)}-(CH-CH)_{1.43}-(CH_2)_y\ CH_3 \\ | \\ CH_2-O-\overset{O}{\underset{\|}{C}}-CH_2-(CH_2)_{(x-1)}-(CH-CH)_{1.43}-(CH_2)_y\ CH_3 \end{array}$$

$$\downarrow 6\ CH_3OH$$

$$3\quad CH_3-(CH_2)_3-(BD)_5-\underset{\underset{\underset{CH_3}{|}}{\underset{(CH_2)_3}{|}}}{\underset{(BD)_5}{\underset{|}{C}}}-CH_2-(CH_2)_{(x-1)}-(CH-CH)_{1.43}-(CH_2)_y\ CH_3$$

$$+\ \underset{\underset{OH}{|}\ \underset{OH}{|}\ \underset{OH}{|}}{CH_2-CH-CH_2}$$

$$+\ 6\ CH_3O^{\ominus}Li^{\oplus}$$

According to the result observed for R = 6, we can assume a similar behaviour for R = 4.3, and suppose the presence of a mixture of a monoaddition with a remaining ketone :

$$CH_3-(CH_2)_3-(BD)_5-\overset{O}{\underset{\|}{C}}-CH_2-(CH_2)_{(x-1)}-(CH-CH)_{1.43}-(CH_2)_y\ CH_3$$

and a biaddition as described before. For R = 10, the degree of coupling estimated by GPC is approaching 3. We also were able by ^{1}H NMR (Figure 12) to estimate 41 % (mol.) of residual epoxide group. In these conditions, we are now in the presence of a mixture of a biaddition as before and a triaddition due to the ring opening reaction of the epoxide group :

$$CH_3-(CH_2)_3-(BD)_5-\underset{\underset{\underset{CH_3}{|}}{\underset{(CH_2)_3}{|}}}{\underset{(BD)_5}{\underset{|}{\overset{OH}{\underset{|}{C}}}}}-CH_2-(CH_2)_{(x-1)}-CH-\underset{\underset{\underset{CH_3}{|}}{\underset{(CH_2)_3}{|}}}{\underset{(BD)_5}{\underset{|}{\overset{OH}{\underset{|}{CH}}}}}-(CH_2)_y\ CH_3$$

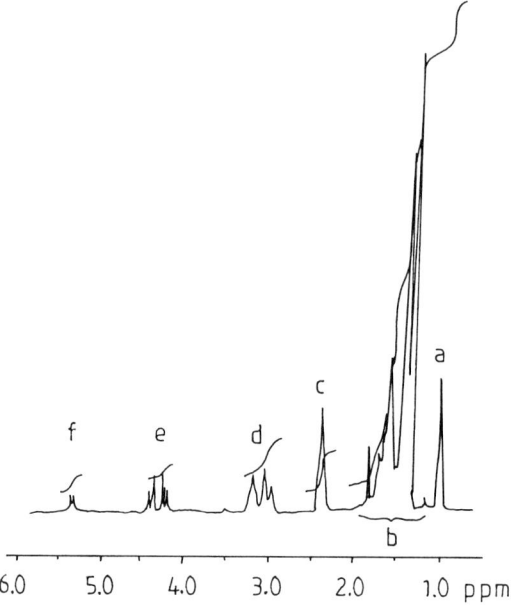

Figure 11. ^{1}H NMR Spectrum in CDCl$_3$ of epoxidized soybean oil.

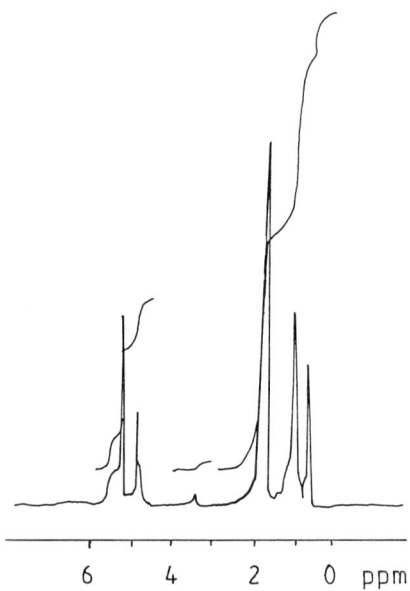

Figure 12. ^{1}H NMR Spectrum in CDCl$_3$ of an oligobutadienyllithium coupled with epoxidized soybean oil (R = 10).

Star-branched polymer synthesis.

Basic studies on the stability of polystyryllithium and polybutadienyllithium species have been of great interest to further synthesis of well defined styrene-butadiene copolymers. Subsequent works on the coupling of living poly(styrene-b-butadienyl)lithium with diesters have led to the control of star shaped structures. The molecular characteristics of the last polymers called CLIPS (Clear Impact Poly Styrene) have been checked with regard to their synthesis. Asymmetrical star-shaped copolymers have been developped in a high temperature, high solid content process, according to the following scheme :

First step (temperature 70°C in ethylbenzene 300 g). A first population of polystyryllithium PS(I) is obtained by polymerizing a first charge of styrene (β_1 = 337 g) with a molar amount (α_1 = 0.0042 mole) of n-butyllithium.

Second step (temperature = 70°C). In the same medium (the first population remaining living) a second population PS(II) obtained by polymerizing a second charge of styrene (β_2 = 174 g) with a molar amount (α_2 = 0.0127 mole) of n-butyllithium :

```
                    β₁                           β₁         β₂
first step.  α₁ |————|            α₁  |—————————|——————|
    PS (I)                                   ⎧ |——————|
                                  2nd step. ⎨ |——————|
                                  PS (II)   ⎩ |——————|
```

α_1 - is the molar fraction of n-butyllithium needed to initiate the first population, α_1 is also the average number of "long chains" in the sample.

β_1 - is the weight fraction of styrene used in the first charge, the number average molecular weight Mn (PS(1)) of the first population being the ratio β_1 (gram)/α_1 (mole).

α_2 - is the molar fraction of n-butyllithium needed to initiate the second population, α_2 is also the average number of "short chains" in the sample.

β_2 - is the weight fraction of styrene used in the second charge.

Third step (temperature = 90°C, final solid content = 70 % by weight). Butadiene (189 g, 27 % by weight) was then added to the resulting mixture from step 2 in order to obtain the block copolymer :

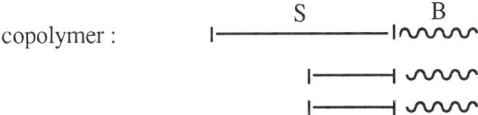

Fourth step (temperature = 90°C). Finaly the resulting block copolymers from step 3, have been linked with a diester (didecyladipate DDA) in a molar ratio R = 3/1 of living chains over DDA.

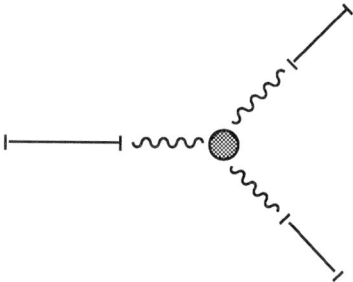

GPC analysis of the samples taken at different steps are reported in table III and (Figure 13).

Table III. Characterization by GPC of the synthesis of a star-branched polymer.

			Theoretical	Experimental
first step	PS(I)	M_n	80,000	82,000
		M_w	-	98,000
		I	-	1.2
2nd step	Bipop. PS	M_n	30,000	28,000
		M_w		78,000
		I		2.8
	M_n (long PS)		90,000	93,000
	M_n (Short PS)		10,000	11,000
3rd step	Bipop. SB	M_n	41,000	45,000
		M_w		81,000
		I		1.8
4th step	linked SB	M_n	92,000	87,000
		M_w		166,000
		I		1.9
		S/B	73/27	73.8/26.2

A good concordance is observed between the theoretical and the experimental values.

Literature Cited

(1) Szwarc, M. "Carbanions living polymers and Electron transfer processes", Intersicence Publishers, New-York, 1968.
(2) Young, R.N., Quirk, R.P., and Fetters, L.J., Adv. Polym. Sci., 1984, 56, 1.
(3) Bywater, S., Adv. Polym. Sci., 1965, 4, 66.
(4) Anderson, J.N., Kern, W.J., Bethea, T.W., and Adams, H.E., J. Appl. Polym. Sci., 1972, 16, 3133.
(5) Nentwig, W., Sinn, H., Makromol. Chem. Rapid Commun., 1980, 1, 59.
(6) Finnegan, R.A., and Kuta, H.W., J. Org. Chem., 1965, 30, 4138.
(7) Colaze, W.H., Lin, J., and Felton, E.G., J. Org. Chem., 1966, 31, 2643.

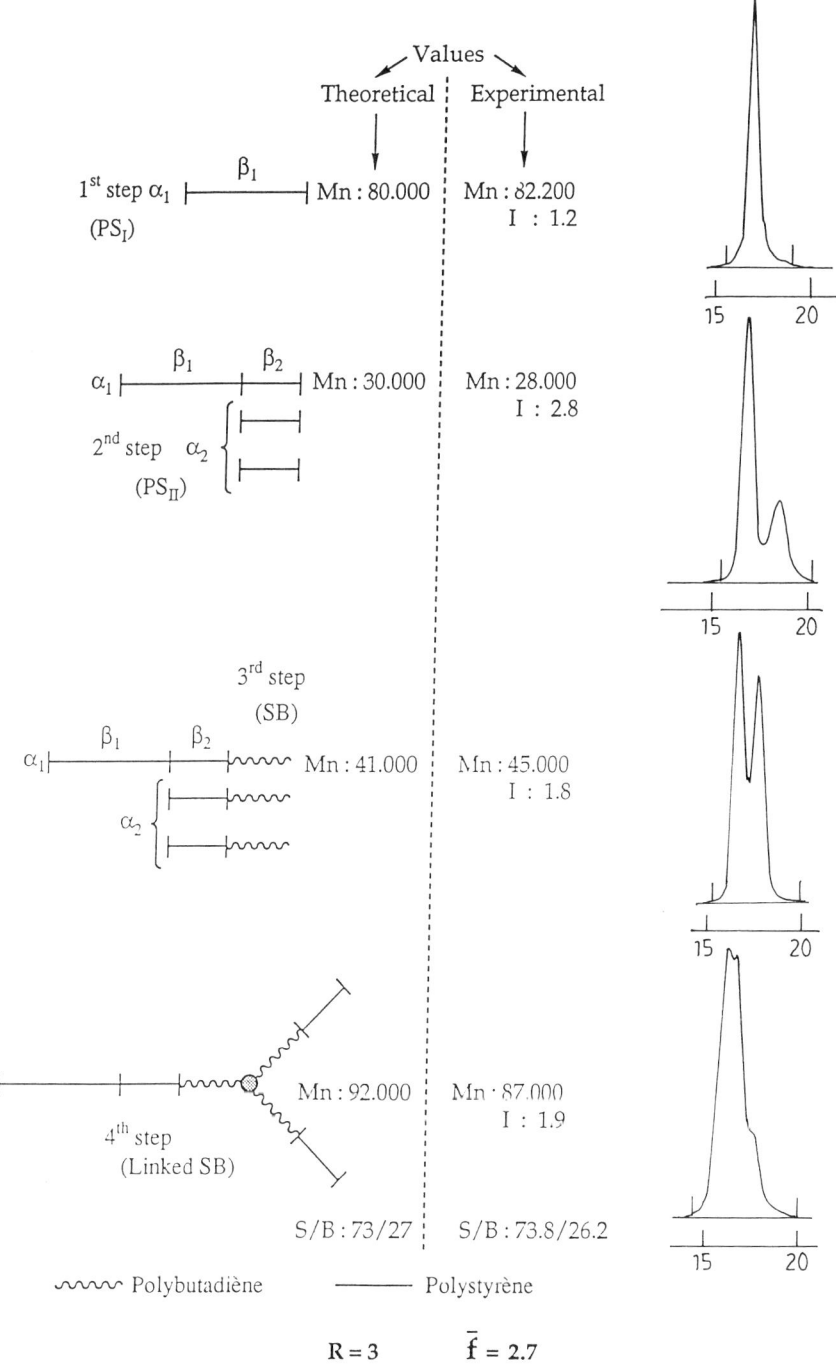

Figure 13. Characterization by GPC of the synthesis of a star-branched polymer.

(8) Colaze, W.H., and Adams, G.M., J. Am. Chem. Soc., 1966, 88, 4653.
(9) Kern, W.J., Anderson, J.N., Adams, H.E., Bouton, T.C., and Bethea, T.W., J. Appl. Polym. Sci., 1972, 16, 3123.
(10) Wang, L.S., Favier, J.C., and Sigwalt, P., Polym. Comm., 1989, 30, 248.
(11) Spach, G., Levy, M., and Szwarc, M., J. Chem. Soc., 1962, 355.
(12) Antkowiak, T.A., Polym. Prep., (Amer. Chem. Soc. Div. Polym. Chem.), 1971, 12, 393.
(13) Hsieh, H.L., and Glaze, W.H., Rubb. Chem. Technol., 1970, 43, 22.
(14) Gehrke, K., Roth, C.H., and Hunerbein, G., Plast. U. Kaut., 1973, 20, 767.
(15) Gatzke, A.L., and Vanzo, E., Chem. Commun., 1967, 1180.
(16) Bywater, S., "Comprehensive chemical kinetics", Amsterdam : Elsevier, 1976, 15, 1.
(17) Morton, M., Helminiak, T.E., Gadkary, S.D., and Bucche, F., J. Polym. Sci., 1962, 57, 471.
(18) Wenger, F., and Yen, S.P.S., Polym. Prep. (Amer. Chem. Soc. Div. Polym. Chem.), 1962, 3, 162.
(19) Altares, T., Wyman, D.P., Allen, V.R., andMeyersen, K., J. Polym. Sci. (Part A), 1965, 3, 4131.
(20) Meunier, J.C., and Leemput, R. van, Makromol. Chem., 1971, 142, 1.
(21) Gervasi, J.A., and Gosnell, A.B., J. Polym. Sci., (Part A1), 1966, 4, 1391.
(22) Hsieh, H.L., Rubb. Chem. Technol., 1976, 49, 1305.
(23) Kitchen, A.G., and Szaila, F.L., U.S. 1972, 3, 639, 517 invs.
(24) Fodor, L.M., Kitchen, A.G., and Biard, C.C., in Deanin, R.D., Ed., "New Industrial Polymers", Am. Chem. Soc. Symposium, Series 4, Am. Chem. Soc., Washington, D.C., 1974, 37.

Chapter 12

Asymmetric Star Block Copolymers: Anionic Synthesis, Characterization, and Pressure-Sensitive Adhesive Performance

J.-J. Ma, M. K. Nestegard, B. D Majumdar, and M. M. Sheridan

3M Adhesive Technologies Center, 3M Center Building 201-4N-01, St. Paul, MN 55144–1000

Asymmetric star block copolymers were prepared using conventional anionic polymerization methodology. The asymmetric star block copolymer contains at least two kinds of arms with varied molecular weight polystyrene (S) endblocks, typically, high molecular weight S endblocks and low molecular S endblocks. When formulated into pressure sensitive adhesives with selected tackifiers, the high molecular weight S blocks provide a necessary physical crosslinking network while the lower molecular weight S blocks provide the system with good deformability and improved lifting resistance.

Poly(styrene-*b*-diene-*b*-styrene) block copolymers have been widely used in fabrication of pressure sensitivie adhesives because of their unique thermoplastic elastomer features (*1*). The conventional block copolymers known as Kraton polymers of poly(styrene-*b*-isoprene-*b*-styrene) (SIS) poly(styrene-*b*-butadiene-*b*-styrene) (SBS) and their hydrogenated versions, however, do not always provide satisfactory performance in removable tape applications where the balanced adhesion and resistance to lifting from the substrate are critical. In this paper, we describe synthesis and characterization of a new kind of asymmetric star block copolymers prepared by sequential addition of styrene monomer and *sec*-butyllithium initiator. The asymmetric star block copolymer contains at least two kinds of arms with varied molecular weight polystyrene (S) endblocks, typically, high molecular weight endblocks of S_{high} =10,000 to 30,000 Dalton and low molecular endblocks of S_{low} =3,000 to 8,000 Dalton (*2*). When formulated into a pressure sensitive adhesive with the selected tackifiers, the high molecular weight S endblocks provide necessary physical crosslinking networks while the lower molecular weight S endblocks provide the system with good deformability and improved lifting resistance. The synthesis, morphology and adhesive performance of the asymmetric star block copolymer will be discussed in this paper.

EXPERIMENTAL

Solvent, initiator, monomer and coupling agent. Cyclohexane, styrene, isoprene and divinylbenzene were purified according to the procedure described by Morton and Fetters (*3*). *sec*-Butyl lithium of 12 wt % in cyclohexane was purchased from FMC Corporation Lithium Division and used without further purification.

Polymerization. Polymerization was conducted in a single neck round bottom flask equipped with a Rotoflo™ stopcock and a magnetic stirring bar under argon atmosphere. The mixed molecular weight polystyrene (S) end blocks were made by the sequential addition of styrene and *sec*-butyl lithium initiator. The living polystyrene anion solution was then added into a flask pre-filled with cyclohexane and isoprene and kept at 55 °C for one to two hours. The living poly(styrene-*b*-isoprene) (SI) anions were then linked into a star architecture by using divinylbenzene as a coupling agent (*4*). The number average molecular weight (M_n) of S endblocks, arms and star block copolymers was determined by a Hewlett-Packard Model 1082B size exclusion chromatograph (SEC) using polystyrene standards.

RESULTS AND DISCUSSION

Synthesis. The synthesis route of the asymmetric star block copolymer is shown in Scheme I. The mixed molecular weight polystyrene endblocks, S_{high} and S_{low}, were prepared by sequential addition of styrene (St) monomer and *sec*-butyllithium (*sec*-BuLi) initiator. The living S endblocks were then allowed to polymerize isoprene (I) to form two kinds of diblocks, I-S_{high} and I-S_{low}. The final star contains a

Scheme I

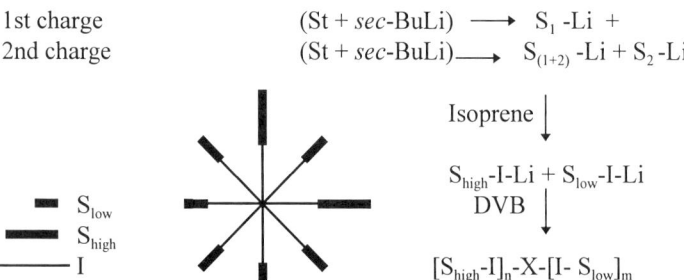

statistical combination of two kinds of arms after coupling reaction using divinylbenzene (DVB) (*4*). The size exclusion chromatography (SEC) curve of the mixed molecular weight S endblocks showed a bimodal molecular weight distribution with peak areas proportional to the ratio of S_{high}/S_{low} (Figure 1). The coupling reaction with DVB typically gave 85-95% coupling efficiency as shown in Figure 1. The SEC results of the asymmetric stars with varied molecular structures are listed in Table I.

Table I. Star block copolymers with mixed molecular weight S endblocks.

Sample No.	Mn (10³ Dalton)[a]					Coupling[b] (%)	Tg (°C)[c] I Block
	S_{high}	S_{low}	(S_{high}/S_{low})	Arm	Star		
A	26	4	0.2/0.8	173	1,270	90	-62
B	15	3	0.4/0.6	135	1,136	99	-64
C	9	2	0.6/0.4	96	775	97	-64
D		4		175	1,166	84	-64
E	13			140	1,109	92	-64

[a] Measured by SEC using polystyrene standards. [b] Calculated from SEC profile. [c] Measured by DSC.

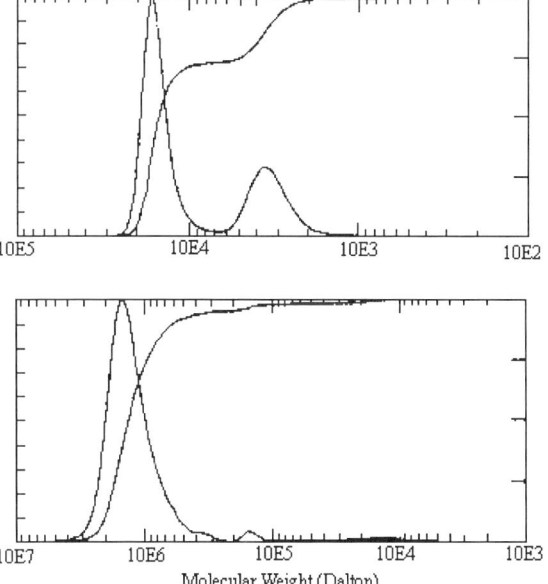

Figure 1. SEC profiles of mixed molecular weight S endblocks (top) and an asymmetric star (bottom, Table I, Sample B).

Morphology and Rheology. It is well known that regular SIS triblock copolymers and $(SI)_n$ star block copolymers have phase separated morphology. The shape of polystyrene domains may be sphere, rod or lamella depending on polystyrene block percentage (St %) (5). Transmission Electron Microscopy (TEM) pictures of the asymmetric star (Table I, Sample B) showed sphere-shaped polystyrene domains evenly distributed in rubber matrix (Figure 2). This is not surprising since the asymmetric star block copolymers listed in Table I have relatively low St %, about 6-10%. The polystyrene domains of the asymmetric have a comparable size to the regular $(SI)_n$ star (Table I, Sample E). This seems to indicate that the endblock domain size of the asymmetric star mainly depends on the high molecular weight endblocks, S_{high}; both S_{high} and S_{low} blocks stay in the same domain.

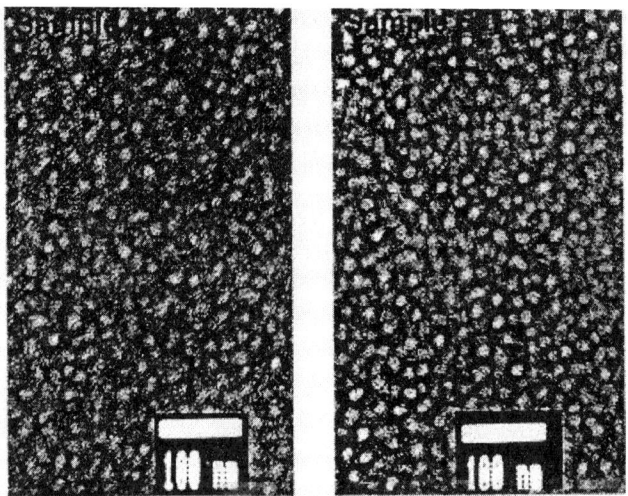

Figure 2. TEM pictures of an asymmetric star block copolymer (Left, Table I, Sample B) in comparison to a regular star block copolymer (Right, Table I, Sample E).

The asymmetric star block copolymer showed sloped plateau on Log G' vs. Temperature plot (Figure 3, Sample A and B). The star with S_{high}=26,000 dalton showed no obvious drop off point (Figure 3, Sample A) within the temperature range tested. Apparently, the rheological behavior of the asymmetric star block copolymer is different from that of the regular star, which generally shows a flat plateau region with a clearly defined drop off point (Figure 3, Sample E). The Log G" vs. Temperature plot of the asymmetric star block copolymer showed an obvious damping peak in the temperature range of 20-70 °C (Figure 3, Sample A and B). The additional energy dissipation peak might be associated with relaxation of the low molecular weight endblocks, S_{low}, from the physically crosslinked network. It has clearly been shown that the rheological behavior of the asymmetric star block copolymer can be manipulated by the molecular structure parameters of this kind of star polymer.

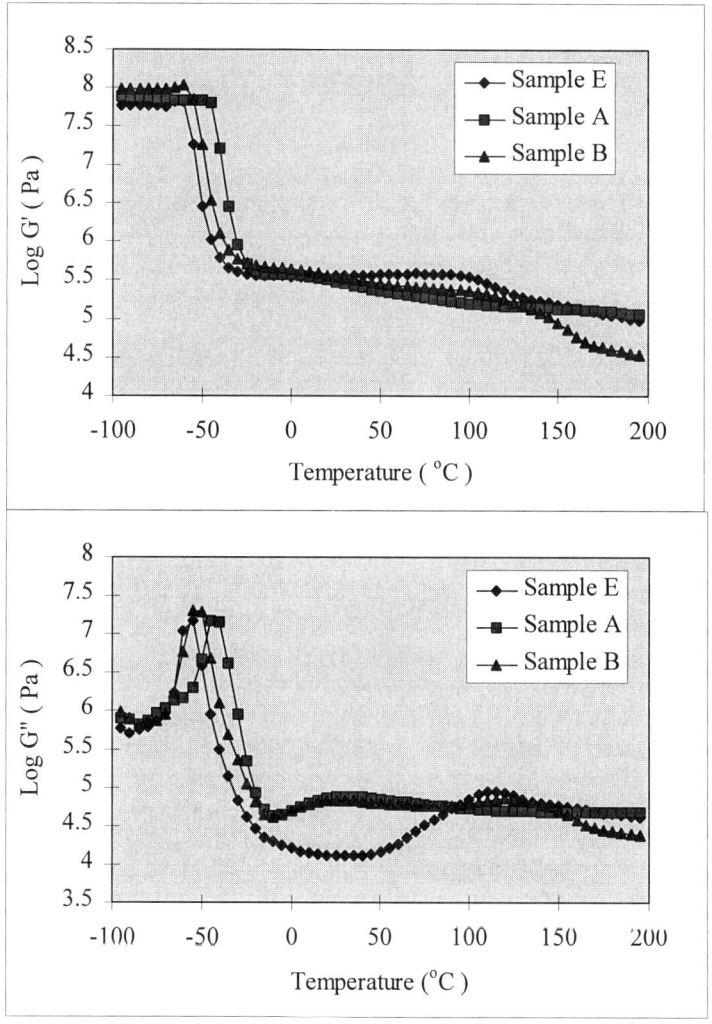

Figure 3. Log G' and Log G'' vs. Temperature plots (0.3 HZ) of asymmetric star block copolymers (Table I, Samples A and B) in comparison to a regular star block copolymer (Table I, Sample E).

Mechanical Properties. Table II shows the tensile behavior of an asymmetric star block copolymer (Table I, Sample A), a regular star with S endblock molecular weight of 13,000 Dalton (Table I, Sample E), and a commercial star block copolymer (Kraton D1320, Shell Chemical Company). The asymmetric star sample showed greatly reduced tensile strength and increased elongation at break as compared to the regular and commercial star block copolymers. The asymmetric star block copolymer also showed reduced tensile modulus. These results indicate that the introduction of low molecular weight S endblocks provides the asymmetric star with better deformability under stress compared to the conventional block polymers.

Table II. Tensile properties of star block copolymers with different molecular structures[a]

Molecular Structure	Modulus (MPa)	Stress Max (MPa)	Strain Max
Regular Star (Sample E)[b]	0.74	8.71	12.2
Kraton D1320(Shell Star)[c]	0.58	7.64	13.2
Asym. Star (Sample A)[b]	0.47	3.97	20.3

[a] Tensile properties were tested on an Instron with crosshead speed at 90 in/min.
[b] See Table I for molecular structure of the star block copolymer. [c] Commercial sample from Shell Chemical Company.

Tape Performance. One of the objectives of developing the asymmetric star block copolymer is to provide a base polymer which can be formulated into pressure sensitive adhesives suitable for a variety of tapes including removable tapes. This objective has been difficult to achieve by using regular triblock or star block copolymers. Conventional block copolymers usually lack sufficient resistance to low stress peel when formulated into a range of adhesion desirable for removable tapes. The adhesive performance of asymmetric star block copolymers (Table III, Sample A, B, and C) is listed in comparison to the regular stars with S endblocks of 4,000 Dalton (Table III, Sample D) and 13,000 Dalton (Table III, Sample E) respectively.

The adhesive made from an uncrosslinked block copolymer having only low molecular weight S endblocks showed insufficient resistance to low stress peel (3.2 minutes) and failed in the cohesive mode, leaving adhesive residue on the substrate upon separation (Table III, Row 4,). The tape comprised of an uncrosslinked block copolymer having only high molecular weight styrene endblocks also showed insufficient resistance to low stress peel (18 minutes, Table III, Row 5,). When crosslinked, Sample D showed improved resistance to low stress peel but still failed cohesively, and Sample E had unsatisfactory resistance to low stress peel.

Tapes made from the asymmetric star (Samples A, B, and C) showed good resistance to low stress peel; they resist lifting under light loads. They maintained moderate adhesion, thereby being easy to remove. They were removed cleanly from the substrate without leaving adhesive residue (Table III, Rows 1-3 and 6-9). In addition, crosslinking the adhesive further improved its cohesive strength, and temperature and solvent resistance. This adhesive performance is very useful for making removable tapes such as masking tapes, packaging tapes and labels.

Table III. Adhesive formulation and performance.[a]

Polymer	Dose (Mrads)	Adhesion (oz/in)	(N/m)	LSP[b] (min)	Fail Mode
A	0	44.8	490	193	ADH
B	0	54.4	595	399	ADH
C	0	54.4	595	546	ADH
D	0	43.2	472	3	COH
E	0	44.8	490	18	ADH
A	5	25.6	280	4000	ADH
B	5	35.2	385	6000	ADH
C	5	30.4	332	5000	ADH
D	5	70.4	770	10000	COH
E	5	28.8	315	10	ADH

[a] All samples were formulated with 40 phr tackifier and 30 phr plasticizer.
[b] LSP: low stress peel was measured on 4x0.75 in^2 against stainless steel under 200 gram load at 90 degree. ADH indicates adhesive failure and COH cohesive failure.

CONCLUSION

We demonstrated in this paper that an asymmetric star block copolymer having both low and high molecular weight polystyrene endblocks can be made using anionic polymerization method by sequential addition of styrene monomer and butyllithium initiator. Mechanical and rheological behaviors of asymmetric star block copolymers depend on the molecular structure. The properties of the asymmetric star block copolymer can be adjusted by simply tuning the molecular parameters, such as the molecular weight of S_{high} and S_{low} and the ratio of S_{high}/S_{low}. The presence of both high and low molecular weight S endblocks on the same molecule is essential to provide adhesives with desired removable adhesion and good resistance to lifting with or without crosslinking.

REFERENCES

1. Ewins, E. E.; St. Clair, Jr. D.; Erickson, J. J. R.; Korcz, W. H. in *Handbook of Pressure Sensitive Adhesive Technology*; Editor, Satas, D.,Ed.; 2nd Ed., Van Nostrand Reinhold Book: New York, **1989**; p 317.
2. Nestegard, M. K.; Ma, J.-J. *US Patent,* **1994**, 5296547; *US patent,* **1995**, 5393787.
3. Morton, M.; Fetters, L. J. *Rubber Chem. Technol.*, **1975**, *vol*, **48**, p 359.
4. Bi, L.-K.; Fetters, L. J. *Macromoleculaes*, 1976, *vol*, **9**, p 732.
5. Arggarwal, S. L. *Polymer*, **1976**, *vol*, **17**, p 938.

Chapter 13

Preparation of Branchef Polystyrenes with Known Architecture

J. L. Hahnfeld, W. C. Pike, D. E. Kirkpatrick, and T. G. Bee

STYRON® Resins Research and Development, The Dow Chemical Company, 438 Building, Midland, MI 48667

Three-arm and randomly branched polystyrene samples with well characterized architecture have been prepared for comparison with linear polymers. In order to investigate the influence of polymer architecture on the properties of atactic polystyrene, we have prepared branched polystyrene samples using anionic synthesis. Randomly branched polystyrenes were synthesized by preparing polystyryl dianions of specified molecular weight and coupling these dianion macromers using α,α',α''-trihalomesitylene and α,α''-dihaloxylene. Equations to accurately calculate the branch density and molecular weight were developed. By controlling the ratio of dianion to di- and tri-functional coupling agent, the branch density and molecular weight of the polystyrene could be predicted and controlled. Polystyrene samples with branch densities of zero to 0.75 branches per 500 monomer units were prepared.

Atactic general purpose polystyrene is a multi-million ton per year polymer which was commercialized more than 55 years ago. The commercial success of polystyrene is largely due to transparency, lack of color, ease of fabrication, thermal stability, low specific gravity, relatively high modulus, excellent electrical properties, and in particular, its low cost. Today, commercial polystyrene is virtually all produced via free-radical polymerization.(*1*) General purpose polystyrene is a relatively pure and simple product. The primary variables which differentiate products are molecular weight, molecular weight distribution, and additives. Due to the polymerization chemistry, commercial polystyrene's architecture is essentially linear.

The purpose of this study was to determine the effect of polymer architecture on the properties and processability of polystyrene. In particular, this paper will describe the preparation of polystyrene samples with well characterized architecture. It is particularly important to be able to control and predict the architecture of these model polymers in order to accurately assess the effects of molecular architecture. It should be noted that as a commercial supplier of polystyrene, we are most interested in the effect of polymer architecture on polymer properties. We are interested whether changing the architecture of the polymer will fundamentally change rheology of the polymer such that processing advantages can be obtained. There are numerous studies which have shown that polymer architecture can significantly affect the relationship between molecular weight and viscosity. We consider this effect to be relatively unimportant for this study and are interested in the effects of polymer architecture on rheology irrespective of molecular weight. Due to these priorities, we are most interested in investigating the effects of random branching in polystyrene.

Synthetic Approach

There are many examples of the preparation of star branched polystyrene.(2,3) The number of branches in these polymers can be varied almost infinitely. Comb polymers are also well known in the literature.(3,4) By comparison, there are relatively few examples of the preparation of well characterized, randomly branched polymers.(5) We will describe the preparation of randomly branched polystyrenes in which the number of branches and the molecular weight can be controlled and predicted. Our study of the effects of branching also included the preparation of polymers containing a single branch point (symmetrical three-arm stars). Three-arm stars are, in fact, the most simple branched polymer structure and result in arms with the highest available molecular weight for entanglements. Similar synthetic approaches were used for the preparation of the three-arm stars and the randomly branched samples. Figure 1 shows the general scheme for the preparation of these branched polymers. Both of the synthetic schemes rely on the coupling of polystyryl anions with tri-halo mesitylene to form branch sites. In the case of 3-arm stars we used polystyryl mono-anions and in the case of the randomly branched polymers we coupled polystyryl dianions. α, α'-Dihaloxylenes are also used to extend chains in the syntheses of randomly branched polymers.

Sodium naphthalene was chosen as the difunctional initiator for the preparation of polystyryl dianion. Thus, in the randomly branched samples, the counterion was sodium. A number of initiator systems exist which form similar dianions, many with lithium counterions. Sodium naphthalene was chosen for its high yield of dianion, convenience, and reproducibility.

Model Reactions The reaction conditions were chosen to optimize the coupling efficiency of the polystyryl anions. Figure 2 shows the possible reaction paths when coupling is attempted.(6) Reaction path (a) is the desired nucleophilic substitution. The side reactions shown in paths (b) and (c) are proton abstraction and electron transfer to radical respectively. Proton abstraction, path (b), leads to chain transfer resulting in uncoupled polymer. Electron transfer, path(c), leads to radical coupling which is uncontrolled. Experiments were designed to help determine the conditions

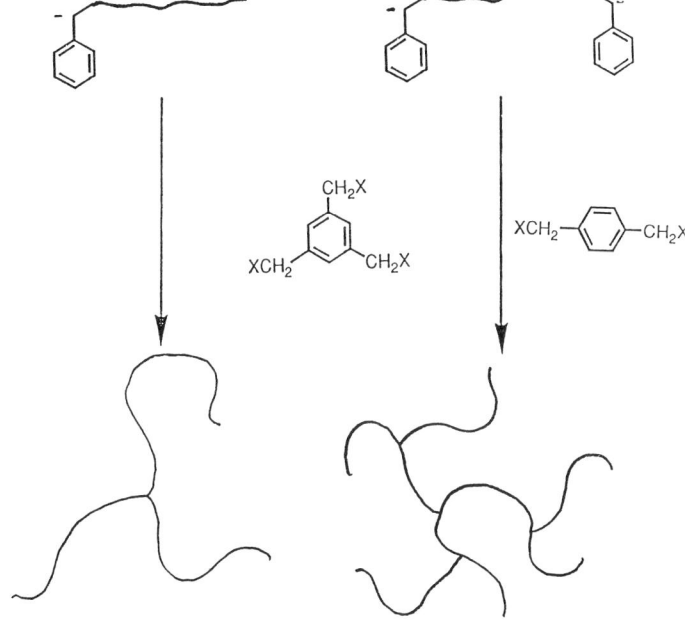

Figure 1. General reaction scheme for the preparation of three-arm and randomly branched polystyrenes.

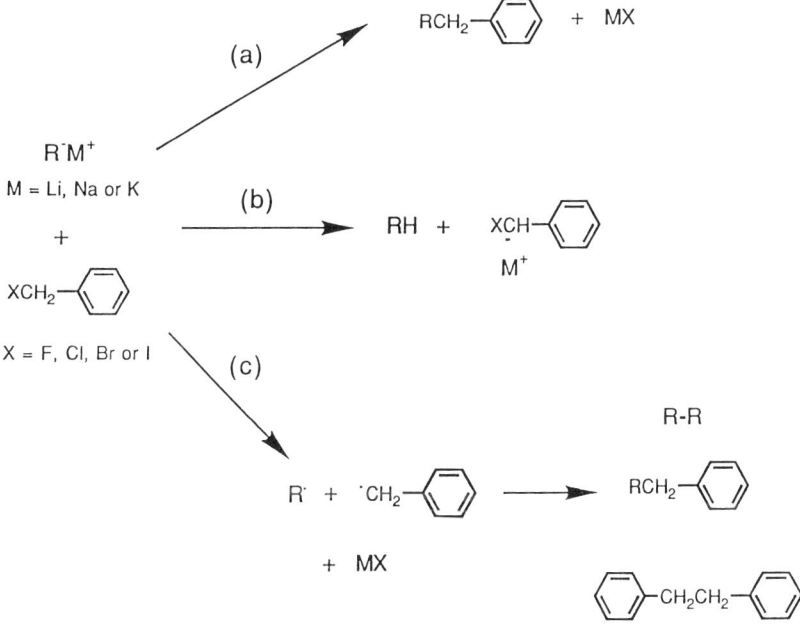

Figure 2. Reactions of alkyl metals with benzyl halides.

leading to exclusive nucleophilic substitution. Polystyryl dianions were reacted with benzyl halide and dihaloxylene in separate reactions. Figure 3 shows that, if nucleophilic substitution is taking place, the reaction with benzyl halide results in polystyrene with no molecular weight change but reaction with dihaloxylene results in a large increase in molecular weight. The results of proton abstraction are shown in Figure 4. In this case, the reaction with benzyl halide as well as the reaction with dihaloxylene will result in polymer showing no change in molecular weight. Finally, Figure 5 shows that electron transfer with radical coupling will result in an increase in molecular weight in reactions with benzyl halide as well as dihaloxylene. The amount of proton abstraction is expected to increase in the coupling agent order of $\Phi\text{-}CH_2I < \Phi\text{-}CH_2Br < \Phi CH_2Cl < \Phi\text{-}CH_2F$ and as the counterion ranges from lithium to potassium. However, the amount of electron transfer is expected to follow the inverse order in both halide and counterion. Also, additives which complex the counterion and low termperatures suppress both of these side reactions.

Reactions in model systems were carried out in order to optimize the amount of nucleophilic substitution. Figure 6 illustrates the reactions of polystyryl dianion with benzyl chloride and benzyl bromide in benzene. Nucleophilic substitution should result in virtually no increase in molecular weight when the dianion is capped with the benzyl halides. The GPC curves in Figure 6 indicate that in the reaction with benzyl bromide significant molecular weight increase is seen, indicating electron transfer and radical recombination is taking place. This pathway is suppressed in the reaction with benzyl chloride. The relatively small amount of molecular weight increase indicates that nucleophilic substitution or proton abstraction predominate. Figure 7 shows the results of the reaction of polystyryl dianion with dihalozylenes in benzene. With nucleophilic substitution we would expect a large increase in molecular weight. The GPC molecular weight results in Figure 7 show that reaction of the dianion with both α,α'-dibromo-p-xylene and α,α'-dichloro-p-xylene result in significant polymer remaining which shows no molecular weight increase. The large amount of polymer showing no molecular weight increase indicates relatively low coupling efficiencies, indicating proton abstraction is interfering.

The poor coupling efficiencies can be improved by changing the solvent to a 1:1 mixture of benzene/THF and using the chloride coupling agents. Figure 8 shows the reaction of polystyryl dianion with α,α'-dichloro-p-xylene and benzyl chloride in benzene/THF. The reaction with α,α'-dichloro-p-xylene results in a very nice molecular weight increase with very little of the original molecular weight material remaining. This indicates that proton abstraction has been efficiently suppressed. The reaction with benzyl chloride results in almost all of the polymer not changing molecular weight. There is a very small peak at twice the original molecular weight which could result from a small amount of radical transfer or contamination with a small amount of oxygen which also results in coupling. Evaluation of these reactions indicate that the coupling efficiency of this system is 95%.

Preparation of branched polymers using this strategy requires the reaction of three macromers at the same branch point. To determine if steric hindrance significantly

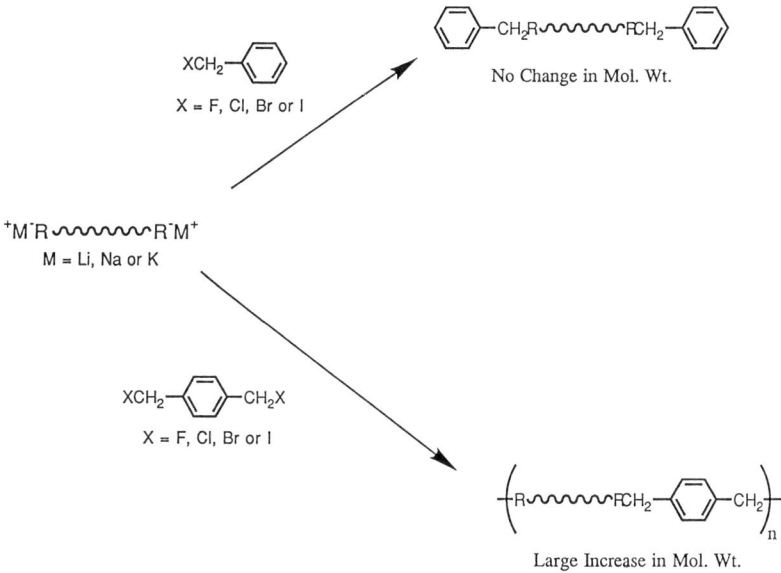

Figure 3. Products expected from nucleophilic substitution in the reaction of polystyryl dianion with benzyl halide and dihaloxylene.

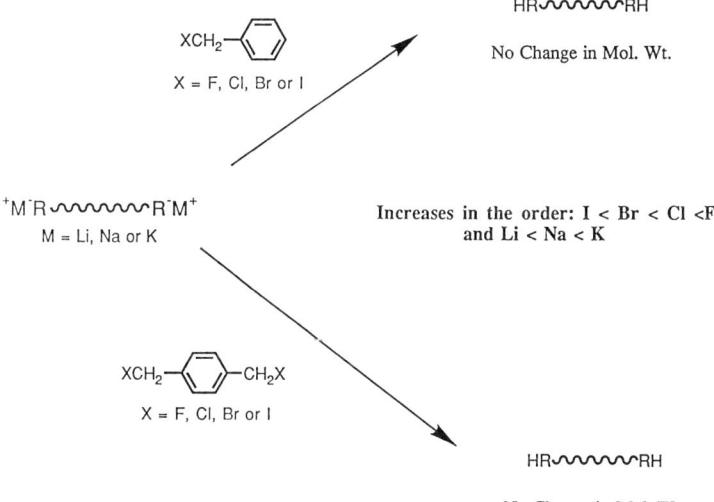

Figure 4. Products expected from proton abstraction in the reaction of polystyryl dianion with benzyl halide and dihaloxylene.

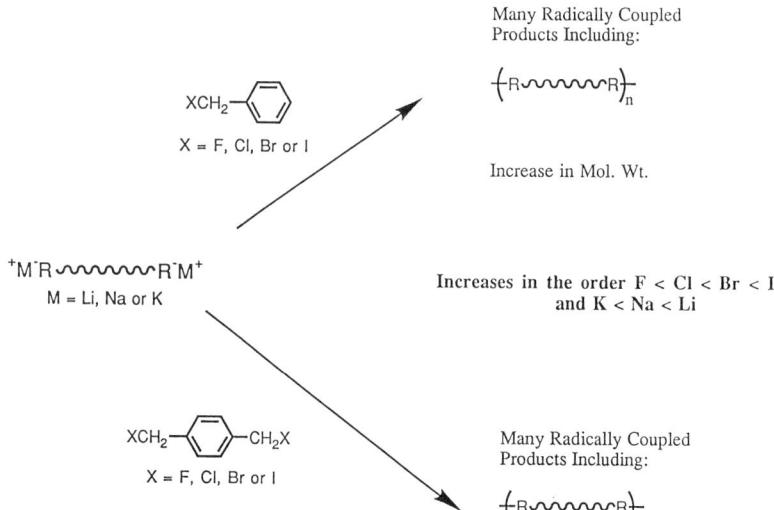

Figure 5. Products expected from electron transfer and radical coupling in the reaction of polystyryl dianion with benzyl halide and dihaloxylene.

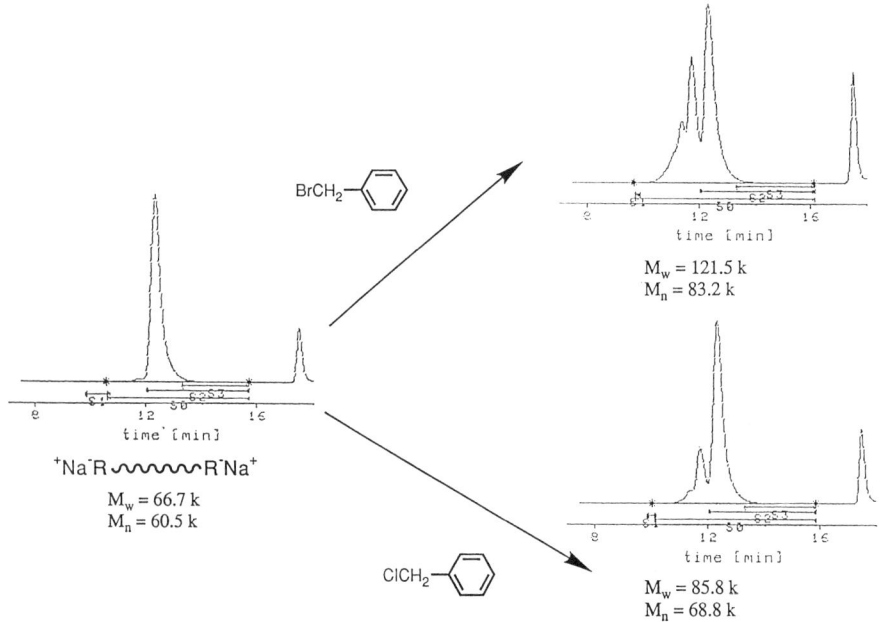

Figure 6. GPC traces of the reaction products of polystyryl dianion with benzyl bromide and benzyl chloride in benzene solvent.

173

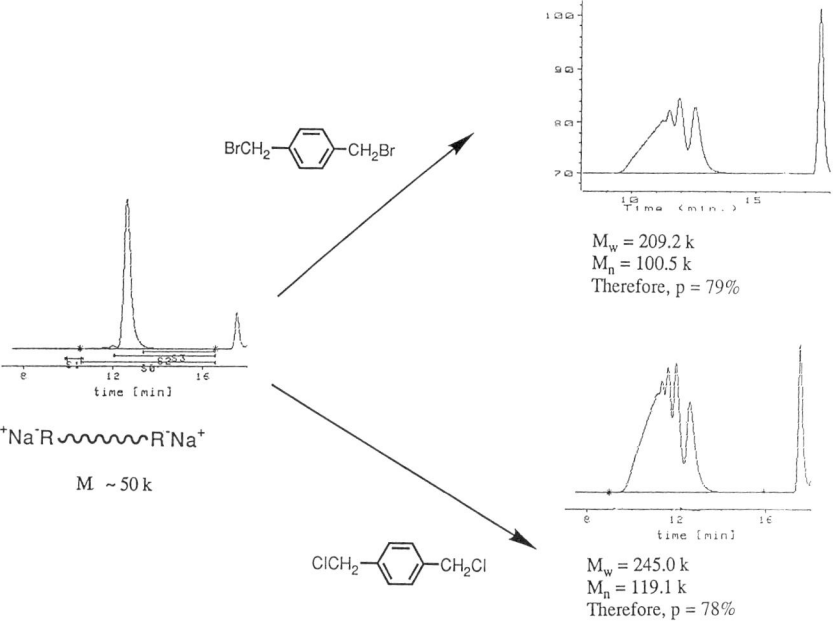

Figure 7. GPC traces of the reaction products of polystyryl dianion with α,α'-dibromo-*p*-xylene and α,α'-dichloro-*p*-xylene in benzene solvent.

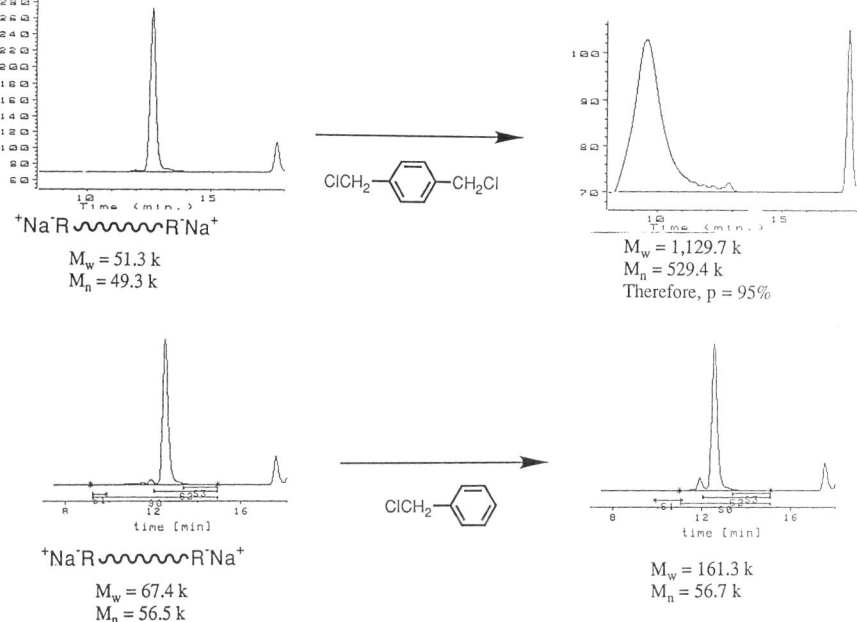

Figure 8. GPC traces of polystyrenes prepared by the reactions of polystyryl dianion with α,α'-dichloro-*p*-xylene and benzyl chloride in benzene/THF solvent (1:1).

lowers the coupling efficiency at these branch points, mono-functional polystyryl lithium can be reacted with α,α',α"-trichloromesitylene to form three-arm star polymers. Figure 9 shows the synthetic route and the GPC molecular weight data for the preparation of three-arm star polystyrene. In this synthesis the polystyryl anion is generated using n-butyl lithium in benzene/THF. Bryce and coworkers have performed this experiment and determined the coupling efficiency, p, to be 92%.(7) We obtained similar results in our laboratory.

Synthesis and Purification of Three-Arm Polystyrene   In the synthesis of three-arm polymers, the molecular weight is most accurately determined by removing a small amount of the reaction solution for molecular weight analysis after the generation of the polystyryl anion. The coupling reaction is then carried out by adding the α,α',α"-trichloromesitylene to the anion solution. The best results are obtained when the coupling agent is the limiting reagent. In this case the three-arm star contains a small amount of uncoupled material. If the polystyryl anion is the limiting reagent the final product contains three-arm, two-arm, and uncoupled material. We have prepared materials with molecular weights of 120,000, 200,000, 300,000, and 450,000. Each of these materials was contaminated with a small amount of uncoupled material. The uncoupled material can be effectively removed by molecular weight fractionation via selective precipitation.(8) The benzene/THF solution (1 part) was mixed with 8 parts acetone and 1 part methanol to selectively precipitate the desired product. These results are shown in Figure 10. The solution contains essentially all of the uncoupled product and a small amount of the three-arm polymer. A very small amount of two-arm product is also seen in this solution. The precipitated polymer is essentially all three-arm polymer. Each of the three-arm polymer samples was purified in this manner.

Synthesis of Randomly Branched Polystyrene   The reaction scheme for preparing the randomly branched polymer is shown in Figure 11. In this reaction scheme, polystyryl dianion of a specified molecular weight is prepared using sodium naphthalene as the initiator. The polystyryl dianion is coupled using α,α'-dichloro-p-xylene and α,α',α"-trichloromesitylene as the coupling agents. The branching density depends on the ratio of tri-functional coupling agent to polystyryl dianion. The molecular weight of the final polymer depends on the molecular weight of the polystyryl dianion and the molar ratio of difunctional coupling agent, tri-functional coupling agent and polystyryl dianion. Choosing the molecular weight of the polystyryl dianion fixes the minimum distance between branch points. Thus, these polymers are not truly random in the sense that two branches cannot be closer than this minimum segment molecular weight. They are random from the step growth polymerization point of view. However, given the low concentration of branches, we do not expect the rheology of these polymers to vary significantly from a randomly branched polymer where each styrene monomer could potentially be a branch point.

Calculation of Molecular Weight and Branch Density   While the synthetic scheme provides a pathway to the preparation of branched polymers, the utility is realized only when the branching and molecular weight of the polymers can be

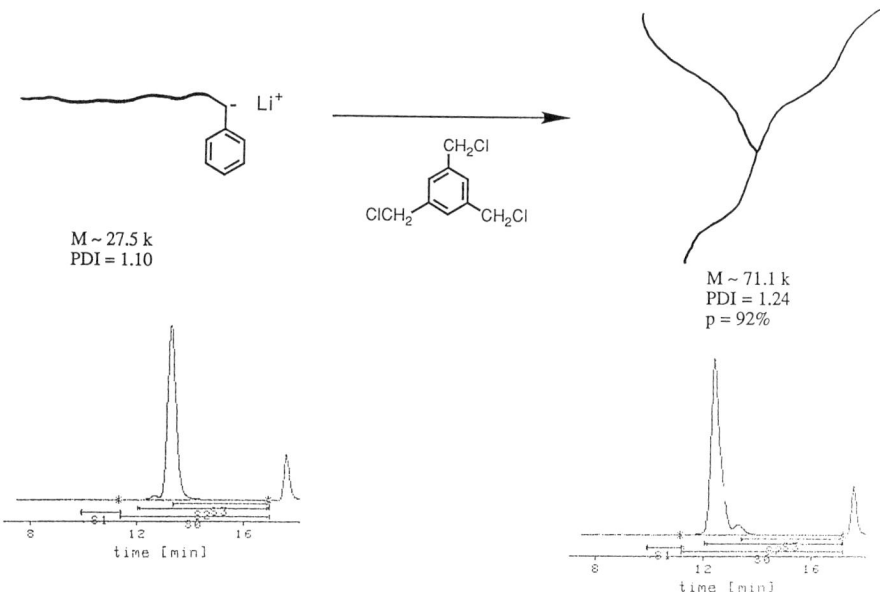

Figure 9. Synthetic route and GPC data for the preparation of three-arm star polystyrene from polystyryl anion.

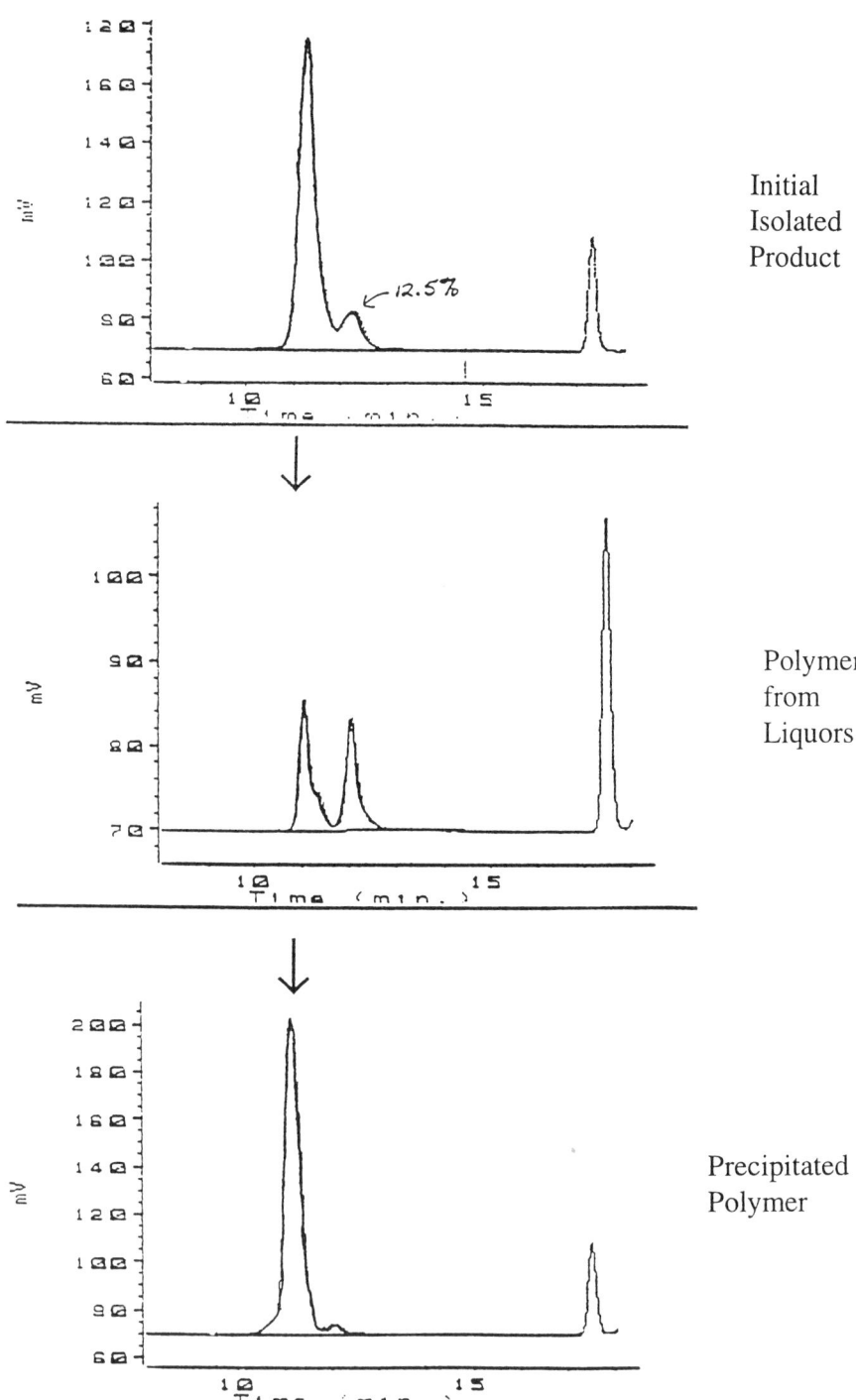

Figure 10. GPC traces of products from the selective precipitation of three-arm polystyrene for the removal of linear material.

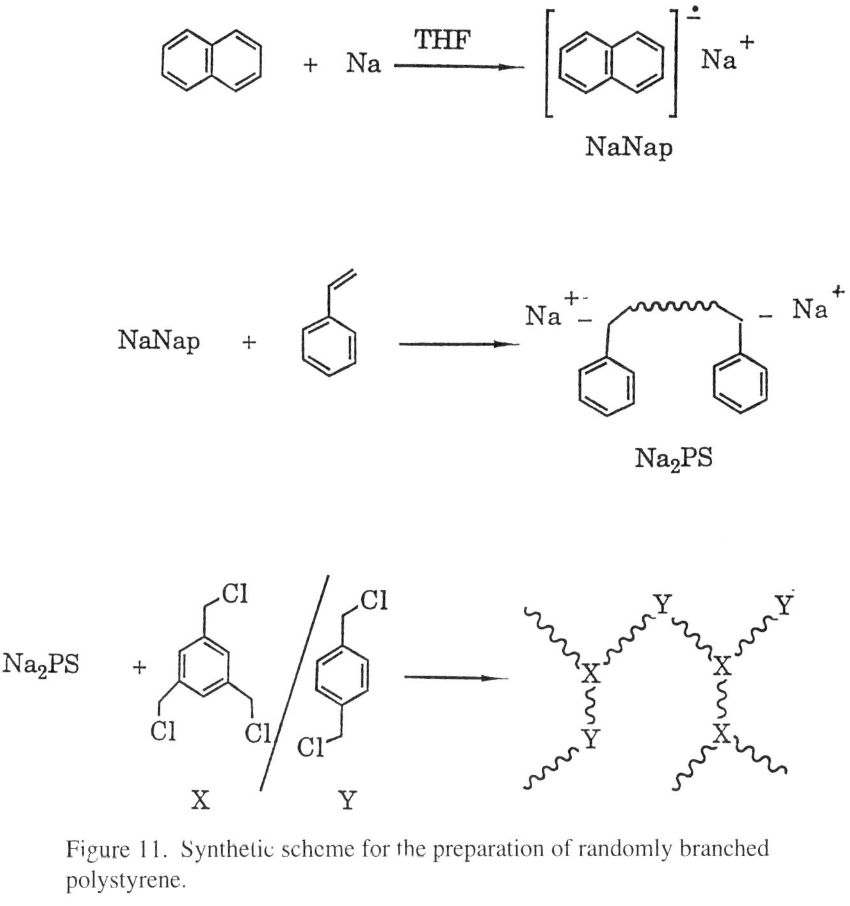

Figure 11. Synthetic scheme for the preparation of randomly branched polystyrene.

controlled and predicted. These variables can be predicted if one knows the molecular weight of the polystyryl dianion macromer and the ratios of dianions and difunctional and tri-functional coupling agents. These calculations are similar to those used in some condensation polymerizations. The molecular weight of the branched polystyrene can be predicted using the equations below. These equations were derived using a Macosko-Miller approach.(9)

$$M_w = 1 + \frac{6pr + \dfrac{6r(1+p(F-1))(p+p^2r)}{1-p^2r(F-1)}}{(5r+3-rF)} \times 104.15$$

$$M_n = \frac{-5r+rF-3}{-5r+rF-3+6pr} \times 104.15$$

wherein:

104.15 = molecular weight of styrene monomer
p = extent of reaction, probability that chloride group has reacted (0.95).
r = stoichiometric ratio of chloride groups to anion groups.
F = average functionality of chloride groups as defined by:

$$\frac{(3 \times Cl \text{ from trihalo coupling agent}) + (2 \times Cl \text{ from dihalo coupling agent})}{\text{total Cl groups on dihalo and trihalo coupling agents}}$$

The branch density of the randomly branched polymers is a function of the ratio of tri-functional coupling agent to anion groups and the molecular weight of the polystyryl dianion. Knowing these variables we can also calculate the branch density of our randomly branched polymers using the Flory approach.(10) We express the branch density as the number of branches per 1000 carbon atoms in the polymer backbone. This is also the number of branches per 500 monomer units. The branch density is predicted by the following formula:

$$Br/1000C = \frac{(\text{mole fraction trichloride coupling agent})(p^3)}{(\text{mole fraction polystyryl dianion})(M_w \text{ dianion}/104.15)}$$

$$= \frac{\dfrac{(4r-2rF)}{(-5R+rF-3)} p^3}{500 \left(\dfrac{-3}{-5r+rF-3}\right)(M_w \text{ dianion}/104.15)}$$

Wherein:

Mw dianion = Molecular weight of the polystyryl dianion before coupling

Synthesis Thus, using the equations above, polymers with given molecular weights and branch densities can be reliably synthesized. In a typical synthesis, freshly prepared sodium naphthalene initiator is used to initiate the anionic polymerization of styrene to a chosen molecular weight. The samples reported here used a polystyryl dianion molecular weight of 20,000. A small aliquot of the reaction mixture is removed to accurately determine the molecular weight of the polystyryl dianion. Using the equations above, the correct amounts of α,α'-dichloro-p-xylene and α,α',α''-trichloromesitylene are chosen via iterative calculation. A dilute solution of the coupling agents is slowly added to the polystyryl dianion macromer with vigorous mixing. After the addition is complete, the reaction is terminated via the addition of degassed methanol. The reactions are typically run using 1 mole of monomer in 1500 ml of solvent.

There are several reaction variables which are important for the success of this polymerization. The macromer synthesis is subject to the normal constraints of an anionic polymerization. The coupling reaction is extremely rapid and care must be taken to ensure a completely homogeneous reaction. The styrene concentration must be low enough that viscosity does not begin to limit coupling efficiency. However, the styrene concentration is kept high enough to ensure production of a useful quantity of polymer. Mixing during the coupling reaction is very important. If the coupling reaction is not completely homogeneous, the result is the formation of gel as well as macromer which has not coupled. Therefore the coupling agent is slowly added as a dilute solution with vigorous mixing. Gel formation is certainly one indictor of inhomogeneous coupling. The amount of uncoupled macromer is another very good indicator of coupling efficiency. Due to the stoichiometry of macromer to coupling agents, there is always some uncoupled macromer left at the end of the reaction. By comparing the amount of uncoupled macromer with that predicted from the reaction stoichiometry, the coupling efficiency can be analyzed. Analysis of the amount of uncoupled macromer was used to optimize the reaction conditions. The temperature is also held below 50°C to limit side reactions and anion degradation. The amount of THF is also important: too little gives poor coupling efficiencies while too much can lead to reaction with the THF.

Polymer Analysis GPC analysis using LALLS was also used to check the molecular weight and the branch density vs. those predicted from the equations above. All samples used for further study had uncoupled macromer levels which were in good agreement with the predicted levels. The molecular weight of the samples is an extremely sensitive test of the proper architecture. Due to the high molecular weights of the components being coupled, small differences in stoichiometry can lead to large molecular weight shifts. Typical results are shown in Table I.

Table I. Comparison of Predicted and Measured Molecular Weights and Branch Densities in Randomly Branched Polystyrene Samples

SAMPLE	BRANCHES/1000C		Mw		M_z/M_w
	Aim	LALLS	Aim	LALLS	
A	0.10	0.10	300K	390K	3.0
B	0.25	0.13	300K	295K	4.2
C	0.50	0.28	300K	369K	8.5
D	0.50	0.32	500K	446K	8.1
E	0.75	0.60	500K	589K	10.9

These results indicate that branched polystyrene can be prepared with an architecture which is accurately predicted. The agreement of molecular weight measured vs. predicted is very good considering the sensitivity of this parameter to the extent of coupling. The amount of branching measured is lower in each case than would be predicted. This shortfall is consistent in all polymerizations. It is important to note, however, that analytical standards are not available for branching density. Therefore, the accuracy of the analytical method is not known.

Separation of Product from Linear Polymer   Due to the stoichiometry of the coupling reaction, the product polymer always contains some uncoupled macromer (mono-product) as well as some product at twice the molecular weight of the starting macromer (di-product). Since the mono-product and di-product consist of linear material it is desirable to remove this material from the final product. As in the case of the three-arm stars, this can be done very efficiently using selective precipitation.(8) Figure 12 shows the GPC traces from initial isolated product as well as the products separated via selective precipitation. Figure 13 shows the effect of solvent polarity in the selective precipitation. In Figure 13, the GPC traces of products precipitated with varying solvent ratios are seen. In each case one part of reaction solution (benzene/THF) is combined with eight parts of acetone and from zero to two parts of methanol. When eight parts of acetone and two parts of methanol are used to precipitate the product, virtually all of the polymer, including the linear polymer, is precipitated. Figure 12 illustrates that fairly fine molecular weight fractionation can be accomplished using this method.

Conclusions

We have shown that symmetrical three-arm polystyrene and randomly branched polystyrene with well characterized structure can be efficiently prepared. Furthermore, the synthetic routes illustrated here are reproducible and reliable. Using calculations developed for condensation polymerizations, the branch densities and molecular weights of the polymers can be accurately predicted. We have also shown that the linear polymer, which is a by-product of these polymerizations, can be effectively removed via selective precipitation of the desired product.

181

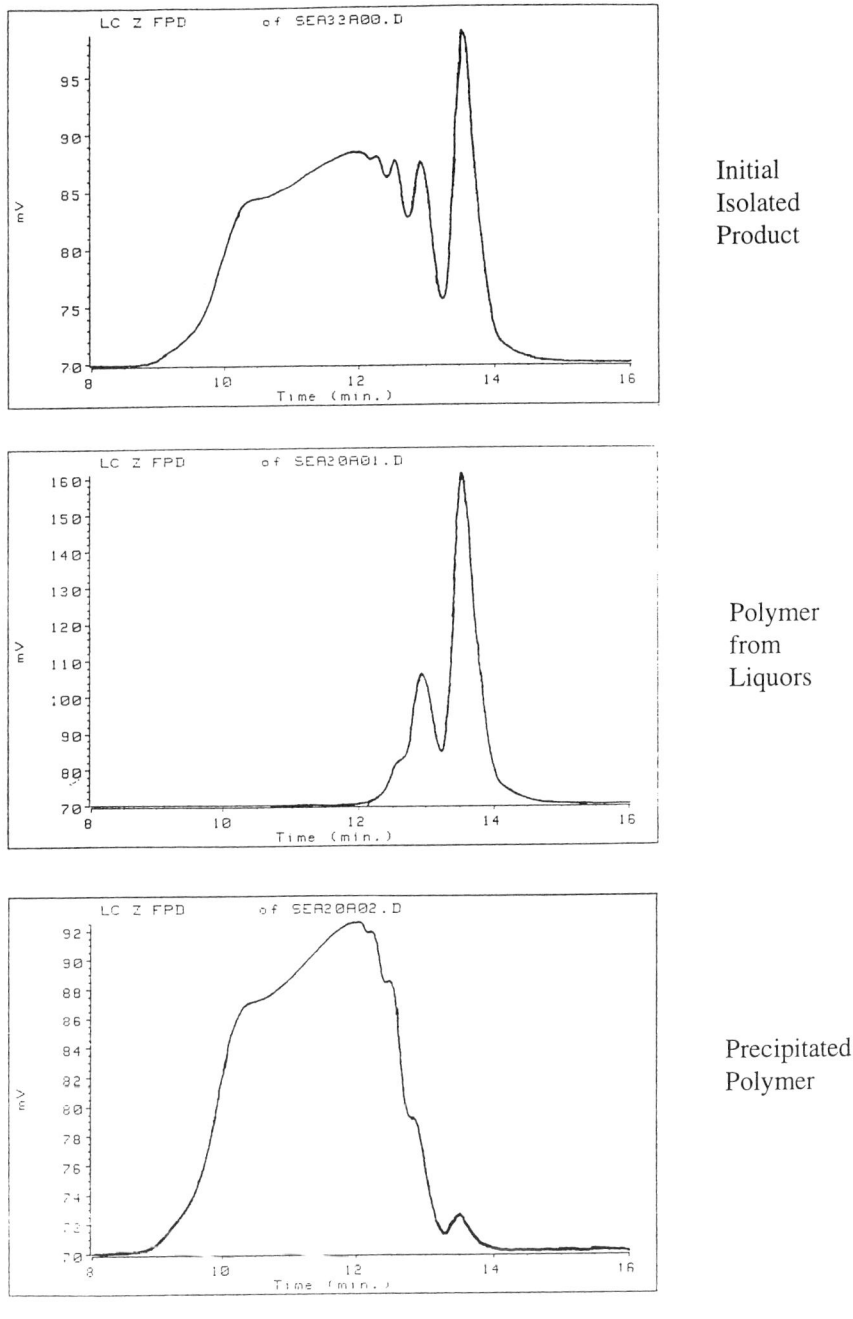

Figure 12. GPC traces of polystyrene isolated from the selective precipitation of randomly branched polystyrene for the separation of linear polymer.

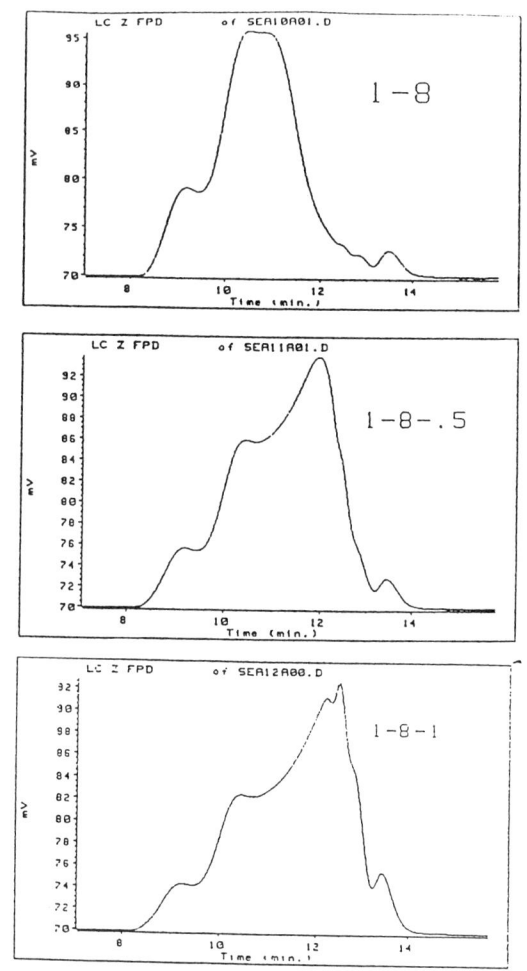

Figure 13. GPC traces showing the effect of solvent polarity on the selective precipitation of randomly branched polystyrene. Solvent composition in parts (1:1) benzene/THF-parts acetone-parts methanol.

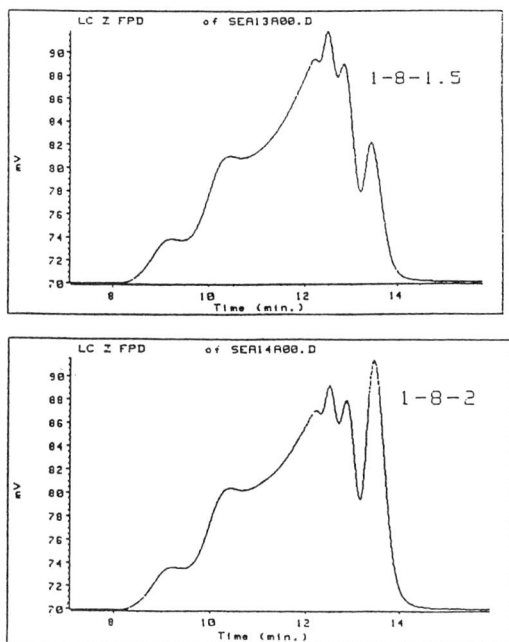

Figure 13. *Continued*

Acknowledgments

The authors thank Professor Bob Howell of Central Michigan University for the preparation of α,α',α"-trichloromesitylene. Molecular weight and branching analyses were provided by Greg Kormelink, Phil Kuch and Rose Nelson of The Dow Chemical Company. Bernie Meister of The Dow Chemical Company provided many helpful discussions.

Literature Cited

1. "Styrene Polymers" in *Encyclopedia of Polymer Science and Engineering, 2nd Ed.,* John Wiley, N.Y., Vol. 16, pp. 1-246.
2. Roovers, J.; Hadjichristidis, N.; Fetters, L.J.; *Macromolecules*, **1983**, *16*, 214.
3. Hsieh, H.L. and Quirk, R.P., *Anionic Polymerization Principles and Applications*, Marcel Dekker, N.Y., 1996, Chapter IV.
4. Tsukahara, Y.; Inoue, J.; Ohta, Y.; Kohijiya, S.; *Polymer*, **1994**, *35*, 5785.
5. Kasehagen, L.J.; Macosko, C.W.; Trowgridge, D.; Magnus, F. *J Rheol.*, 1966, *40*, 689.
6. Takaki, M.; Asami, R.; Kuwata, Y. *Macromolecules,* **1979**, *12*, 378.
7. Bryce, W.A.J.; McGibbon, G.; Meldrum, I.G. *Polymer*, **1970**, *11*, 394.
8. Wyman, D.P.; Elyash, L.J.; Frazer, F.J. *J. Polym. Sci., Part A,* **1965**, *3*, 681.
9. Macosko, C.W.; Miller, D.R., *Macromolecules, 1976, 9*, 199.
10. Flory, P.J., *Principles of Polymer Chemistry,* Cornell Univ. Press., NY, 1953, Chapter IX, "Molecular Weight Distributions in Nonlinear Polymers and Theory of Gelation."

Diene Polymers

Chapter 14

New Isoprene Polymers

Makoto Nishikawa, Mizuho Maeda, Hiromichi Nakata, Hideo Takamatsu, and Masao Ishii

Isoprene Chemicals Research and Development Department, Kuraray Company, Ltd., 36 Towada Kamisu-cho Kashima-gun Ibaraki 314-02, Japan

We have been developing various kinds of isoprene-based polymers by using synthesized isoprene monomer. We have newly developed and commercialized different types of isoprene rubbers with anionic living polymerization. The first one is liquid isoprene rubber (L-IR) with finely controlled low molecular weight, which acts as a co-curable plasticizer for solid rubbers. The second one is polystyrene and hydrogenated polyisoprene block copolymer (S-EP-S). It has a well designed molecular structure and shows good elasticity and adhesion properties. The third one is block copolymer of polystyrene and vinyl bonded polyisoprene (S-VI-S). It has not only a high damping effect but also thermoplasticity. Furthermore, the hydrogenated type of S-VI-S has good miscibility with polypropylene and it gives good flexibility and transparency to its polypropylene blends. Begin Abstract here.

We have developed isoprene monomer production not based on an extraction method but based on a synthetic method from C4 fraction (*1*). The purity of isoprene monomer, which is synthesized from isobutylene and formaldehyde by using phosphoric acid catalyst, is higher than that obtained by the extraction method. We have been developing various kinds of isoprene polymers synthesized from this high purity isoprene monomer (Figure 1). Before now, generally we have discovered two ways of polymerization. Ziegler polymerization and anionic living polymerization are commonly known in order to synthesize isoprene polymers. Solid isoprene rubbers with high <u>cis</u>-1,4-bonding and <u>trans</u>-1,4 bonding are our first products using Ziegler polymerization technology. Ziegler polymerization has the advantage of producing poly-1,4-isoprene having high stereoregularity. But it is difficult to strictly control the molecular weight and microstructure of polymer molecules by using Ziegler polymerization technology. On the other hand, anionic living polymerization technology is able to not only control the molecular weight and microstructure of polymer molecules but also synthesize block copolymers. We have already commercialized three kinds of novel polyisoprene rubbers by using anionic living polymerization technology. One is liquid isoprene rubber (L-IR) of which the molecular weight (2) is controlled and others are styrenic thermoplastic rubbers. We have two kinds of styrenic thermoplastic rubber. One is polystyrene and hydrogenated polyisoprene block copolymer (S-EP-S) which is a high performance thermoplastic rubber and another is polystyrene and vinyl-bonded polyisoprene block copolymer (S-VI-S) in which microstructure is controlled.

Characteristics and Applications of Liquid Isoprene Rubber (L-IR)

In the rubber industry, addition of process oil is necessary in order to improve the processability. However, the process oil cannot be curable, so the problem of oil bleeding occurs after the cure. The general remarks about the liquid rubber have already published (3). We presumed that low molecular weight polyisoprene would act as a co-curable plasticizer. We investigated the molecular weight range of polyisoprene from two points of view; one was co-curability with diene rubbers and the other was the same processability as oil. This polyisoprene (Liquid polyisoprene, abbreviation for it is LIR) is prepared by anionic living polymerization with n-butyllithium (NBL) initiator in hexane as solvent. The ratio of 1.4-bonding/3.4-bonding is 92/8 (mol ratio) and the ratio of cis-bonding/trans-bonding in the 1.4-bonded units is about 65/35 (mol ratio). The relationship between melt viscosity at 38°C and number average molecular weight (Mn) by GPC is shown in Figure 2.

Figure 3 shows that the relationship between the molecular weight of polyisoprene and crosslinking rate of it. Crosslinking rate was measured by curelastometer. The measurement temperature of curelastometer was 145°C. The polyisoprenes with molecular weights more than 30,000 are almost co-curable with natural rubber. Figure 4 shows the improvement effect on the processability using low molecular weight polyisoprene. The polyisoprenes with molecular weights of 50,000 and 28,000 have almost the same effect as with process oil. Finally, we decided that the useful molecular weight range of polyisoprene for these applications was from 30,000 to 50,000 and we call this controlled polymer liquid isoprene rubber (L-IR). The applications of L-IR are as a co-curable plasticizer for vulcanized rubber compounds. L-IR can reduce the stage or shorten the time of the mixing of rubber without an influence on the mechanical properties. Moreover, L-IR gives other advantages (high extrusion rate, smooth surface in calendering and good adhesion to each calendered sheet) to the vulcanized rubber compounds.

Characteristics and Applications of Hydrogenated Styrene Isoprene Block Copolymers (S-EP-S)

By anionic living polymerization, polystyrene and polydiene block copolymers (S-I-S,

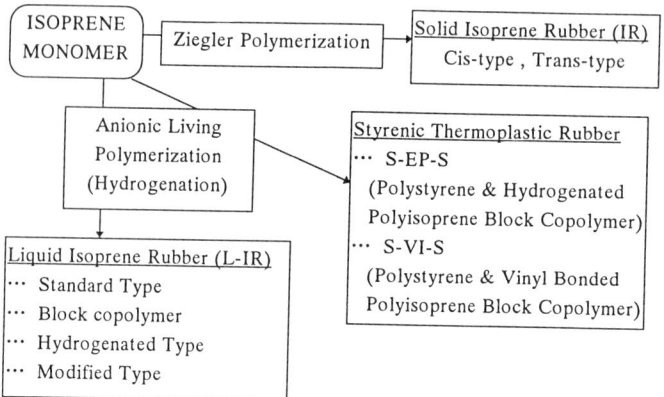

Figure 1. KURARAY's isoprene polymers.

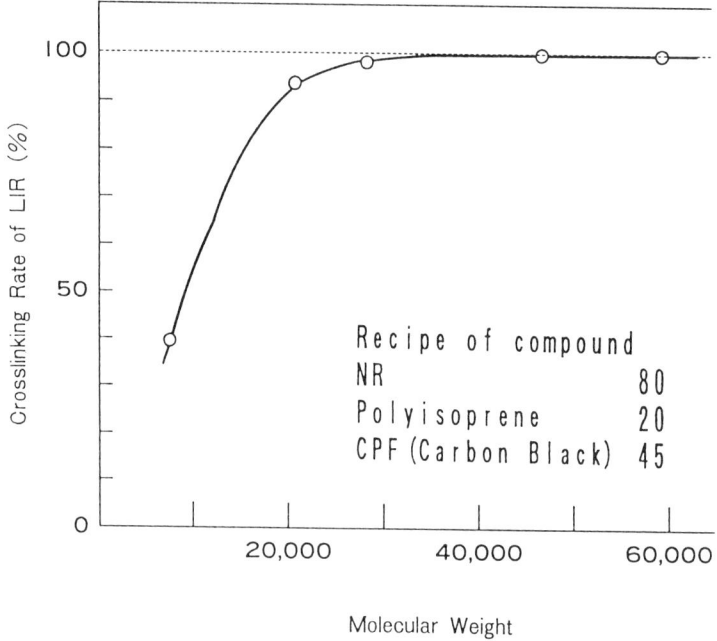

Figure 2. Relationship between melt-viscosity at 38 °C and number-average molecular weight by GPC

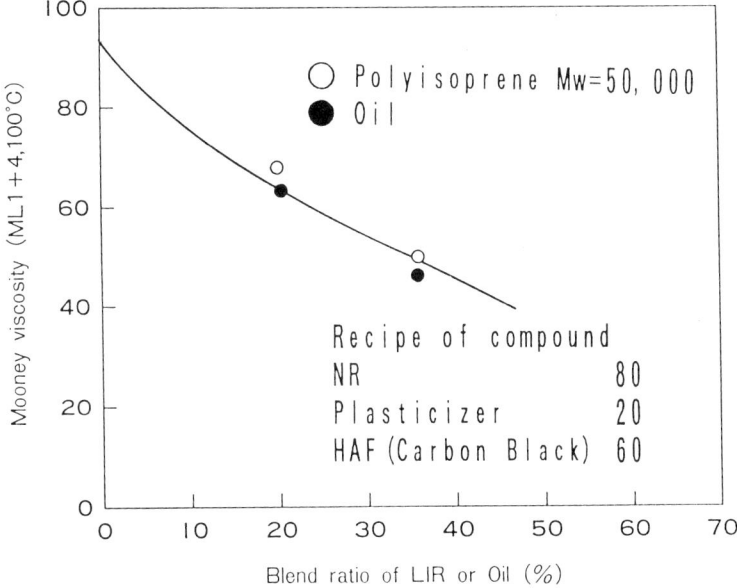

Figure 3. The relationship between molecular weight and crosslinking rate of polyisoprene.

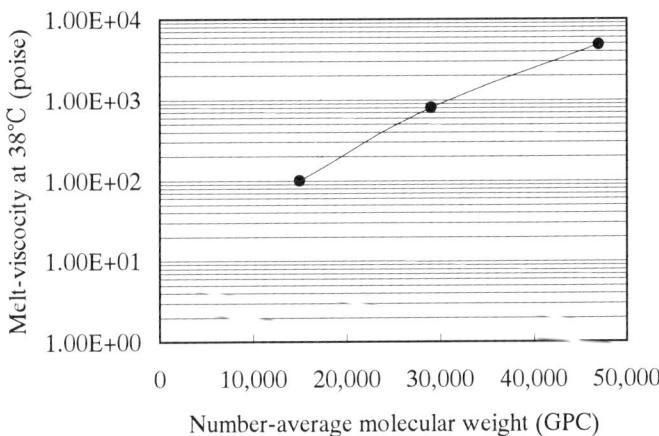

Figure 4. The relationship between blend ratio of plasticizer and Mooney viscosity.

S-B-S and S-EB-S) are produced by several companies. We have commercialized polystyrene and hydrogenated polyisoprene block copolymers (S-EP-S). The polymerization method of S-EP-S is the anionic living polymerization with sec-butyllithium initiator (SBL) in hydrocarbon solvent such as pentane, hexane, heptane or cyclohexane by sequential monomer addition.

S-EP-S has polystyrene end blocks and a center poly(ethylene/propylene) block. The range of styrene content of S-EP-S is from 13 wt% to 65 wt%. S-EP-S has better heat and weather resistance than those of SIS and SBS, because it has the saturated rubber block. S-EP-S is obtained by hydrogenation of SIS. Hydrogenation was accomplished by using a metal catalyst like as Pd, Pt, Ti and the combination of Co/Al, Ni/Al etc. (4). The degree of hydrogenation of these S-EP-S copolymers was above 98 %. The rubber block of S-EP-S has primarily an ethylene propylene alternating structure. On the other hand, S-EB-S is obtained by hydrogenation of SBS with 40 to 50 mol% 1,2-bonded structure. The rubber block of S-EB-S has an ethylene and butylene random structure. We compared the behavior of the orientation crystallinity (Table I) and the mechanical properties (Table II) of S-EP-S with those of S-EB-S, which has the same molecular weight and styrene content.

The intensity of WAXS of S-EP-S exhibits no change when the orientation ratio is varied from 1 to 5. We confirmed that S-EP-S has no crystallinity at any orientation by DSC measurements. This means that S-EP-S has no crystallinity. On the other hand, S-EB-S has some crystallinity even with no orientation as measured by DSC and the intensity of WAXS; the crystallinity increased as the orientation ratio increased up to 5. These results show that the rubber block of S-EP-S is more randomized than that of S-EB-S. For these reasons, we thought that the EP block is softer than EB block. As for mechanical properties, S-EP-S has lower modulus, lower tensile strength and higher elongation compared to S-EB-S. These properties of S-EP-S give better adhesion in their adhesive compounds.

We thought that the softness of S-EP-S and diblock S-EP would provide a good base polymer for pressure sensitive adhesives. Table III shows the effect of the addition of S-EP to S-EP-S for pressure sensitive adhesives. By adding S-EP to S-EP-S, better tack and better adhesion, especially adhesion to polyethylene, can be obtained. Blending of S-EP to S-EP-S is effective for use as the base polymer of a pressure sensitive adhesive.

Another application of S-EP-S/S-EP blends is as impact-modifier of polypropylene, polyphenylene oxide, polystyrene, etc. In this case, the softness of these blends is used extensively. By controlling the molecular weight, styrene content and block structure that is triblock or diblock, by anionic living polymerization and controlling the diblock polymer content, these polymers can have various kinds of properties to meet many applications. Applications include their use as adhesives, sealants, coating materials, rubber- like compounds (substitute of vulcanized rubber), plastics modifiers, compatibilizers, thermosetting resin modifiers, anti-shrinking agents for unsaturated polyesters asphalt modifiers and so on.

Characteristics and Applications of Polystyrene Vinyl-Bonded Polyisoprene Block Copolymer (S-VI-S)

In general, energy loss by polymers occurs near the glass transition temperature (Tg) of

Table Ⅰ. Comparison of the orientation crystallinity between SEPS and SEBS

Orientated ratio of rubber	Intensity of WAXS			
	×0	×1	×2	×5
S-EP-S	0.69	0.69	0.68	0.66
S-EB-S	0.94	0.93	1.03	1.28
S-EP-S	Styrene content = 30 wt. % MW = 70,400			
S-EB-S	Styrene content = 29 wt. % MW = 72,000			

Table Ⅱ. Comparison of the mechanical properties between SEPS and SEBS

	Modulus 100 % (MPa)	Tensile strength (MPa)	Elongation (%)	Hardness (JIS A) (-)
S-EP-S	2.3	24	720	74
S-EB-S	3.0	30	590	74
S-EP-S	Styrene content = 30 wt. % MW = 70,400			
S-EB-S	Styrene content = 29 wt. % MW = 72,000			

Table Ⅲ. The effect of addition of S-EP to S-EP-S for pressure sensitive adhesives

Run Number		1	2	3
(Formulation) (PHR)				
S-EP-S		100	75	50
S-EP		0	25	50
ARKON P-85[1]		100	100	100
IRGANOX 1010		1	1	1
(Properties)				
Tack: Rolling ball-tack[2]	(Ball No.)	7	9	12
Cohesion: Creep test at 40°C				
Holding Power	(min)	>240	>240	>240
Slippage, 240 min	(mm)	0.08	0.11	0.14
Adhesion: to stainless	(g/cm)	520	690	790
to PE	(g/cm)	80	150	390

(1) Alicyclic hydrocarbon resin   ARAKAWA CHEMICAL
(2) Rolling ball-tack   Larger No. means better tack
  Creep test   25mm × 25mm   1Kg Load
  180° peel test   Rate of peel=300mm/min at 25°C

polymers. We thought that the existence of the Tg at room temperature would lead to energy loss at room temperature.

Control of Microstructure of Polyisoprene. By anionic polymerization with Lewis bases such as diethyl ether, tetrahydrofuran, triethylamine , N,N,N´,N´-tetramethyl ethylenediamine and so on, the amount of vinyl-bonded units of polyisoprene can be controlled. Figure 5 shows that the Tg of polyisoprene increases by increasing the vinyl content. Polyisoprene with more than 70 % vinyl microstructure has a Tg around room temperature. So high vinyl-bonded polyisoprene delivers high energy loss at room temperature. By use of this characteristic, we have commercialized the polystyrene and vinyl-bonded polyisoprene block copolymer. The polystyrene blocks act as physical crosslinks and provide thermoplasticity; the vinyl-bonded polyisoprene block acts as rubber and provides vibration damping.

Vibration Damping Effect. S-VI-S gives the vibration damping effect to various kinds of polymers (PE, PP, EVA, PSt, SMC and so on). Table IV shows that vibration damping effect and mechanical properties of a high density polyethylene (HDPE) and S-VI-S blend. Twice the damping effect can be obtained by blending of only 10 wt% S-VI-S into HDPE. We confirmed the existence of the same $\tan\delta$ peak of vinyl-bonded polyisoprene in this mixture around room temperature by measuring the dynamic modulus. The existence of the isolated vinyl-bonded polyisoprene phase in the mixture causes the increase of the loss factor of the blend.

Polypropylene/Hydrogenated Vinyl-Bonded Polyisoprene Blends. Vinyl-bonded polyisoprene has another unique characteristic. Hydrogenated vinyl-bonded polyisoprene has miscibility with polypropylene at the molecular level. Table V. shows the miscibility of PP and this hydrogenated diene polymer. The hydrogenated low vinyl-bonded polyisoprene has good compatibility with polypropylene but has no miscibility (as shown by scanning electron microscopy). However, hydrogenated polyisoprene with more than 40 % vinyl-bonded microstructure has miscibility with polypropylene in all compositions.

We compared the miscibility of S-EP-S with polypropylene and of H-S-VI-S with it by Scanning Electron Microscopy(SEM). Photograph 1 shows the blend of PP/S-EP-S (=70/30 weight ratio). The void parts are the S-EP-S portion extracted by toluene treatment. Photograph 2 shows the blend of PP/H-S-VI-S (=70/30 weight ratio) after extraction treatment by toluene.

S-EP-S and H-S-VI-S have the same styrene content (20 wt%) and molecular weight (Mn about 120,000). In the case of PP/S-EP-S blend, S-EP-S dispersed into 1 to 5 μm sized particles. On the other hand, in the case of PP/H-S-VI-S blend , a separate H-S-VI-S phase could not be detected by this method.

We have commercialized the hydrogenated type of the polystyrene vinyl-bonded polyisoprene block copolymer (H-S-VI-S). Table VI shows the properties of polypropylene and the hydrogenated type of the polystyrene vinyl-bonded polyisoprene block copolymer blend. This polymer gives flexibility and transparency to blends with polypropylene in spite of maintaining the same degree of crystallinity. As for

Table IV. Vibration damping effect and mechanical properties of HDPE/S-VI-S

(Formulation)				
HDPE	(wt. %)	70	90	100
S-VI-S	(wt. %)	30	10	-
(Damping properties)			10	
tan δ at 25°C	(-)	0.140	0.093	0.072
Loss factor at 25°C	(-)	0.155	0.072	0.036
(Mechanical properties)				
Tensile modulus	(MPa)	343	598	853
Tensile Strength	(MPa)	13	20	25
Elongation	(%)	910	1170	1655

Loss factor was measured by resonance method with cantilever beam

Table V. The miscibility of PP/hydrogenated vinyl bonded polyisoprene

Compositions Rubber/PP	H-PIp $V = 6\%$	H-PIp $V = 40\%$	H-PIp $V = 78\%$	H-PBd $V = 38\%$	EPR
1/9	I	M	M	I	I
3/7	I	M	M	I	I
5/5	I	M	M	I	I
7/3	I	M	M	I	I
9/1	I	M	M	I	I

H-Ip	:Hydrogenated polyisoprene	MW=100,000
H-Bd	:Hydrogenated polybutadiene	MW=100,000
EPR	:Ethylene-propylene rubber	PP = 48 wt. %

V : Vinyl microstructure (mol%)　M : miscible　I : immiscible

Table VI. Characteristics of PP/Hydrogenated type of S-VI-S blend

(Formulation)				
PP (Homo)	(wt. %)	50	70	100
H-S-VI-S	(wt. %)	50	30	-
Transmittance[1]	(%)	85	88	81
(Thermal properties)				
Melting point	(°C)	161	162	164
Heat of fusion[2]	(mJ/mg)	134	127	131
(Mechanical properties)				
Flexural modulus	(MPa)	123	470	1290
Tensile Strength	(MPa)	26	18	39
Elongation	(%)	830	520	40

1):Transmittance was measured by reflectance transmittance meter
2):Heat of fusion was calibrated according to PP ratio

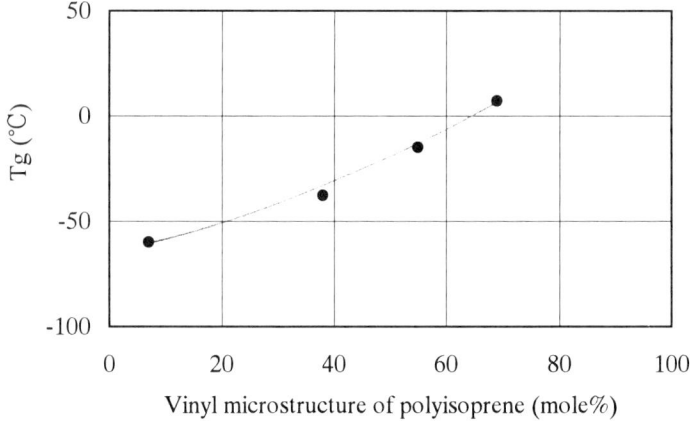

Figure 5. Relationship between vinyl microstructure and Tg of polyisoprene.

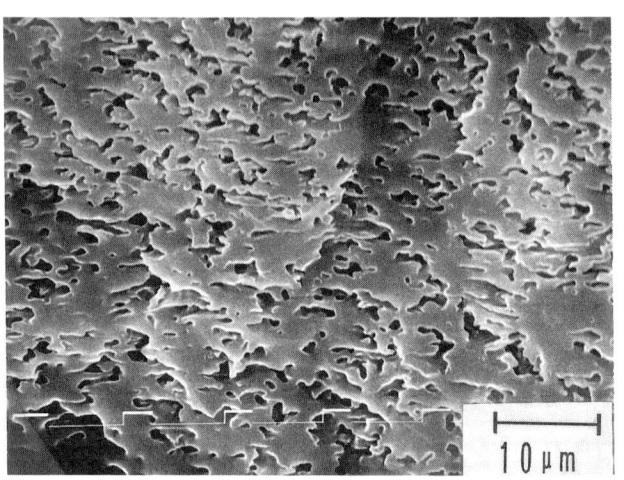

PP/S−EP−S = 70/30 (wt/wt)

Photograph 1. Electron micrograph of PP/S-EP-S blend (70/30, wt/wt) after toluene extraction.

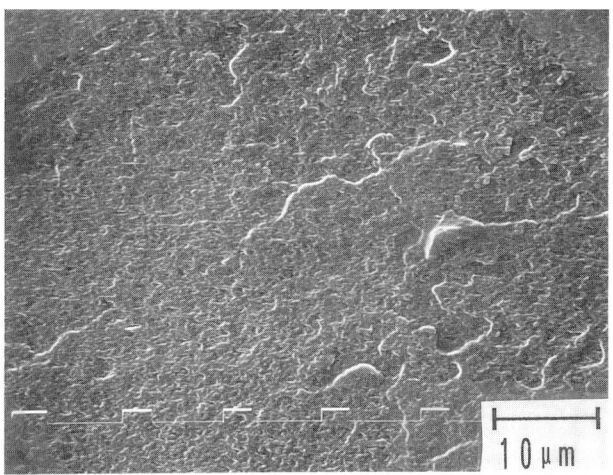

Photograph 2. Electron micrograph of PP/H-S-VI-S blend (70/30, wt/wt) after toluene extraction.

transparency, we presume that the molecular chain of hydrogenated vinyl bonded polyisoprene, which is miscible with polypropylene, prevents the formation of large crystallites of polypropylene in the cooling process from the melt to the solid. As for flexibility, we presume that the tan δ peak of polypropylene shifts to low temperature because of the miscibility with hydrogenated vinyl-bonded polyisoprene. These compounds are good for film, sheet, tube applications and so on.

The applications of S-VI-S include use as a vibration-damping material by using energy loss at room temperature and as a transparent and soft material, for example, as a replacement of soft PVC by using miscibility with polypropylene.

Conclusion

As applications of anionic living polymerization, we have commercialized three kinds polyisoprene polymers. L-IR has controlled molecular weight. S-EP-S has controlled molecular weight and controlled block copolymer structure. S-VI-S has controlled molecular weight, controlled block copolymer structure and controlled microstructure. Anionic living polymerization can strictly control the molecular structure to meet many industrial applications. We believe that anionic living polymerization technology provides many possibilities for producing new polymers.

References

1. K. Kushida, S. Kyo, O. Yamada, *Petrotech* **1992**, 15, 64.
2. S. Minatono, H. Takamastu, Y. Inukai, J. Yamauchi, Koubunshi ronbunshu **1978**, 35, 599.
3. J. P. Kennedy, E.G.M. Törnquist, "Liquid Rubber" in "Polymer Chemistry of Synthetic Elastomers (II)", Interscience, New York **(1969)**
4. Robert H. Schwaar, *SRI Report*, Report No. 104 **(1976)**

Chapter 15

Commercial Production of 1,2-Polybutadiene

A. A. Arest-Yakubovich[1], I. P. Golberg[2], V. L. Zolotarev[2,4], V. I. Aksenov[2], I. I. Ermakova[3], and V. S. Ryakhovsky[2]

[1]Karpov Institute of Physical Chemistry, Vorontsovo pole 10, Moscow 103064, Russia
[2]Efremov Synthetic Rubber Plant, Stroitelei St. 2, Efremov, Tula Region 301860, Russia
[3]Synthetic Rubber Research Institute, Gapsalskaya St. 1, St. Petersburg 198035, Russia

> The technology of the synthesis, properties, and fields of application of high and low-molecular-weight 1,2-polybutadienes are described. The polymerization is performed as a continuous solution process with the use of n-BuLi - diglyme catalytic system. Certain advantages of anionic process over a standard solution polymerization with Ziegler-Natta catalysts are demonstrated. The improvements in the process based on the use of sodium or potassium alkoxides or new organosodium initiators are also described.

One of the main practical advantages of anionic polymerization processes is their unique flexibility. They can be adapted relatively easily to various technological schemes; besides, they give a wide possibility for controlling micro and macro structure of polymers (the structure of monomer units, molar masses, MWD, degree of branching, etc.) and, consequently, their properties. These advantages are clearly demonstrated by the experience of the development of the industrial synthesis of 1,2-polybutadiene rubber (1,2-PB) via a solution process.

As is well known, polybutadiene of high (50-60%) 1,2-unit content obtained via heterogeneous polymerization with the use of sodium metal was the first type of commercial synthetic rubber (1,2). After World War II it was completely forced out from tire industry by emulsion SBR and stereoregular polymers. However, due to its unique set of properties, 1,2-PB has found many industrial applications for non-tire goods. Nowadays, several types of high- and low-molecular-weight high vinyl polybutadienes are produced via continuous solution polymerization. This process was developed in the middle 1970s in application to already functioning technology of the synthesis of solution cis-1,4-polybutadiene rubber (3).

[4]Current address: Syntezkautschuk Company, Ltd., Myasnitskaya Street, 20, Moscow 101851, Russia.

High-molecular-weight 1,2-Polybutadiene

Manufacturing. High-molecular-weight 1,2-polybutadiene is produced by means of a *n*-butyllithium - diglyme catalytic system in toluene with the use of the standard scheme of the synthesis of solution *cis*-1,4-polybutadiene (2). After unreacted monomer and high boiling substances are separated and the water is removed by azeotropic distillation, recycled toluene undergoes rectification and, finally, is dried with aluminum oxide. As a result, the content of residual water together with other reactive impurities in the incoming solution does not exceed 5-10 ppm and of oxygen (in the vapour phase) 1-2 ppm. The amount of water and other impurities in toluene is controlled by a on-line chromatograph.

The polymerization step is performed in a series of 5 standard reactors, 16,6 m^3 each, equipped with a stirrer and a cooling jacket; the polymer formed is recovered by the usual procedure when the solvent and unreacted monomer are steam-stripped. The capacity of a single line varies from 2 to 5 tons per hour depending on the type of the rubber. Thus, the production of high-molecular-weight 1,2-PB requires no extra investments as compared to other processes of solution polymerization.

From the chemical point of view, the simplest task is the control of polymer microstructure. As can be seen from Figure 1, by changing the ratio diglyme/BuLi, it is possible to vary the 1,2-unit content in the polymer within a wide interval. Similarly, the average molecular weight of the polymer can be controlled via the ratio monomer/initiator.

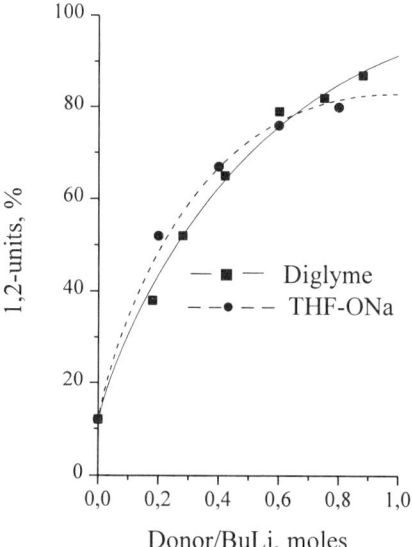

Figure 1. The effect of electron donors on the polybutadiene structure (toluene, 30 ^{0}C).

The control of the MWD and the architecture of macromolecules is a more complicated problem. However, these parameters are of a great importance for the technology of manufacturing and processing the polymer.

It is known that the polymer formed in a series of continuous stirred tank reactors has a fairly narrow MWD ($\overline{M}_w/\overline{M}_n \leq 2$) when initiator and monomer are fed only in the first reactor (4). However, a many-years experience has shown that a much wider MWD is required for most fields of application. One of possible ways to broaden the MWD is the fractional addition of the monomer and/or the initiator into several sites of the series.

The macrostructure of polymers is of a great importance for the separation of the rubber from reaction solution. It was observed that the cold flow of linear polybutadienes with 1,2-unit content higher than 50% is very low due to its high molar cohesion (5). However, at the temperatures higher than 40-50 °C some kind of melting takes place, and the polymer flow increases sharply (Figure 2). This shortcoming was overcome by means of the formation of a certain amount of branchings in the macromolecules with the use of a cross-linking agent, e.g., divinylbenzene (DVB). Due to the importance of this problem, an elaborate study of DVB-butadiene copolymerization was performed (3,6,7). The reactivity of the first double bond of DVB in reaction with polybutadienyl lithium growing chain end was studied quantitatively (7).

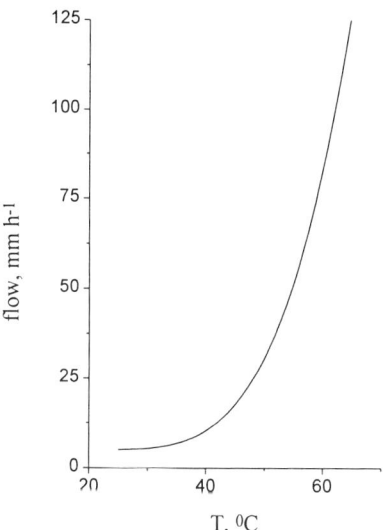

Figure 2. Dependence of the flow of linear polybutadiene on temperature. [η] = 1.58 dl g⁻¹; 65 % 1,2-units.

The value of $1/r_B = k_{BD}/k_{BB}$, where indexes B and D denote butadiene and divinylbenzene, respectively, was calculated from the data on the kinetics of DVB consumption during polymerization of butadiene. It was found that in the presence of diglyme both p- and m-DVB are more reactive than butadiene, as far as the addition of the first double bond of DVB is concerned (Table I). At the same time, the reactivity of the second (pendant) double bond with respect to the addition of the second growing chain was found to be 3 to 4 times lower than that of the first bond and seemed to decrease with increasing chain length. As a result, some amount of pendant double bonds remains unreacted even after complete conversion of butadiene *(6)*.

Table I. The dependence of relative reactivity of the first double bond of DVB ($1/r_B = k_{BD}/k_{BB}$) on the molar ratio diglyme/BuLi for DVB isomers

Diglyme/BuLi, moles	0.05	0.1	0.2
m-DVB	1.02	1.43	1.92
p-DVB	3.57	3.57	3.57

SOURCE: Adapted from ref. 7.

It was found that the best results (from the point of view of technological properties of the polymer) were obtained when DVB was added in the middle of the process, so that tetrafunctional sites of branching with approximately equal branch lengths were formed. The use of DVB, provided that it is introduced in the system in the proper place, was found to give more opportunities to control the MWD, the degree of branching and the ratio between linear and branched macromolecules than the use of other coupling agents such as $SiCl_4$ or $SnCl_4$. As a result of this study, it became possible to reduce the polymer flow at 90 °C to the level similar to that of a standard *cis*-1,4-rubber without loss in solubility and without gel formation in the polymerization reactors (Figure 3).

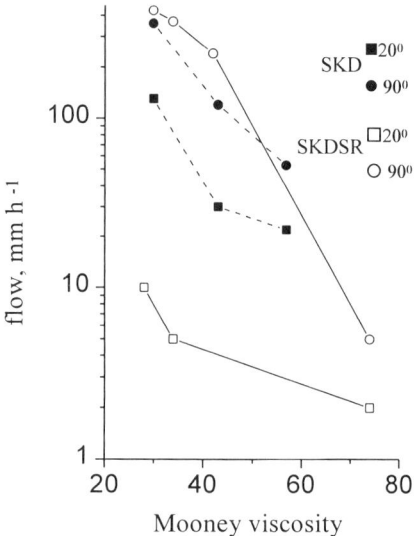

Figure 3. Polymer flow vs. Mooney viscosity for commercial *cis*-1,4 (SKD) and 1,2 (SKDSR) polybutadiene rubbers.

A prolonged experience in the simultaneous exploitation of two processes, namely, the synthesis of cis-1,4-polybutadiene with Ti catalyst and the anionic synthesis of 1,2-PB, with the use of the same standard equipment has confirmed a number of advantages of anionic process. Anionic polymerization is much more flexible and controllable. In anionic process, it is possible to vary the 1,2-unit content from 10 to 80 % and, changing inlet sites of reagents and the amount of DVB, $\overline{M_w}/\overline{M_n}$ ratio from 1.5 - 2 to 8 - 10, thus obtaining unimodal, bimodal and, if necessary, polymodal MWD curves (Figure 4).

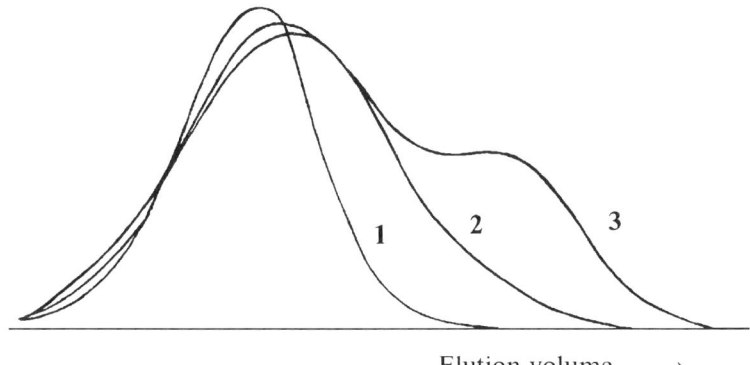

Figure 4. GPC-traces of typical commercial 1,2-polybutadienes.
1-SKDSR-SH, $\overline{M_w}/\overline{M_n}$ = 2.5; 2-SKDSR, $\overline{M_w}/\overline{M_n}$ = 5;
3-SKDLB, $\overline{M_w}/\overline{M_n}$ = 7.5.

Because of the absence of side reactions, even in the presence of DVB no cross-linked polymer or oligomers are formed which are characteristic of many polymerizations with Ziegler-Natta catalysts due to the presence of cationic-active components in the catalytic system. As a result, no gel is formed in anionic polymerization reactors even in the absence of special gel-suppressing agents. Consequently, a continuous operation of the reactors during indefinitely long time is possible, whereas in the synthesis of cis-1,4-polybutadiene the reactors should be cleaned from gel at least once a year. The use of toluene which is a weak chain transfer agent in lithium-initiated anionic polymerization, also seems to be favourable for suppressing gel formation, if the latter is caused by chain transfer to polymer *(4,8)*.

Properties and fields of application. One of the most important properties of 1,2-PB is its high resistance to heat and thermal-oxidative ageing. The main part of double bonds in a 1,2-polybutadiene macromolecule is located in side groups and not in the backbone. That is why, in contrast to other polydiene rubbers, 1,2-polybutadiene does not destruct when heated but forms cross-linked structures without loss in strength (Figure 5). This feature, along with a high filling capacity (up to 900 % with respect to the weight of the rubber) and low abradability, makes this polymer especially useful as a binder for abrasives, e.g. grinding wheels, and asbestos friction articles, such as brake shoes, which can be successfully used at eleva(9)ted temperatures. A lot of grinding wheels and other grinding and cutting tools containing 7-10% 1,2-PB and ~90% abrasive powder is consumed by different branches of engineering industry.

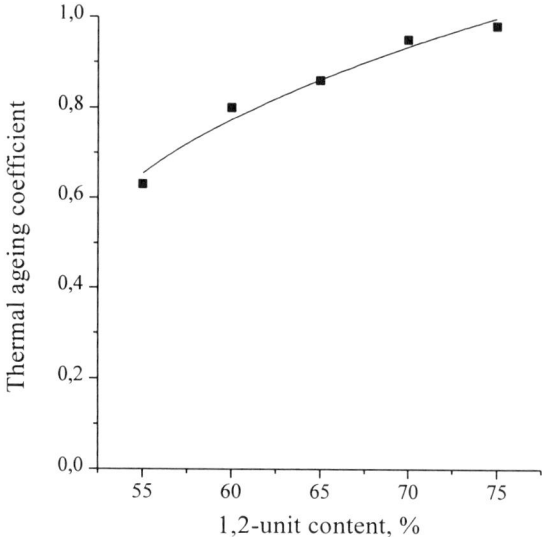

Figure 5. Thermal ageing properties of polybutadienes of different structure. Ageing for 72 h at 100 °C.

Besides, this rubber is characterised by the absence of odor (due to the absence of oligomers), very low cold flow (see Figure 3), good processability, good miscibility with ingredients and relatively low cost. Blends on the basis of 1,2-PB are easily processed by extrusion and rolling and form a smooth surface with satisfactory tackiness. The shrinkage of carbon black-filled blends is considerably less than that of the majority of other rubbers. All this enables to use this rubber in the manufacturing of such articles as linoleum, flexible pipes, rubberized fabric, artificial leather, rubber shoes, boats, etc. It is also used in electric engineering for cable sheaths. Useful additional information the properties of 1,2-PB one can find in a recent review (9).

One of the most important features which determines processability of the rubber is its MWD. As a first approximation, one can say that, at equal Mooney viscosity, the processability of the polymer is the better the broader MWD. For some articles, for example, for grinding wheels and conveyor belts, a special rubber with bimodal MWD (SKDLB) was developed (10) while the main type (SKDSR) has a unimodal MWD. The production of an oil-extended rubber is also possible.

As was mentioned previously, the production of 1,2-PB is oriented mainly towards non-tire consumers. Nevertheless, due to the recent interest of tire manufacturers in medium- and high-vinyl polydienes (9,11,12), a special type of rubber for tires, SKDSR-SH, was developed. This rubber has a somewhat lower 1,2-unit content and a narrower MWD, that improves its freeze-resistance and strength.

This polymer is also well compatible with other types of tire rubbers. Typical properties of commercial high-molecular-weight 1,2-polybutadiene rubbers produced by Efremov Synthetic Rubber Plant are given in Table II. The MWD curves are shown in Figure 4.

Table II. Properties of commercial 1,2-polybutadiene rubbers

	SKDSR 1	SKDSR 2	SKDLB	SKDSR-SH
1,2-unit content, %	≥70	≥70	55-70	60
Mooney viscosity	28-38	38-47	28-37	40-50
Elastic recovery, mm	1.5-1.7	2.3-2.5	1-1.3	2.3-2.5
Plasticity	0.50	0.45	0.54	0.5
Modulus 300 %, MPa	6	9	6-8	8-10
Tensile strength, MPa	16	18	15-17	18-20
Elongation, %	500	550	450-600	500-600

Low-Molecular-Weight 1,2-Polybutadiene

Along with high-molecular-weight rubbers, a non-functional liquid 1,2-PB is also produced. Similar to the high-molecular-weight polymer, this rubber is obtained by a continuous process at 40-50% monomer in solution, and the solvent is removed by the rotary evaporator. The use of lithium catalyst together with various electron donors enables to control the microstructure and molecular weight of the polymers within a wide range according to requirements of consumers. The presence of electron donors, especially of heavier alkali metal alkoxides, greatly intensifies chain transfer to solvent which enables to reduce considerably the consumption of n-BuLi in comparison to the synthesis of non-functional liquid 1,4-polybutadienes. Some properties of main types of commercial liquid 1,2-polybutadienes are presented in Table III.

Table III. Properties of commercial liquid 1,2-polybutadienes

	SKDSN	LKN-S	LKN-1-5
1,2-unit content, %	40-60	40-60	20-50
Dynamic viscosity, Pa·s, at 50 ^{0}C	30-80	60-150	-
at 25 ^{0}C	-	-	1-15
Ash content, % max.	0.1	0.3	0.1
Drying mass loss, % max.	0.7	1.0	0.5-0.8
$\overline{M}_n \cdot 10^{-3}$	5-10	7-15	1-5

Owing to a high film-forming ability and a high metal adhesion, these rubbers are used as a base for anticorrosive sealants and coatings, as plastic lubricants, depressing additives to paraffin-rich oils and petroleum products, for production of some kinds of mechanical goods etc. In contrast to most polymers, 1,2-PB, being subjected to a short-time action of high temperature, does not destruct, but forms a dense dielectric film. That is why liquid 1,2-PB is also used for deposition of electric insulating coatings by impulse burning.

Prospects for Development

A lot of work was performed on purpose to improve economical and ecological characteristics of the process and improvement of the polymer quality by means of varying the catalytic system without changing the technological scheme as a whole. Two directions of the improvement seem to be most promising: the replacement of expensive and non-biodegradable diglyme by a cheaper alkoxide donor and the replacement of n-BuLi by an organosodium initiator.

It was found previously *(13)* that due to the presence of the strong chelate-forming di-oxy-ethylene group, bidentate alkali metal alkoxides of the type

$$Mt\text{-}O\text{-}CH_2\text{-}CH_2\text{-}O\text{-} ,$$

where Mt is an alkali metal, have a considerably higher solvating ability and the solubility in hydrocarbons than alkoxides of aliphatic alcohols. Sodium alkoxide of tetrahydrofurfuryl alcohol (THF-ONa)

$$\begin{array}{cc} CH_2-CH_2 \\ | \quad\quad | \\ CH_2 \quad CH-CH_2ONa \\ \diagdown \quad \diagup \\ O \end{array}$$

is especially promising for the synthesis of high-molecular-weight polymers *(14)*. This alkoxide is readily obtained from a cheap and commercially available alcohol and is soluble in toluene. Its solvating ability is comparable to that of diglyme and at low donor/BuLi ratio even exceeds the latter (see Figure 1, curve 2). However, in contrast to diglyme, the alcohol which is formed after hydrolysis of THF-ONa, is biodegradable and is completely absorbed by waste water thus contaminating neither the rubber nor recycling toluene.

For liquid polymers, THF-OK turned out to be very useful. Due to a higher solvating ability, it gives polymers of higher 1,2-unit content than commonly used aliphatic potassium alkoxides *(14)*. LKN-1-5 polymer presented in Table III is obtained with the use of this compound.

A considerable success was achieved recently in the development of sodium-based initiators. It should be mentioned that there exists a kind of prejudice concerning the possibility of using sodium in homogeneous solution polymerization of dienes. Until recently, in some publications and even textbooks it has been stated that only low-molecular-weight polymers with low yield can be obtained by sodium-initiated diene polymerization in nonpolar solvents *(1,15)*. However, it was shown that this opinion is wrong. The first example was a complex sodium-aluminum system which contained well-known disodium oligo-α-methyl styrene in combination with tri-*iso*-butylaluminum or di-*iso*-butylaluminum hydride *(16-18)*. It was found that the use of alkylaluminum modifiers suppresses chain transfer processes, which are characteristic of organosodium initiators, and thus enables to obtain high-molecular-weight polymer even in toluene solution. However, the necessity to use additional polar solvent for the synthesis of disodium oligomers does not make this system very convenient for large-scale industrial application.

Later on, a hydrocarbon-soluble organosodium initiator was developed. This allows to adapt the "classical" synthesis of sodium-butadiene rubber to the conditions of a continuous solution process (19-21). The initiator, 2-ethylhexyl sodium, can be synthesised from commercially available starting compound, 2-ethylhexanol, via the corresponding chloride with essentially the same technique and equipment as for the standard synthesis of butyllithium. However, in this case, contrary to all previously mentioned processes, aliphatic solvents must be used for synthesis of both initiator and high-molecular-weight polymer because toluene is metallated quickly by organosodium compounds. At the same time, some amount of toluene or other strong chain transfer agent - isobutene - enables to control the molecular weight of the polymer within the wide interval up to the synthesis of liquid rubber (20,22). High- and low-molecular weight styrene-butadiene rubbers with high vinyl group content also can be obtained with this initiator (19,21).

Properties of pilot-plant polymers obtained in the presence of several new initiators are demonstrated in Table IV.

Table IV. Properties of 1,2-polybutadiene, obtained with sodium-containing initiators

Initiator	BuLi-THFONa	RNa[a]
1,2-unit content, %	76	46
Mooney viscosity	48	45
Plasticity	0.23	0.45
Modulus 300%, MPa	8.7	10.3
Tensile strength, MPa	15.8	17.8
Elongation, %	430	445

[a] in aliphatic solvent (21)

Conclusion

The large-scale industrial process for the synthesis of 1,2-polybutadiene rubber by continuous solution polymerization with n-BuLi-diglyme anionic initiator has been developed and successfully exploited for many years. The technology of the synthesis, properties and fields of application of high and low-molecular-weight polybutadienes are described. This rubber is applied for manufacturing various industrial and consumer goods; it is especially useful as a binder for abrasives, asbestos friction articles, electric insulation and other articles the exploitation of which is connected with high heat evolution. A special type of rubber for the tire industry is also manufactured. An experience in the parallel exploitation of anionic synthesis of 1,2-polybutadiene and the synthesis of cis-1,4-polybutadiene with Ti-containing catalyst, with the use of the same standard equipment has confirmed a number of advantages of the anionic process. The improvements in the process based on the use of sodium and potassium alkoxides or new organosodium initiators are also discussed.

Literature cited

1. Tate, D.P.; Bethea, T.B. In *Encyclopedia of Polymer Science and Engineering;* Kroshwitz, J.J., Ed; 2d Ed., J.Wiley & Sons: N.Y., **1985**; Vol. 2, p. 537.
2. Babitskii, B.D.; Krol, V.A. In *Synthetic Rubber;* Garmonov, I.V., Ed.; Khimia: Leningrad, **1983**, p. 134.
3. Diner E.Z.; Zaboristov, V.N.; Ryachovsky, V.S.; Krol, V.A. *Kautchuk i rezina* (Moscow). **1983**, no 6, p. 7.
4. Litvinenko, G.I.; Arest-Yakubovich, A.A.; Zolotarev, V.L. *Vysokomol. Soedin., Ser. A.* **1991**, *33*, 1410.
5. Bojkova, I.N.; Diner, E.Z.; Drozdov, B.T.; Ermakova, I.I.; Krol', V.A. *Promyshlennost' Sinteticheskogo Kauchuka (Synthetic Rubber Industry)* **1976**, no. 10, 7.
6. Shpakov, P.P.; Zak, A.V.; Lavrov, V.A.; Ermakova, I.I. *Vysokomol. Soedin., Ser. A.* **1988**, *30*, 467.
7. Ryakhovsky, V.S.; Ermakova, I.I.; Zaboristov, V.N.; Krol', V.A. In *Anionnaya Polymerizatsia. Voprosy Teorii i Praktiki (Anionic Polymerization. Theory and Practice)* TSNIINeftekhim: Moscow, **1985**, p. 28.
8. Arest-Yakubovich, A.A; Litvinenko, G.I. *Prog. Polym. Sci.* **1996**, *21*, 335.
9. Halasa, A.F., Massie, J. In *Kirk-Othmer Encyclopedia of Chemical Technology*, 4th ed., J.Wiley&Sons: N.Y., **1993**; Vol. 8, p. 1031.
10. Ermakova, I.I.; Mongayt, E.Z.; Eremina, M.A.; Drozdov, B.T. *Proc. of the Conf. "Kautchuk-89";* Moscow, **1990**, p. 30.
11. Nordsiek, K.N. *Kautsch. Gummi. Kunstst.*, **1982**, *35*, 371.
12. Yoshioka, A.; Komuro, K.; Kedo, A.; Watanabe, H.; Akita, S.; Masuta, T.; Nakajima, A. *Pure Appl. Chem.* **1986**, *58*, 1697.
13. Basova, R.V.; Arest-Yakubovich, A.A. *Vysokomol. Soedin., Ser. B* **1978**, *20*, 709.
14. Aksenov, V.I.; Sokolova, A.D.; Afanasjeva, V.V.; Arest-Yakubovich, A.A. *Kautchuk i:Rezina* (Moscow). **1990**, no. 11, 11.
15. Cheng, T.C. *Polymer. Sci. Techn.* **1984**, *24*, 155.
16. Arest-Yakubovich, A.A.; Anosov,V.I.; Basova, R.V.; Zolotarev, V.L.; Kristal'nyi, E.V.; Makortov, D.V.; Nakhmanovich, B.I. *Vysokomol. Soedin., Ser. A.* **1982**, *24*, 1530.
17. Arest-Yakubovich, A.A. *Chem. Revs.* **1994**, *19*, pt. 4, 1.
18. Arest-Yakubovich, A.A. *Macromol. Symp.*, **1994**, *85*, 279.
19. Arest-Yakubovich, A.A.; Litvinenko, G.I.; Basova, R.V. *Polymer Prepr.* **1994**, *35*, no 2, 544.
20. Arest-Yakubovich, A.A.; Pakuro, N.I.; Zolotareva, I.V.; Kristalnyi, E.V.; Basova, R.V. *Polymer Int.* **1995**, *37*, 165.
21. Arest-Yakubovich, A.A.; Zolotareva, I.V.; Pakuro, N.I.; Nakhmanovich, B.I.; Glukhovskoi, V.S. *Chem. Eng. Sci.* **1996,** *51*, 2775.
22. Arest-Yakubovich, A.A.; Kristal'nyi, E.V.; Zhuravleva, I.L. *Vysokomol. Soedin., Ser. B.* **1994**, *36*, 1553.

Other Applications of Controlled Anionic Polymerization

Chapter 16

Practical Applications of Macromonomer Techniques for the Synthesis of Comb-Shaped Copolymers

Sebastian Roos[1], Axel H. E. Müller[1,3], Marita Kaufmann[2], Werner Siol[2], and Clemens Auschra[2]

[1]Institut für Physikalische Chemie, Universität Mainz, Mainz, Germany
[2]Röhm GmbH, Darmstadt, Germany

Acrylic macromonomers (MM) with a wide range of molecular weights and high functionality have been synthesized both by group transfer polymerization (GTP) using functionalized initiators and by radical polymerization using transfer agents. Important parameters controlling the MM reactivity in radical copolymerization with MMA and nBuA have been determined. The MM reactivity decreases with increasing concentration, increasing MM chain length and decreasing spacing of side groups. Various comb-shaped copolymers have been obtained by radical copolymerization of MM's with acrylic monomers like n-butyl acrylate (nBuA) and n-butyl methacrylate (BMA). PnBuA-g-PMMA is a highly transparent, microphase-separated material showing good thermoplastic elastomer properties. Glass transition temperatures and spin diffusion measurements indicate that the PMMA phase contains large amounts of butyl acrylate segments. PBMA-g-poly(alkyl methacrylate) with C_{12}-C_{16} alkyl groups shows better properties as an viscosity index improver than linear poly(alkyl methacrylate).

Comb-shaped (graft) copolymers offer all properties of block copolymers but are much easier to synthesize. Moreover, the branched structure leads to decreased melt viscosities which is an important advantage for processing. Depending on the nature of their backbone and side chains, (hard/soft, water/oil-soluble) they can be used for a wide variety of applications, such as impact-resistant plastics, thermoplastic elastomers, compatibilizers, and polymeric emulsifiers. According to the theory of Erukhimovich,[1] graft copolymers should show better phase separation than triblock copolymers.

[3]Corresponding author.

Up to now, mainly styrene and dienes have been used to build the hard and soft blocks of block copolymers used as impact-resistant plastics and thermoplastic elastomers. Owing to the advantageous properties of the polymers formed, e.g. improved weatherability and optical properties, it is of some interest to replace the hydrocarbon monomers by acrylic ones.

The state-of-the-art technique to synthesize graft copolymers is the macromonomer (MM) technique. It allows the control of both chain length and spacing of the side chains. The former is given by the experimental conditions of the macromonomer synthesis and the latter is determined by the molar ratio of the comonomers and the reactivity ratio of the low-molecular-weight monomer, $r_1 = k_{11}/k_{12}$.

Major criteria for the selection of the macromonomers to be synthesized were the potential product properties as well as ready availability of the educts and easy manageability of the end products. If possible, the products should be offered in the form of pure substances (macromonomers of lauryl methacrylate, liquid; macromonomers of methyl methacrylate, solid), or as highly concentrated solutions (macromonomers of butyl methacrylate). The macromonomers have very low color indices which enable their use even in transparent systems.

Synthesis of Macromonomers and Copolymerization

Synthesis by Group Transfer Polymerization: Methacryloyl-terminated PMMA macromonomers of various molecular weights ($10^3 \leq M_w \leq 2 \cdot 10^4$) and low polydispersity ($M_w/M_n < 1.2$) were prepared by group-transfer polymerization (GTP) using a trimethylsiloxy-functionalized initiator.[2] After polymerization the protecting trimethylsiloxy group was removed and the resultant HO-CH_2-CH_2-terminated PMMA was reacted with methacryloyl chloride in order to introduce the methacryloyl group with a functionality, $f > 0.95$.

Synthesis on an Industrial Scale: On an industrial scale, simple free-radical polymerization in presence of 2-mercaptoethanol is an alternative route to obtaining HO-CH_2-CH_2-S-terminated poly(alkyl methacrylates) with a functionality, $f > 0.98$. Via the amount of chain transfer agent (0.5 to 2%) the chain length was varied between $5 \cdot 10^3 \leq M_w \leq 4 \cdot 10^4$. Subsequent transesterification with methyl methacrylate or butyl acrylate was used to convert the OH-terminated poly(alkyl methacrylate) into the desired macromonomer.[10] A polymer-analogous transesterification in the side chain was not observed in the process. The yield of (meth)acryloyl terminal groups was > 0.95.

Determination of Functionality by Liquid Chromatography under Critical Conditions (LCCC): In order to determine the yield of polymerizable groups, LCCC[3] was applied. This method works at the critical point of the polymer/solvent/non-solvent/stationary phase system where size exclusion and adsorption balance each other in a way that no separation according to molecular weight occurs. In contrast, separation according to the chain-end structure is very efficient (see Fig. 1).

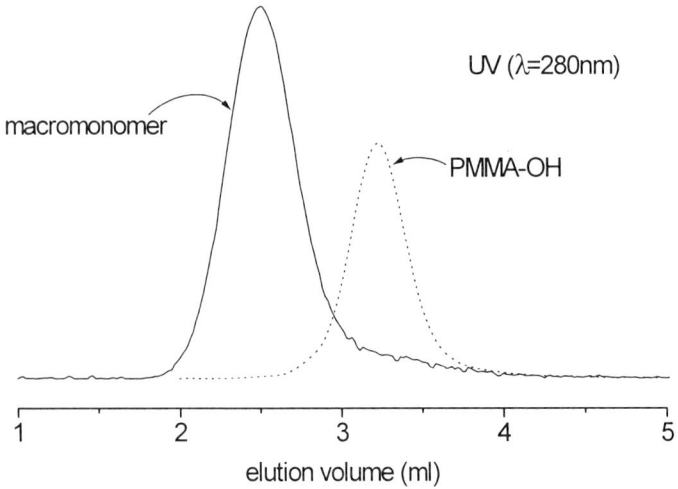

Fig. 1. LCCC eluograms of hydroxyl- and methacryloyl-terminated PMMA

Synthesis of the graft copolymers: The copolymerization of methacryloyl-terminated PMMA with nBuA was investigated in toluene at 60°C using *tert.*-butylperoxyneodecanoate as an initiator. Ca. every sixth MM was labeled with naphtylcarboxyethyl methacrylate in order to facilitate the determination of the MM conversion by GPC analysis of the copolymer.[4] The conversion of nBuA was determined by gas chromatography.

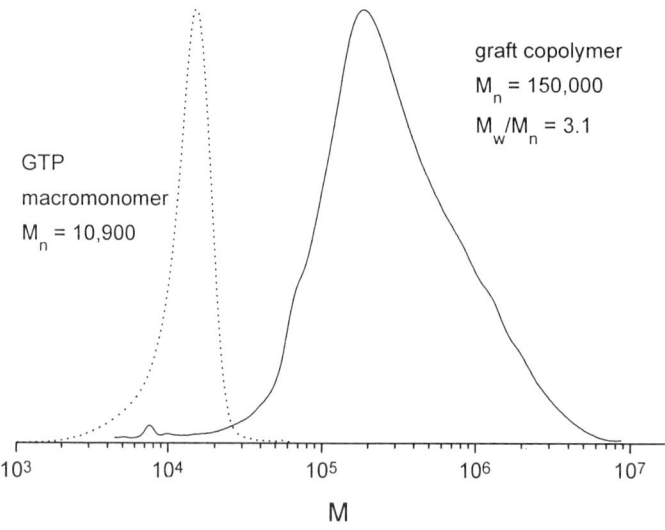

Fig. 2. GPC traces of the MM and the resulting graft copolymer.

Determination of Macromonomer Reactivities in Radical Copolymerization

The macromonomers obtained by GTP were used in order to determine their relative reactivities in radical copolymerization, measured as $1/r_1 = k_{12}/k_{11}$, as a function of concentration, chain length, and spacing. For a system where incompatibility between backbone and incoming macromonomer is excluded, i.e. the copolymerization of ω-methacryloyl-PMMA and methyl methacrylate (MMA, index 1)[2,5] three conclusions could be drawn:

- the reactivity of macromonomers (MM) strongly decreases with increasing MM weight concentration, due to the increasing viscosity which limits the mobility of the macromonomer;
- the reactivity of MM decreases with decreasing molar ratio $[MMA]_0/[MM]_0$, i.e. with decreasing spacing between the side chains, due to the increasing segment density in the vicinity of the propagating center which hinders the approach of the MM;
- under undisturbed conditions, i.e. at low polymer concentration and large spacing ($[MMA]_0/[MM]_0 \geq 100$), the reactivity of MM does not depend on the MM chain length.

In a real system, i.e. the copolymerization of ω-methacryloyl-PMMA (index 2) with butyl acrylate (nBuA; index 1) the dependences of the relative MM reactivity on the molar ratio of the monomers, $[MM]_0/[nBuA]_0$, and on the total monomer concentration, are more complicated.[4,6] In all these experiments the MM reactivity, $1/r_1$, is always lower than that of the low-molecular-weight analogue, MMA, but higher than that of nBuA. This has the advantage that the MM is completely consumed.

The results indicate that the copolymerization is controlled by two additional effects:

- the kinetic excluded-volume effect: at these low concentrations the MM and the graft copolymer molecules try to avoid each other. At higher concentrations overlap of polymer coils cannot be avoided, leading to an initial increase in reactivity with concentration.
- preferential solvation ("bootstrap effect"[7]): the chain end is preferentially surrounded by monomers of its own kind, leading to an increased concentration as compared to stoichiometry.

In contrast to copolymerizations with MMA, there is now a distinct decrease in MM reactivity with increasing molecular weight of the MM (Fig. 3).

Microphase Separation in Bulk

Microphase separation is indicated by the DSC diagrams (Fig. 4) which show two glass transition temperatures, one at $T_{g1} = -47$ °C (extrapolated to zero heating rate) for the PnBuA main chains and one at $T_{g2} = 76$ °C for the PMMA side chains. The value for PMMA is significantly lower than that of the macromonomer which is $T_g = 100$ °C. This is a first indication that the PnBuA segments penetrate into the PMMA microphase acting as a softener. Xie and Zhou found Tg's at -45 °C and 94 °C for the different phases in PnBuA-g-PMMA copolymers.[8]

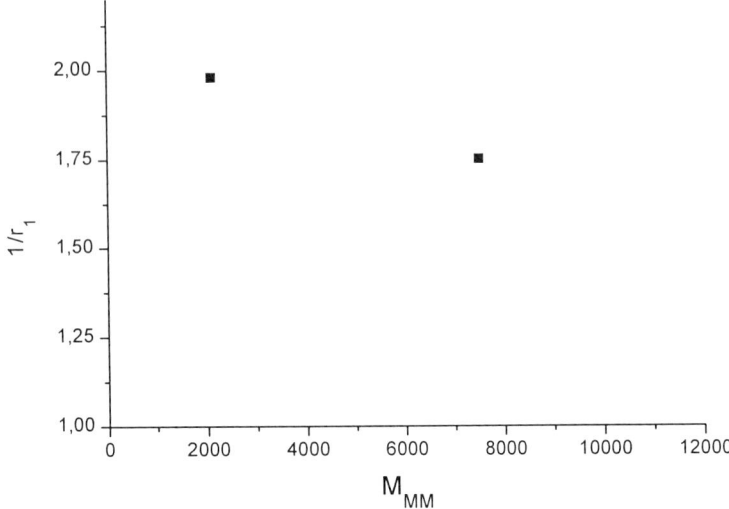

Fig. 3. Effect of molecular weight of macromonomer on MM reactivity. AIBN, butyl acetate, T = 60 °C; $[nBuA]_0 + [MM]_0 = 1.05$ mol/l; $[nBuA]_0/[MM]_0 = 170$.

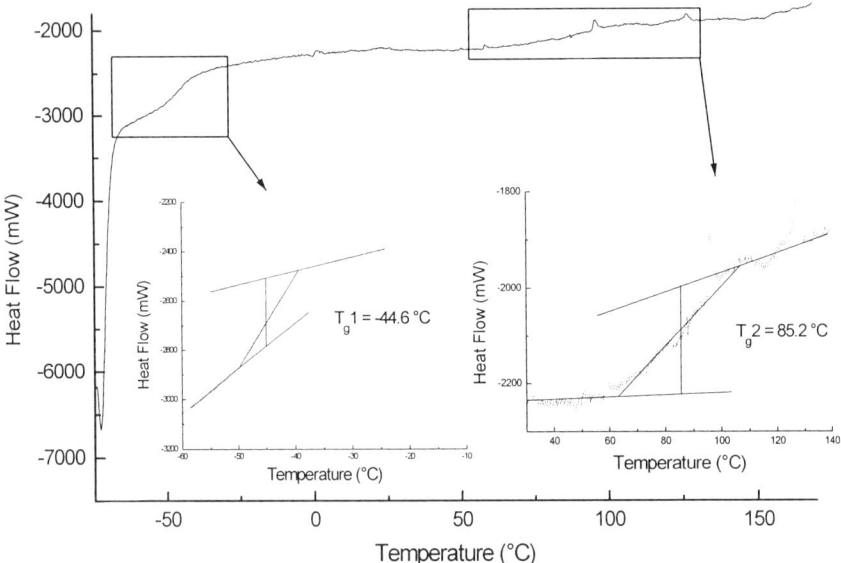

Fig. 4: DSC diagram of PnBuA-g-PMMA at a heating rate of 10 K/min. Reproduced with permission from Ref. 4

Morphological studies confirm microphase separation with very small domain sizes (< 10 nm). NMR spin diffusion measurements[9] corroborate the DSC and TEM findings. They show that the amount of PnBuA in the PMMA phase increases with increasing weight ratio of nBuA to MMA. At a weight ratio of nBuA:MMA = 1 this amount can reach nearly 50%.

Properties and Practical Applications of Comb-Shaped Copolymers

Depending on the macromonomers and comonomers chosen, polymer structures with the most varied properties can be obtained. Hard/soft systems (e.g. PnBuA-*g*-PMMA) can be used as thermoplastic elastomers or impact modifiers, amphiphilic systems (e.g. poly(acrylic acid)-*g*-poly(lauryl methacrylate)) can be used as polymeric emulsifiers; polar/non-polar or amorphous/crystalline systems may be used as compatibilizers for polymer blends, and oil-soluble/insoluble systems (e.g. poly(butyl methacrylate)-*g*-poly(alkyl methacrylate) make better viscosity index improvers for motor oils.

Thermoplastic Elastomers: For the preparation of the films the graft copolymers formed by copolymerization of PMMA macromonomers with nBuA were dissolved in butyl acetate and poured into a mould of 50 mm length and 15 mm width. After the solvent was evaporated the films were placed in a vacuum oven for 2 days at room temperature and one day at 50 °C. The resulting films are 1-1.5 mm thick and crystal-clear. They were elongated at a rate of 50 mm/min and show typical behavior of thermoplastic elastomers as demonstrated by the stress-strain diagram in Fig. 5.

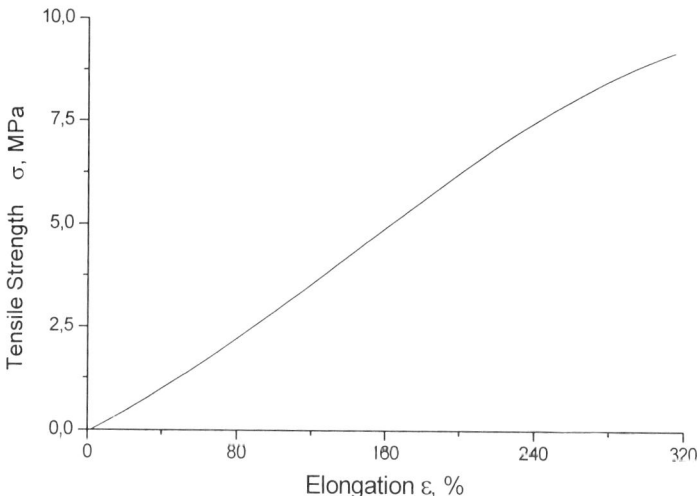

Fig. 5: Typical stress-strain diagram of PnBuA-*g*-PMMA thermoplastic elastomer.

The elastomeric properties strongly depend on the structure of the graft copolymers, as defined by composition, spacing and chain length of grafts, and molecular weight of the copolymer. The polymers investigated in Tables 1,3, and 4 were obtained from macromonomers made by radical polymerization and those in Table 2 from macromonomers made by GTP.

Table 1 shows that the tensile strength, σ, strongly increases with increasing weight fraction of PMMA. Analogously, the maximum elongation decreases. Thus copolymers with > 40 wt.-% of PMMA are impact-resistant plastics rather than thermoplastic elastomers. A similar effect is observed when increasing the length of the PMMA side chains and simultaneously decreasing the spacing in order to maintain a constant fraction of PMMA (w_{MMA} = 33%) in the copolymer. Table 2 shows that longer side chains (and correspondingly, larger spacing) lead to harder materials. This may be explained by more efficient formation of the domains with increasing the M_w of the macromonomer in agreement with results of Xie and Zhou.[8]

Table 1: Effect of composition on the elastomeric properties of PnBuA-g-PMMA. DP_n(PMMA-MM) = 40, DP_w = 70, M_w^{app}(copolymer) ≈ 750,000

Spacing, $[nBuA]_0/[MM]_0$	w_{PMMA}, %	σ, MPa	ε, %
177	15	3.4	720
125	20	5.3	600
73	30	10	435
47	40	15	375

Table 2: Effect of length of side chain on the elastomeric properties. Graft copolymers containing 33 wt.-% PMMA; M_w/M_n of MM ≈ 1.1, M_w^{app} (copolymer) ≈ 450,000

DP_n of MM	Spacing	σ, MPa	ε, %
109	170	9.1	170
55	100	6.1	400
21	43	1.8	970

Table 3 shows that the elastomeric properties improve for larger molecular weights of the copolymers due to a larger average number of PMMA side chains per molecule. For low molecular weights of the copolymer there is a significant fraction of polymers which carry no or only one side chain making them ineffective for crosslinking the soft backbones.

Table 3: Effect of molecular weight of graft copolymer on the elastomeric properties. Graft copolymers containing 30 wt.-% PMMA; DP_n (MM) = 40, DP_w = 70.

$10^5 M_w^{app}$	σ, MPa	ε, %
4.8	5.7	598
3.3	5.2	491
2.3	4.7	482
1.2	2.9	385

Finally, Table 4 demonstrates that only polymerization in solution leads to the desired thermoplastic elastomers, whereas in bulk polymerizations there is just a homopolymerization of nBuA leading to polymer blends only. This is due to the incompatibility of the macromonomers with the backbone formed from nBuA. In addition, it was shown in the kinetic studies that the (apparent) macromonomer reactivity decreases with increasing concentration of polymer due to the restricted mobility of macromonomers as compared to the low molecular weight monomers.

Table 4: Effect of MM concentration in copolymerization. 15 wt.-% PMMA-MM

Conditions of polymerization	σ, MPa	ε, %	Appearance of polymer films
Bulk;15 wt.-% MM in nBuA	0.5	276	white
in butyl acetate; 5 wt.-% MM; 25 wt.-% nBuA	3.4	720	transparent

Viscosity Index Improvers: Linear polymers, e.g. copolymers of MMA and alkyl methacrylates (C_{12}-C_{16}) (PAMA) are used in order to decrease the viscosity index (VI) of motor oils, thus giving these oils a larger span of operating temperatures. In a first approximation the viscosity index (VI) of a motor oil corresponds to the slope of a plot of log(log ν) versus temperature, where $\nu = \eta/\varrho$ is the kinematic viscosity (ASTM D 2270). Compared to the linear PAMA's, graft copolymers with poly(n-butyl methacrylate) (PBMA) backbone and PAMA side chains have greatly enhanced properties. At low temperatures their backbone is hardly soluble in mineral oil leading to a collapsed structure with low viscosity. At high temperatures the solubility of PBMA increases leading to an expanded structure with high viscosity. As a consequence, the viscosity decrease of the base oil with increasing temperature is much better compensated than with linear polymers having the same permanent shear stability index, PSSI (see Fig. 6 and Table 5). The graft copolymer also efficiently decreases the crystallization of the base oil at low temperatures as is indicated by relative viscosities below unity for T < 0 °C. PSSI is a technical quantity to define the shear stability of lubricants. It is given as the difference of the viscosity of used and fresh oils divided by the difference of the viscosity of fresh oil and base oil (ASTM D3945-93; DIN 51382).

Table 5: Properties of VI improvers in base oil 150N. Concentrations were chosen in order to obtain a constant viscosity of 14 mm^2s^{-1} at 100 °C

product	concentration, wt.-%	PSSI[a] %	VI[b]
Viscoplex® 8-500[c]	5,5	24	193
PBMA-g-P(AMA-co-MMA) 55:45 wt-%	5,5	6,7	308
same, but higher M_w	3,6	26	302

[a] permanent shear stability index, [b] viscosity index, [c] linear P(AMA-co-MMA). AMA = alkyl methacrylate (C_{12}-C_{16}), BMA = n-butyl methacrylate.

Scheme 1: Chain dimensions of graft copolymer in a motor oil at different temperatures.

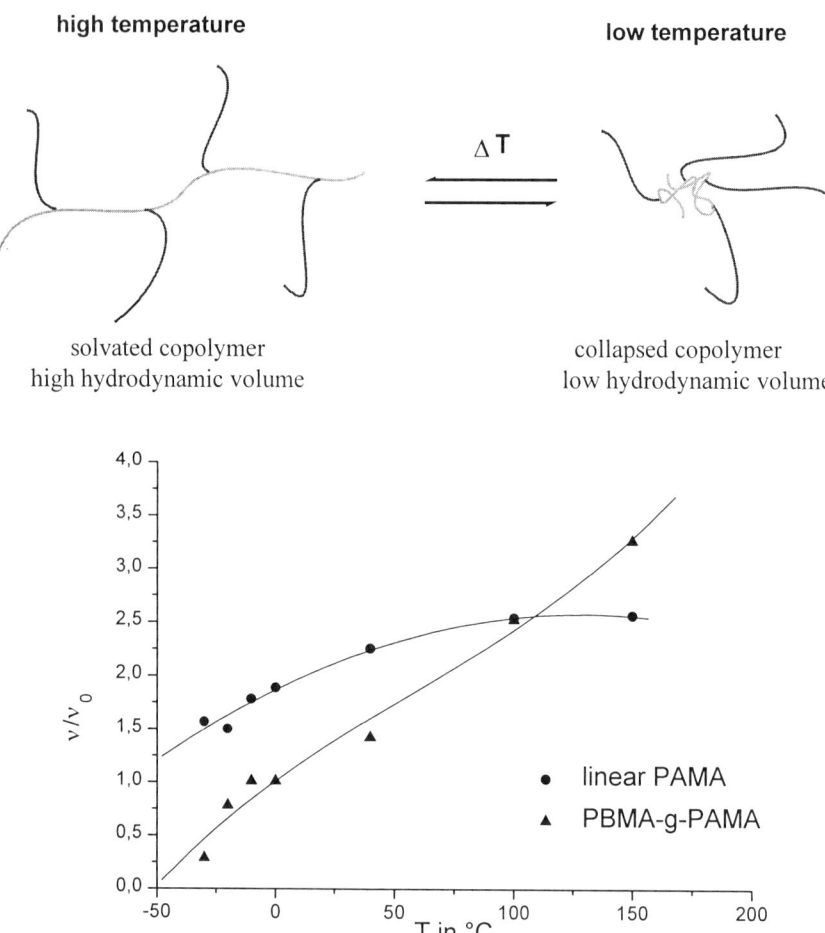

Fig. 6: Relative kinematic viscosities of linear and grafted methacrylates in base oil 150N (see Table 5).

Acknowledgment: This work was supported by the *German Minister of Research and Technology* (BMFT).

References
1) Dobrynin, A.V.; Erukhimovich, I.Y. *Macromolecules* **1993**, *26*, 276
2) Radke, W.; Müller, A.H.E. *Makromol. Chem., Macromol. Symp.* **1992**, *54/55*, 583
3) Entelis, S.G.; Evreinov, V.V.; Gorshkov, A.V. *Adv. Polym. Sci.* **1980**, *76*, 129
4) Radke, W.; Roos, S.; Stein, H.M.; Müller, A.H.E. *Macromol. Symp.* **1996**, *101*, 19
5) Radke, W.; Müller, A.H.E. *Polym. Prepr. (Am. Chem. Soc., Div. Polym. Chem.)* **1991**, *32(1)*, 567
6) Radke, W.; Stein, H.M.; Müller, A.H.E. *Polym. Prepr. (Am. Chem. Soc., Div. Polym. Chem.)* **1993**, *34(2)*, 652
7) Harwood, J. *Makromol. Chem., Macromol. Symp.* **1987**, *10/11*, 331-354
8) Xie, H.-Q.; Zhou, S.-B. *J. Macromol. Sci., Chem.* **1990**, *27*, 491
9) Landfester, K.; Roos, S.; Müller, A.H.E. in preparation
10) US Patent 5,254,632 (1993) to Röhm GmbH

Chapter 17

Preparation of Lithographic Resist Polymers by Anionic Polymerization

Hiroshi Ito

IBM Research Division, Almaden Research Center, 650 Harry Road, San Jose, CA 95120–6099

Anionic polymerization can play a unique role in preparation of resist polymers, although radical polymerization is most commonly employed in syntheses of polymers for lithographic use. In some cases anionic initiation is the only feasible mechanism to prepare a desired polymer. Structural modification to improve radiation sensitivity can sometimes result in reduction of radical homopolymerizability. Methyl α-trifluoromethylacrylate and isopropenyl t-butyl ketone are such examples. In the field of chemical amplification resists, anionic polymerization has been utilized to prepare polyphthalaldehydes as dry-developable resists and poly(4-t-butoxycarbonyloxystyrene) and poly(4-hydroxy-α-methylstyrene) for investigation of polymer end group effects.

Radiation-sensitive polymeric materials called "resists" play an extremely important role in semiconductor device manufacture by microlithography (*1*). The lithographic process as practiced in the semiconductor industry is schematically illustrated in Figure 1. A radiation-induced chemistry makes the exposed region more (positive) or less (negative) soluble in a developer solvent. The microelectronics revolution we are witnessing today has been brought about by the astounding progress of the lithographic imaging technology, in which the resists are the major component. In addition to a film-forming property, resist polymers must satisfy many stringent requirements. In some cases, resist polymers themselves are sensitive to radiation, in other cases completely inert, and yet in some other cases reactive to radiation-generated species. As the miniaturization of electronic devices continues, the increasingly demanding requirements placed on resist materials have necessitated syntheses of resist resins as specialty polymers.
Although radical polymerization is generally considered more practical and economical and therefore most commonly employed in preparation of resist polymers, in some cases desired polymers are not accessible by this mechanism. Poly(methyl α-

trifluoromethylacrylate) (polyMTFMA) and poly(isopropenyl *t*-butyl ketone) (polyIPTBK), which undergo efficient main chain scission upon electron-beam and UV irradiation, are accessible only by anionic polymerization. Another example of anionic polymerization employed in the resist synthesis is cyclopolymerization of phthaladehydes. Polyphthaladehyde derivatives are useful in the design of self-developing or thermally-developable chemical amplification resists (2). The ability of "living" anionic polymerization to produce well-controlled polymer structures has allowed us to study the effect of polymer end groups on chemically-amplified lithographic imaging. In this paper are discussed the above-mentioned applications of anionic polymerization to preparation of resist polymers.

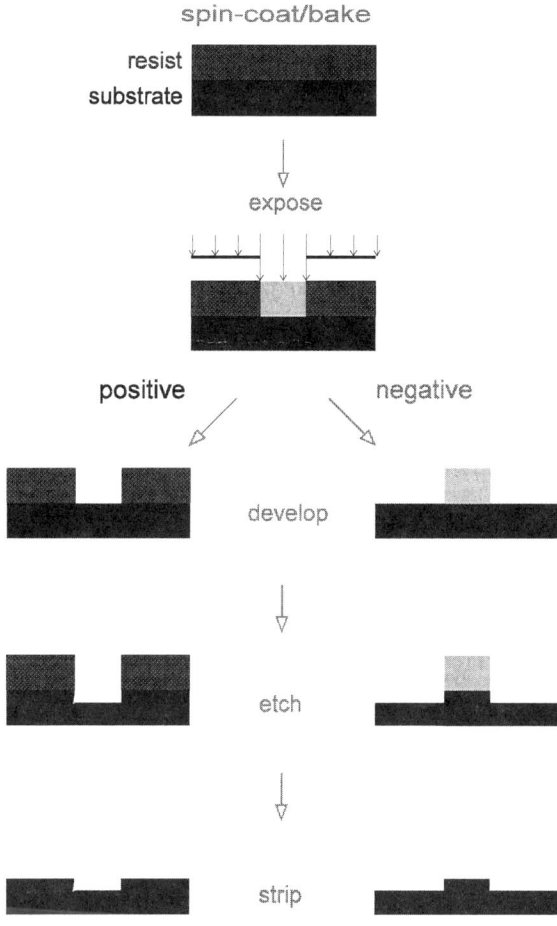

Figure 1 Lithographic imaging process

Resist Polymers Accessible only by Anionic Polymerization

Radical polymerization is generally considered the most economical and therefore the most preferred technique to prepare resist and other polymers. However, there are instances where structural modification to improve sensitivity results in alteration of reactivity in such a fashion that radical polymerization is no longer feasible. There are also instances where a desired resist polymer can be obtained only by ionic polymerization.

Poly(methyl α-Trifluoromethylacrylate). Poly(methyl methacrylate) (PMMA) undergoes main chain scission upon exposure to deep UV (< 300 nm), electron-beam, or x-ray radiation and has been known for its high resolution capability and low sensitivity. Serious efforts were directed toward sensitivity enhancement of the methacrylate resists and many analogs and copolymers were synthesized and evaluated as positive resists (*3-5*). Introduction of an electron-withdrawing group into the α-position was reported to increase the main chain scission susceptibility (*6*). Poly(methyl α-fluoroacrylate) degrades more efficiently than PMMA upon γ-irradiation but suffers from concomitant crosslinking essentially due to HF elimination to form C=C double bonds in the backbone (*6*).

We reasoned that replacement of the α-methyl group of PMMA with a strongly electron-withdrawing CF_3 should increase the scission susceptibility without introducing a crosslinking component (*7*). Methyl α-trifluoromethylacrylate (MTFMA) is similar to methyl methacrylate (MMA) in structure. However, we have found that this fluoro monomer is completely different from MMA in reactivity (*8*). MTFMA undergoes radical homopolymerization very sluggishly even in bulk (*8*). We have carried out radical copolymerization of MTFMA (M_1) with MMA (M_2) to study the radical reactivity of MTFMA. The copolymer composition curve is presented in Figure 2. The reactivity ratio values determined are $r_1=0$ and $r_2=2.36$ (*8*), indicating that MTFMA does not self-propagate. Thus, radical polymerization cannot provide polyMTFMA in a practical sense under normal polymerization conditions. Although the Alfrey-Price Q and e parameters cannot be calculated from reactivity ratios when one of the values is zero, combination of the reactivity ratios and the mercury method developed by Giese (*9*) allowed us to calculate Q and e for MTFMA as 0.74 and 2.5, respectively (*10*). Thus, MTFMA is extremely electron-deficient.

Electron-deficient monomers are expected to undergo anionic polymerization readily. However, anionic initiators such as butyllithium and Grignard reagents typically used for MMA failed to polymerize MTFMA due to S_N-2' addition-elimination side reactions as outlined in Figure 3 (*11*).

Owing to the low electron density on the β-carbon, however, MTFMA undergoes facile anionic polymerization with amines or organic and inorganic salts in the presence of 18-crown-6 (*12*) as shown in Table I and Figure 4. Although the polymerization is slow, polymers with $M_n=20,000$ and $M_w=30,000$-$40,000$ can be prepared in a good to excellent yield. Formation of a charge transfer complex through an electron transfer has been suggested as the initiation mechanism with pyridine (*12*).

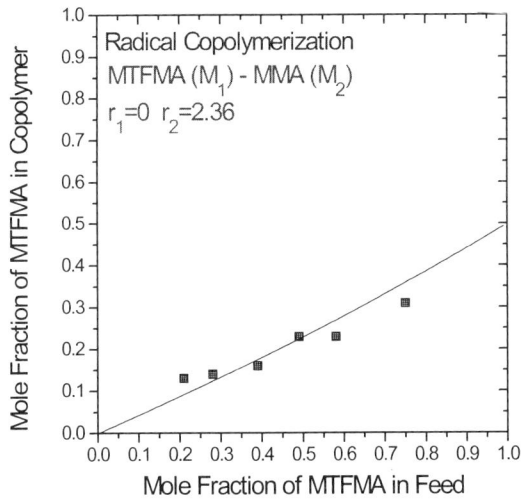

Figure 2 Radical copolymerization of MTFMA with MMA

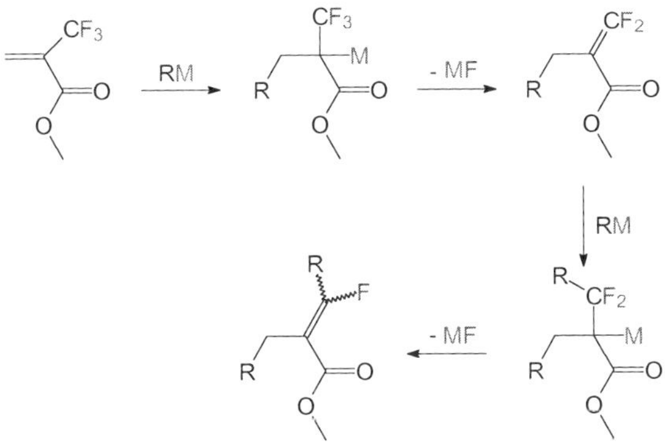

Figure 3 Addition-elimination sequences in anionic polymerization of MTFMA with alkyl alkali

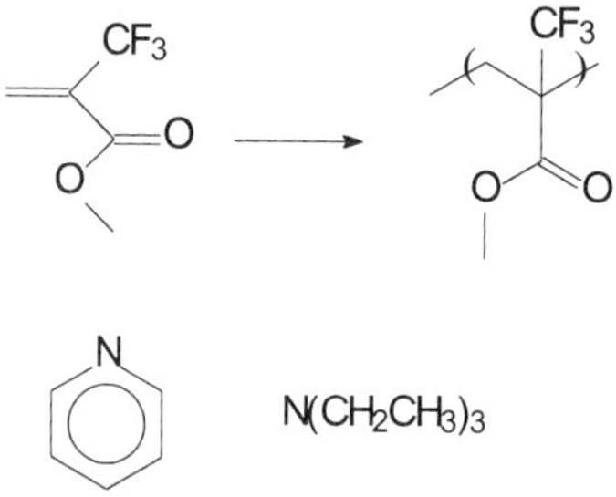

KOH KCN K_2CO_3 KOtBu KF NaOAc NaSCN LiF
(with 18-crown-6)

Figure 4 Anionic polymerization of MTFMA

Table I Anionic Polymerization of MTFMA in Bulk at Room Temperature with Amines and Salts*

Initiator (mol%)	Time (hr)	Yield (%)	M_n	M_w
pyridine (1.3)	24	94	21,300	30,600
crosslinked poly(4-vinylpyridine) (1.3)	72	85	10,100	28,000
Et_3N (2.0)	24	76	16,900	24,800
KOH (2.0)	24	64	8,200	11,400
KCN (2.0)	24	82	11,600	21,300
K_2CO_3 $1^1/_2H_2O$ (2.0)	24	84	13,300	20,200
KOCOPh (2.0)	24	70	11,700	16,200
KOtBu (2.0)	15	76	20,300	41,600
KF $2H_2O$ (2.0)	17	50	19,200	52,400
$NaOCOCH_3$ (2.0)	72	44	18,700	36,900
NaSCN (2.0)	113	43	14,700	21,600
LiF (2.0)	24	0		

*with 18-crown-6

PolyMTFMA thus prepared exhibited a G_s value (the number of main chain scission events per 100 eV of absorbed dose) of 2.5 upon ^{60}Co γ-radiolysis (*7*), while the value for PMMA is 1.3.

Poly(isopropenyl *t*-Butyl Ketone). Polyketones are another class of polymers that undergo main chain scission upon UV irradiation. In order to increase the sensitivity of poly(methyl isopropenyl ketone) (polyMIPK), we wanted to replace the methyl group with a *t*-butyl group because the Norrish type I photolysis (Figure 5) involving homolysis of the C-CO bond is much more favored in the solid state than the type II pathway which requires chain mobility to form a cyclic intermediate. In polyMIPK photolysis occurs only on the bond between the carbonyl carbon and the α-carbon to produce a stable *tertiary* radical. In the target polymer, poly(isopropenyl *t*-butyl ketone) (polyIPTBK), cleavage of either bond linking with the carbonyl carbon produces stable *tertiary* radicals. Thus, polyIPTBK should undergo much more facile photolysis in the solid state through the enhanced Norrish type I mechanism.

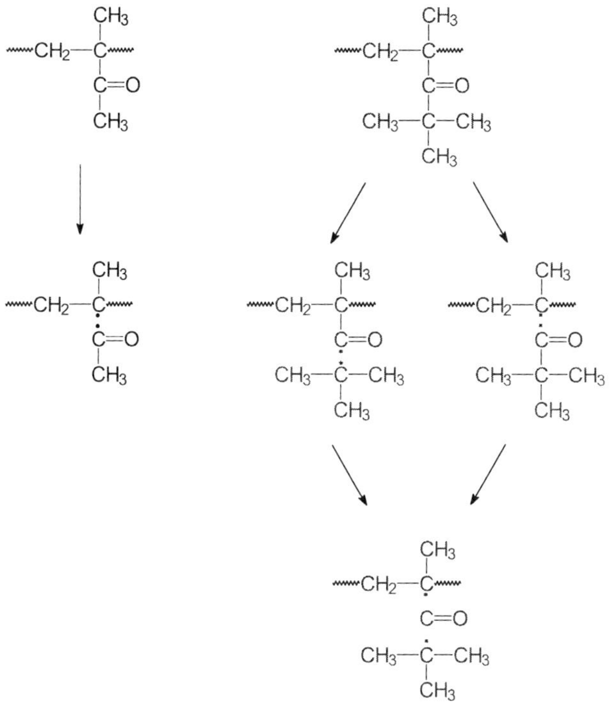

Figure 5 Norrish type I photolysis of polyMIPK and polyIPTBK

Although vinyl ketones are generally reactive monomers, IPTBK did not undergo radical homopolymerization under typical conditions due to steric hindrance (*13*). The C=C and C=O double bonds of this enone is not coplanar (Figure 6) and the electron density on the β-carbon of IPTBK is much higher than that of other vinyl ketone monomers, which is reflected in ^{13}C NMR, IR, and UV spectra (*13*). In consequence, anionic polymerization of IPTBK is a slow equilibrium process with a low ceiling temperature ($T_c < 0$ °C).

Based on our detailed investigation on the reactions between IPTBK and several anionic initiators, we have selected *t*-butyllithium (tBuLi) for the polymerization of IPTBK (*14*). tBuLi adds cleanly and rapidly to the β-carbon of IPTBK at 25 °C without contamination with propagation or addition to C=O (*14*). IPTBK and tBuLi were mixed in tetrahydrofuran (THF) at room temperature to promote clean and fast initiation and then cooled to a polymerization temperature (typically -78 °C) to promote propagation by shifting the equilibrium to the polymer side (*14*). Grignard reagents failed to polymerize IPTBK presumably due to the steric hindrance. The group transfer polymerization of IPTBK resulted in only dimer formation (*14*).

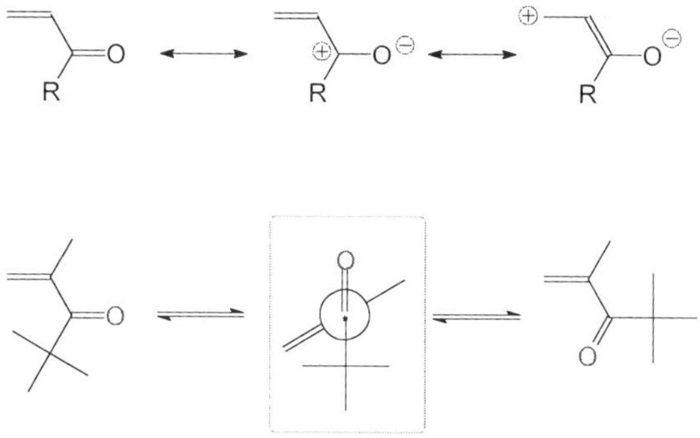

Figure 6 Conjugation in vinyl ketones and conformation of IPTBK

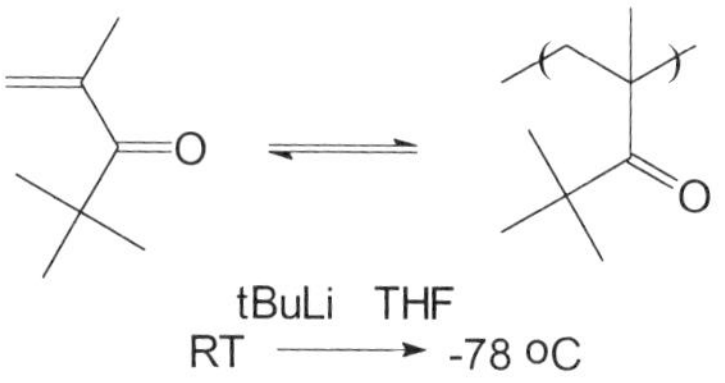

Figure 7 Anionic polymerization of IPTBK with tBuLi

Anionic polymerization of IPTBK with tBuLi is summarized in Table II. When the monomer concentration was low (entries 1 and 2), the polymerization was slow and the molecular weight was low (M_n~3,000) but the molecular weight distribution was narrow (~1.2). The bulk polymerization (entry 4) produced a high molecular weight polymer (M_n=78,600 and M_w=221,600, M_w/M_n=2.8) in a poor yield. Accordingly, we have selected the IPTBK/THF ratio of 1/1. By decreasing the initiator concentration (entries 5, 6, 9, and 10), the molecular weight increased from M_n=6,900 to 223,500 with the polydispersity maintained at a relatively small value of 1.4-1.6. When the polymerization was carried out at -55 °C, the yield and molecular weights were lower (entry 5 vs 6), suggesting the ceiling temperature effect. In one experiment (entry 8), the polymerization mixture was kept at -78 °C for one day, resulting in high viscosity, warmed to -20 °C and maintained at this temperature for one day, which resulted in loss of the high viscosity due to depolymerization, and then cooled again to -78 °C. After one day at -78 °C, a polymer comparable to entry 7 (straight polymerization at -78 °C) was obtained, indicating the equilibrium nature of the polymerization and the high stability of the growing anion. 1,1-Diphenylhexyllithium (DPHLi, entry 11) and s-butyllithium (sBuLi, entry 12) also produced good polymers in good yield.

As tabulated in Table III, while the quantum yield of main chain scission (Φ_s) is essentially the same in solution for polyIPTBK and polyMIPK (0.45), the former degrades > 10 times more efficiently than the latter in the solid state (0.29 vs 0.024) (15). In addition to the more facile Norrish type I cleavage, the steric hindrance, the lack of radical homopolymerizability, and the low ceiling temperature contribute to the high chain scission efficiency.

Table II Anionic Polymerization of IPTBK in THF

Entry	Initiator (mol%)	THF/IPTBK (mL/mL)	Temp (°C)	Time (days)	Yield (%)	M_n	M_w
1	tBuLi (5.0)	5	-78	16	60	3,400	4,100
2	(2.0)	4	-78	3.8	44	2,800	3,300
3	(2.0)	1	-78	0.92	70	5,900	8,100
4	(2.0)	0	-78	1.92	10	78,600	221,600
5	(2.0)	1	-78	1.88	84	6,900	10,900
6	(2.0)	1	-55	2.71	69	3,900	5,200
7	(1.5)	1	-78	1.21	48	9,800	14,400
8	(1.5)	1	-78 -20 -78	1.0 1.0 1.0	66	10,200	14,100
9	(1.0)	1	-78	1.75	85	17,300	28,100
10	(0.5)	1	-78	4.46	52	223,500	324,600
11	DPHLi (2.0)	1	-78	1.83	89	9,600	13,400
12	sBuLi (2.0)	1	-78	1.92	80	14,400	22,200

Table III Quantum Yield of Main Chain Scission

Polymer	Φ_s (solid state)	Φ_s (solution)
polyIPTBK	0.29	0.45
polyMIPK	0.024	0.45

Polyphthalaldehydes. Aldehydes are not compatible with radical polymerization but undergo anionic and cationic polymerizations in equilibrium reactions that are characterized with low T_c. Cyclopolymerization of phthalaldehyde (Figure 8) (*16*) produces a non-crystalline amorphous polymer in contrast to aliphatic aldehydes. Polyphthalaldehyde properly end-capped with an acetyl group is stable thermally to ~200 °C but cleanly depolymerizes to the starting monomer by reaction with a photochemically-generated acid, providing self-development in chemically-amplified lithographic imaging (*17*).

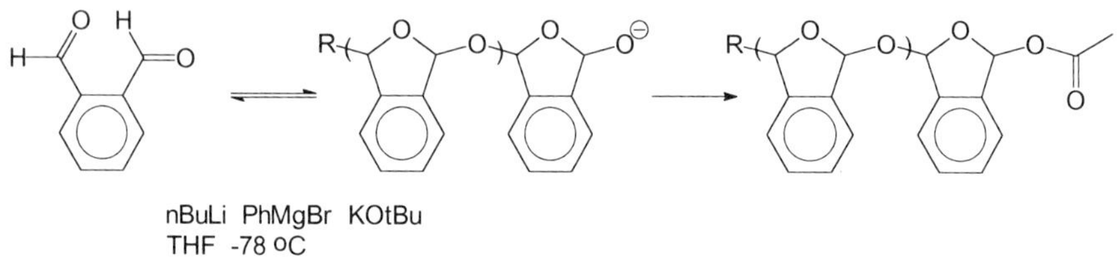

Figure 8 Anionic cyclopolymerization of phthalaldehyde

Anionic polymerization of 4-trimethylsilylphthalaldehyde (SPA) produces a more thermally stable polymer than cationic polymerization presumably due to more efficient end-capping (*18*). The polymerization must be carried out at cryogenic temperatures due to the low T_c. The monomer concentration was kept low (SPA/THF=1/10 g/mL) to prevent the monomer from crystallizing out during the polymerization at -78 °C, which resulted in slow polymerization (Table IV). Nevertheless, high molecular weight polymers (M_n>20,000) with a narrow polydispersity of 1.18-1.32 were obtained with n-butyllithium in good yield of 84 %.

Table IV Anionic Cyclopolymerization of SPA with nBuLi in THF at -78 °C

Initiator mol%	Time (hr)	Yield (%)	M_n	M_w
2.0	48	84	21,100	24,900
1.0	84	84	29,800	39,300

The polySPA containing triphenylsulfonium triflate as a photochemical acid generator functions as a positive tone thermally developable resist, which serves as an oxygen reactive ion etching (RIE) mask during etching of an underlying organic polymer layer in all-dry bilayer lithography (Figure 9) (*18, 19*).

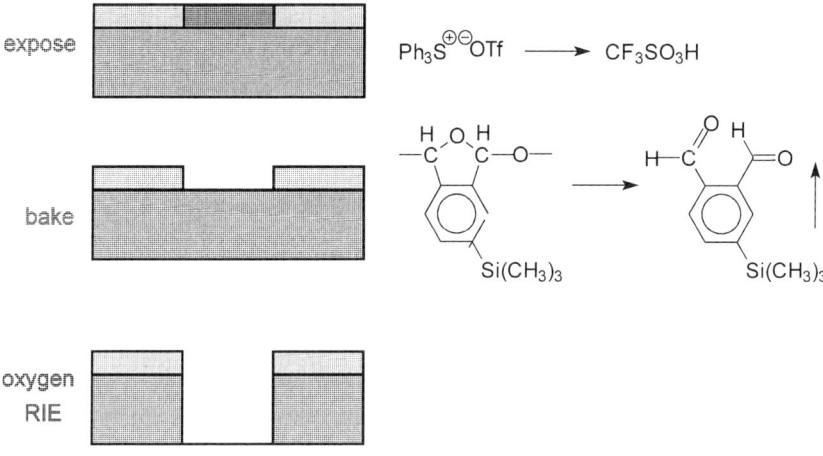

Figure 9 All-dry bilayer lithography with polySPA
(thermal development/oxygen RIE pattern transfer)

Living Anionic Polymerization for Investigation of End Group Effect in Chemically Amplified Resists

As the minimum feature size of microelectronics devices continues to shrink, understanding of the resist chemistry and physics and tailoring resist materials at the molecular level have become indispensable. "Living" anionic polymerization is capable of producing well-controlled polymer architectures in terms of molecular weight and end groups and of providing interesting structures such as block copolymers. Thus, lithographic resist materials could benefit from living anionic polymerization. In fact, it has been reported that narrower polydispersity provides a higher contrast in the case of polystyrene-based negative resist systems (20).

Influence of Polymer End Group on Chemically-Amplified Resist Sensitivity. The chemically-amplified dual-tone tBOC resist consists of poly(4-*t*-butoxycarbonyloxystyrene) (polyBOCST) and a photochemical acid generator (21). Acid-catalyzed deprotection converts the lipophilic polyBOCST to hydrophilic poly(4-hydroxystyrene) (polyHOST) and this polarity change from a nonpolar to polar state allows dual-tone imaging, depending on the polarity of the developer solvent (Figure 10).

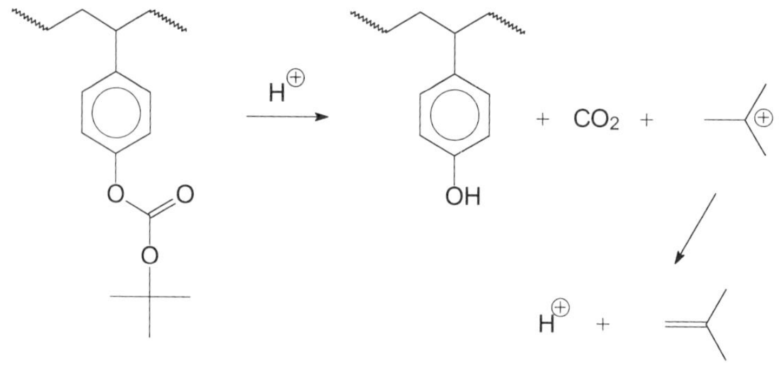

Figure 10 tBOC resist imaging chemistry (acid-catalyzed deprotection)

Although the sensitivity of such resists was expected to be independent of molecular weight, the tBOC resist formulated with a lower molecular weight polymer exhibited a lower sensitivity. The CN end group derived from 2,2'-azobis(isobutyronitrile) (AIBN) used as the initiator was suspected to interfere with the desired acid-catalyzed deprotection (22). In fact, incorporation of a small amount of methacrylonitrile as a comonomer resulted in a lower resist sensitivity (22). To further prove the end group poisoning, we prepared polyBOCST possessing an end group which does not have affinity toward a photochemically-generated acid according to Figure 11 (22).

The *t*-butyl(dimethyl)silyl protecting group has been shown to survive anionic polymerization conditions and to be compatible with living polymerization at cryogenic temperatures in THF (23). We have found that use of cyclohexane as the solvent permits living anionic polymerization with sBuLi to proceed rapidly even at room temperature, providing narrow dispersity polymers in quantitative yield in 10 min (24). No high vacuum break-seal technique nor drying of the reagents was needed to produce narrow polydispersity polymers (M_w/M_n=~1.05) even at a high molecular weight of M_n=21,000. Clean desilylation with hydrogen chloride and subsequent re-protection with di-*t*-butyl dicarbonate provided monodispersed polyBOCST.

Resist formulations were made by mixing 4.75 wt% of triphenylsulfonium hexafluoroantimonate with high (M_n=49,000 and M_w=85,100) and low (M_n=8,400 and M_w=14,500) molecular weight radical polymers made with AIBN as well as the monodispersed low molecular weight (M_n=5,290 and M_w=5,570) polymer prepared *via* living anionic polymerization. Since carbon dioxide and isobutene are released as the deprotection reaction proceeds, the thinning that occurs upon postexposure bake in the exposed resist film is a good measure of the degree of deprotection. Figure 12 presents sensitivity curves for the three resist systems. As mentioned earlier, the higher molecular weight polymer provides a significantly higher sensitivity when AIBN is the initiator. The polymer made *via* living anionic polymerization is as sensitive as the high molecular weight radical polymer in spite of its more than ten times smaller molecular weight (22). Thus, living anionic polymerization followed by polymer reactions could provide PBOCST with no poisoning end groups, which could give rise to high sensitivities even at low molecular weights.

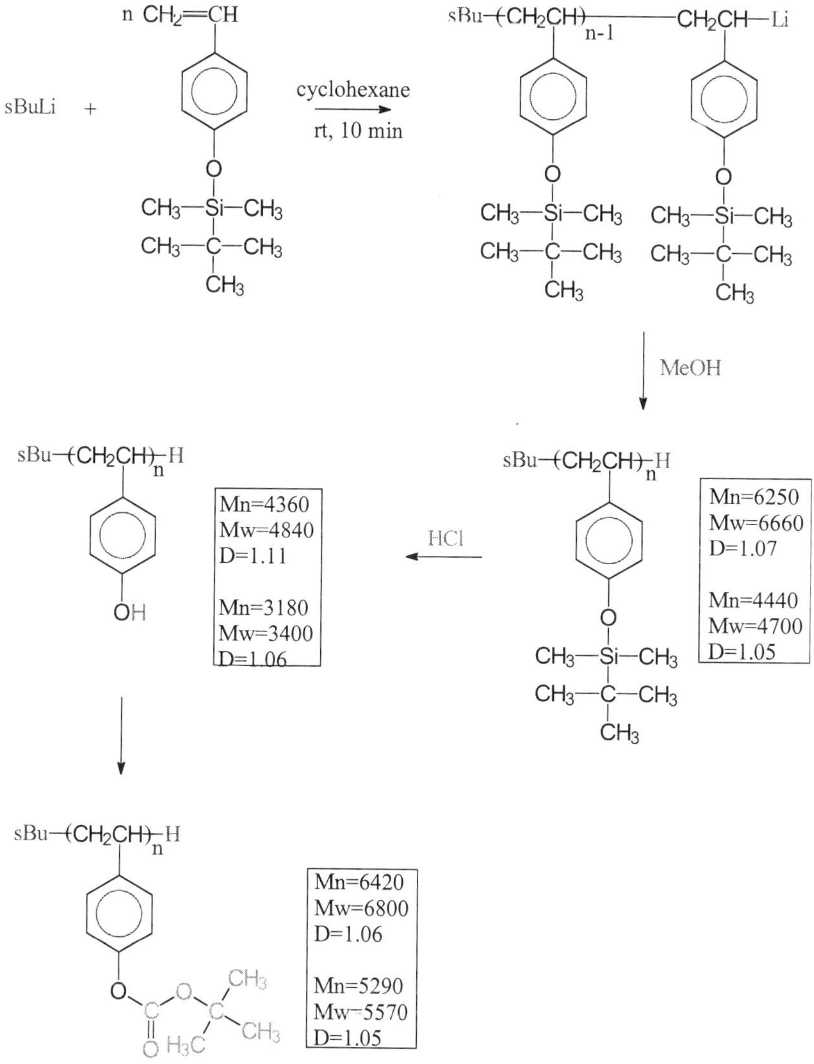

Figure 11 Synthesis of polyBOCST *via* living anionic polymerization

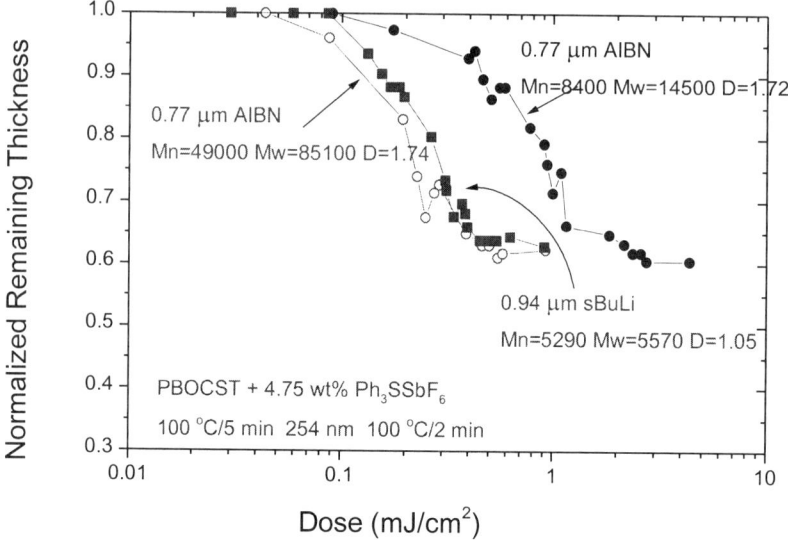

Figure 12 Sensitivity curves of tBOC resists consisting of polyBOCST and 4.75 wt% Ph_3SSbF_6; the resists were postapply-baked at 100 °C, exposed to 254 nm radiation, and postexposure-baked at 100 °C

Furthermore, a chemical amplification positive resist based on partially-protected PHOST with a narrow polydispersity has been reported to provide a higher contrast and higher resolution (25). In the negative-tone chemical amplification systems, use of PHOST with a narrow polydispersity and low molecular weight reportedly reduces edge roughness of developed fine images (< 100 nm) (26).

Elucidation of Acid-Catalyzed Depolymerization Mechanism of Poly(4-hydroxy-α-methylstyrene). Another type of end group effects was observed in acid-catalyzed depolymerization of poly(4-hydroxy-α-methylstyrene) (poly4HOMST) (22,27,28). 4-t-Butoxycarbonyloxy-α-methylstyrene undergoes cationic polymerization with boron trifluoride etherate with the acid-labile tBOC group intact when liquid sulfur dioxide is used as the solvent while use of dichloromethane results in extensive deprotection (27). The resulting polymer can be converted cleanly to poly4HOMST by heating the polymer powder to ~200 °C (Figure 13) whereas acidolysis in solution results in significant main chain scission (27). Cleavage of the polymer end group was suspected to be the cause of the depolymerization. To prove this unzipping mechanism from the polymer end of the cationically-obtained polymer, we prepared the phenolic polymer through living anionic polymerization of the *t*-butyl(dimethyl)silyl-protected monomer (Figure 13) (*22,28*).

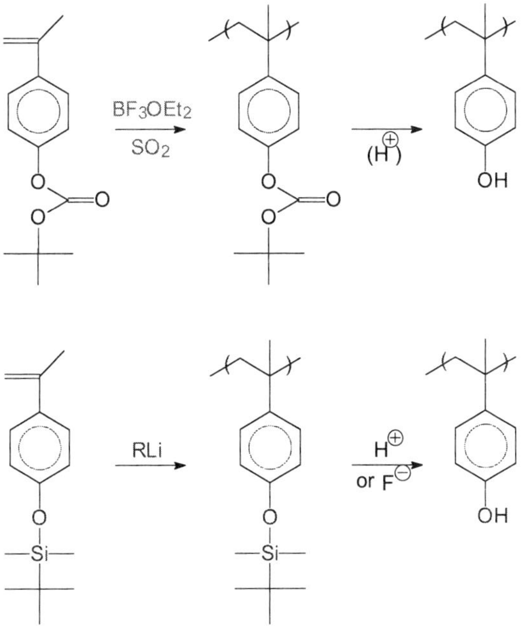

**Figure 13 Syntheses of poly4HOMST
via cationic and living anionic polymerization**

The anionic polymerization of the silyl-protected monomer was carried out using sBuLi as the initiator in THF at -65~-70 °C, considering the low T_c. A polymer with M_n=18,100 and M_w=20,600 (M_w/M_n=1.14) was obtained in 98 % yield in 18 hr. Desilylation with hydrogen chloride produced P4HOMST with M_n=13,300 and M_w=15,400 (M_w/M_n=1.16).

The phenolic resins made by desilylation of the anionic polymer and by thermolysis of the cationically-prepared tBOC-protected polymers were mixed with 2.63 mol% of triphenylsulfonium hexafluoroantimonate. After exposure to 254 nm radiation, the resist films were baked at 130 °C for 2 min and the film thickness measured as shown in Figure 14. The thickness reduction results from depolymerization. The cationically-obtained polymers undergo very efficient depolymerization upon generation of an acid and heating at such a low dose as < 1 mJ/cm^2. When the molecular weight of the phenolic resin is lower, the sensitivity is higher. As expected, however, the anionically-prepared polymer is very inert to acidolysis even at an extremely high dose of 100 mJ/cm^2 with the film shrinkage amounting to less than 10 %. Houlihan et al. proposed a mechanism for depolymerization, which involves addition of acid to the aromatic ring followed by cleavage of the bond between the α-carbon and the aromatic C1 (29). This mechanism completely fails to explain our observations that the anionic polymers are highly resistant to acid-catalyzed depolymerization. We believe that depolymerization propagates from the terminal end of the polymer chain.

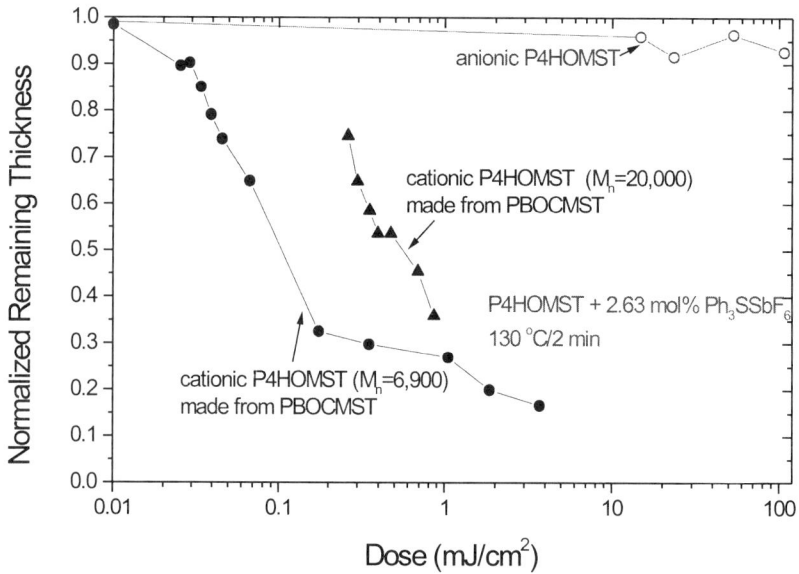

Figure 14 Acid-catalyzed depolymerization of poly4HOMST; the films were exposed to 254 nm radiation and baked at 130 °C

The anionic polymers are expected to have very stable end groups such as alkyl moieties coming from alkyllithium and hydrogen atoms introduced by protonation of the living growing anion (Figure 15). Thus, acids will not react with such structures. In contrast, the cationic polymers are expected to contain end groups that can readily react with acids to form a stable terminal carbocation (*tertiary* benzylic) (Figure 15). Due to the low T_c, depolymerization occurs from the polymer end upon heating. The electron-donating *p*-OH group strongly stabilizes the carbocation, rendering the depolymerization very facile.

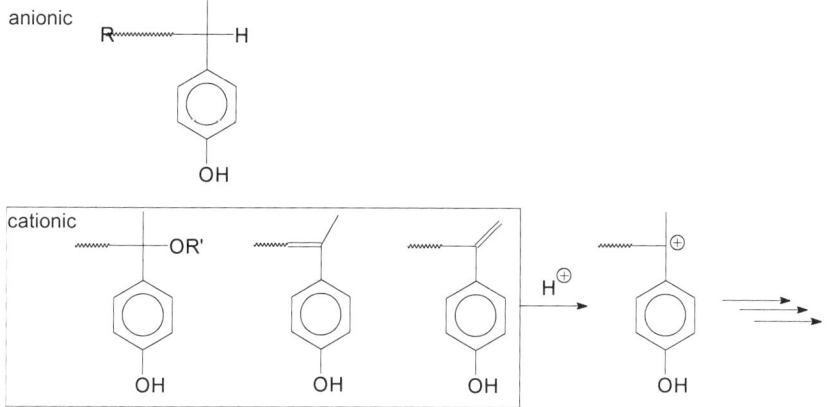

Figure 15 Acid-catalyzed depolymerization of poly4HOMST from polymer end

Acknowledgments

The author thanks his coworkers whose names appear as the co-authors in the cited literature for their contribution.

Literature Cited

1. Moreau, W. M. *Semiconductor Lithography; Principles, Practices, and Materials,* Plenum Press, New York, 1988.
2. Ito, H. in *The Polymeric Materials Encyclopedia: Synthesis, Properties, and Applications* Salamone, J. C., Eds.; CRC Press, Inc: Boca Raton, FL, 1996, Vol. 2C, p. 1146.
3. Willson, C. G. in *Introduction to Microlithography;* Thompson, L. F.; Willson, C. G.; Bowden, M. J., Eds.; Symposium Series 219; American Chemical Society: Washington, D. C., 1983, p. 87.
4. Iwayanagi, T.; Ueno, T.; Nonogaki, S.; Ito, H.; Willson, C. G. in *Electronic and Photonic Applications of Polymers*; Bowden, M. J.; Turner, S. R., Eds.; Advances in Chemistry Series 218; American Chemical Society: Washington, D. C.,1988, p. 107.
5. Ito, H. in *Radiation Curing in Polymer Science and Technology*; Fouassier, J. P.; Rabek, J. E., Eds.; Elsevier: London, 1993, Vol. IV, Chapter 11.
6. Pittman, C.; Chen, C.; Kwiatkowski, J.; Helbert, J. *J. Electrochem. Soc.* **1981**, *128*, 1758.
7. Willson, C. G.; Ito, H.; Miller, D. C.; Tessier, T. G. *Polym. Eng. Sci.* **1983**, *23*, 1000.
8. Ito, H.; Miller, D. C.; Willson, C. G. *Macromolecules* **1982**, *15*, 915.

9. Giese, B.; Meixner, J. *Angew. Chem. Int. Ed. Engl.* **1980**, *19*, 206.
10. Ito, H.; Giese, B.; Engelbrecht, R. *Macromolecules* **1984**, *17*, 2204.
11. Ito, H.; Renaldo, A. F.; Ueda, M. *Macromolecules* **1989**, *22*, 45.
12. Ito, H.; Schwalm, R. in *Recent Advances in Anionic Polymerization;* Hogen-Esch, T. E.; Smid, J., Eds.; Elsevier: New York, N. Y., 1987, p. 421.
13. Ito, H.; MacDonald, S. A.; Willson, C. G.; Moore, J. W.; Gharapetian, H. M.; Guillet, J. E. *Macromolecules* **1986**, *19*, 1839.
14. Ito, H.; Renaldo, A. F. *J. Polym. Sci., Part A, Polym. Chem.,* **1991**, *29*, 1001.
15. MacDonald, S. A.; Ito, H.; Willson, C. G.; Moore, J. W.; Gharapetian, H. M.; Guillet, J. E. in *Materials for Microlithography;* Thompson, L. F.; Willson, C. G.; Fréchet, J. M. J., Eds.; Symposium Series 266; American Chemical Society: Washington, D. C., 1984, p. 179.
16. Aso, C.; Tagami, S.; Kunitake, T. *J. Polym. Sci., Part A-1* **1969**, *7*, 497.
17. Ito, H.; Willson, C. G. *Polym. Eng. Sci.* **1983**, *23*, 1012.
18. Ito, H.; Ueda, M.; Renaldo, A. F. *J. Electrochem. Soc.* **1989**, *136*, 245.
19. Ito, H. *J. Photopolym. Sci. Technol.* **1992**, *5*, 123.
20. Imamura, S.; Sugawara, S. *Jpn. J. Appl. Phys.* **1982**, *21*, 776.
21. Ito, H.; Willson, C. G. in *Polymers in Electronics*; Davidson, T., Ed.; Symposium Series 242; American Chemical Society: Washington, D. C., 1984, p. 11.
22. Ito, H.; England, W. P.; Lundmark, S. B. *Proc. SPIE* **1992**, *1672*, 2.
23. Hirao, A.; Takenaka, K.; Packirisamy, S.; Yamaguchi, K.; Nakahama, S. *Makromol. Chem.* **1985**, *186*, 1157.
24. Ito, H.; Knebelkamp, A.; Lundmark, S. B. *Proc. Polym. Mater. Sci. Eng.* **1993**, *68*, 12.
25. Kawai, Y.; Tanaka, A.; Matsuda, T. *Jpn. J. Appl. Phys.* **1992**, *31*, 4316.
26. Shiraishi, H.; Yoshimura, T.; Sakamizu, T.; Ueno, T.; Okazaki, S. *J. Vac. Sci. Technol.* **1994**, *B12(6)*, 3895.
27. Ito, H.; Willson, C. G.; Fréchet, J. M. J.; Farrall, M. J.; Eichler, E. *Macromolecules* **1983**, *16*, 510.
28. Ito, H.; England, W. P.; Ueda, M. *Makromol. Chem., Macromol. Symp.* **1992**, *53*, 139.
29. Houlihan, F. M.; Reichmanis, E.; Tarascon, R. G.; Taylor, G. N.; Hellman, M. Y.; Thompson, L. F. *Macromolecules* **1989**, *22*, 2999.

Polymerization of Polar and Inorganic Monomers

Chapter 18

Poly(ethylene oxide) Homologs: From Oligomers to Polymer Networks

Christo B. Tsvetanov, Ivaylo Dimitrov, Maria Doytcheva, Elisaveta Petrova, Dobrinka Dotcheva, and Rayna Stamenova

Institute of Polymers, Bulgarian Academy of Sciences, 1113, Sofia, Bulgaria

The most important technological methods for the synthesis of ethylene oxide polymers are discussed. A series of poly(ethylene oxide) networks with different properties is obtained. UV irradiation of anionically prepared poly(ethylene oxide) resins has significant advantages over other methods: it is easy, relatively safe and inexpensive. Another fruitful idea is the synthesis of new materials by the UV crosslinking of PEO composites which offers new possibilities for the synthesis of polymer membranes and smart materials.

Ethylene oxide polymers are materials whose applications include a wide molecular weight range extending from dimers to several millions. The polymeric derivatives of ethylene oxide (EO) can be divided into three classes:
1. Low molecular weight oligomers and polymers (MW between 200-20000) or poly(ethylene glycols) (PEG)s.
2. High molecular weight derivatives (MW between 1.10^5 - 8.10^6) or poly(ethylene oxide) resins (PEO).
3. Poly(ethylene oxide) networks (PEON).
    The ring-opening polymerization of EO is readily affected by a variety of ionic reagents. The synthesis of EO polymers requires the application of different methods, the most important are anionic polymerization in bulk or in solution (low MW range) and anionic suspension polymerization (for synthesis of PEO resins).   This paper reveals the progress of our stadies in the field of EO polymerization and development of new materials based on PEO.

Anionic Polymerization in Solution. Poly(ethylene glycols)

During the 1930s, PEGs were made commercially by the base-catalyzed addition of ethylene oxide to ethylene glycol (*1*). They were first marketed in 1939 by Union Carbide Corporation under the trademark Carbowax. Today PEGs are produced by many companies worldwide. The standard method of producing PEGs is based on the controlled addition of EO to a starting material (water or alcohols) in the presence of alkaline catalysts. During the reaction with EO, the hydroxyl species are in rapid equilibrium with the alkoxides.

The anionic polymerization of EO in solution was studied in detail in the laboratories of Kazanskii (*2-4*), Boileau and Sigwalt (*5,6*), and Panayotov (*8-10*). It proceeds without termination and chain transfer by a living mechanism. We will mention only the most important specific features of EO polymerization in solution. The oxygen atom on the charged end of the growing chain, contributes to a considerable localization of the negative charge, and in this way substantially increases the *tightness of the ion pair*. As a result, the role of the latter in the process of anionic polymerization of EO in solution becomes very significant. *The growing ends are highly associated.* The presence of electron-rich oxygen atoms in the monomer and in the polymer backbone, makes it possible for both of them to compete with the bulk solvent and the growing ends of other living PEO chains in the solvation of the ion-pair cation. The mobility of the formed PEO segments favors *self-solvation* (*4*) *and formation of triple ion pairs* (*7-9*). The ability of solvation and the activity of the growing ends is a function of the increasing number of EO units added (*penultimate effect*). Thus, the electrical conductivity of short-chain living polymers R-$(CH_2CH_2O)_{n-1}$-$CH_2CH_2O^-$,Mt^+ in THF is a function of the chain length and tends to reach a steady value for n between 3 and 7 (*8,9*). The calculation shows that in this case different types of triple ion-pairs are formed. An increase in the polymerization rate in the initial reaction stages was also observed (*2*). This as well as the fact that the rate of propagation is almost insensitive to the influence of the solvent is explained by the self-solvation "shielding" effect of EO units neighbouring to the growing end. The interaction of the cation with the EO monomer was also proposed (*10,11*).

The activation energy for the EO addition to the growing anion is 74.5 kJ/mol (17.8 kcal/mol). An upper limit (up to 28000 - 47000) for the MW has been reported due to a side reaction of a transfer nature (*3*).

The main factors for an effective polymerization process are:
1. The use of alkoxide-enriched starting systems
2. Higher-pressure polymerization
3. Use of catalysts based on potassium, cesium or calcium, since their alkoxides are more active and less associated.

Anionic Suspension Polymerization. Poly(ethylene oxide) Resins

The preparation of PEO resins requires the development of catalyst systems that function in a different mechanism than the one observed in the base-catalyzed synthesis of PEG: alias coordinated anionic polymerization. The coordination has two effects: activating the monomer toward polymerization, and providing an orientation of the reacting molecules leading to a stereospecific polymerization as in the case of alkylene oxides other than EO. Effective initiators are derivatives of divalent and trivalent metals mainly: alkaline-earth compounds (e.g. carbonates, oxides, amides or amide-alkoxides), organometallic compounds (e.g. Ca-diphenyl, Al, Mg, Zn, Sn, or Fe alkyls or alkoxide), commonly with added modifiers, and binary or ternary bimetallic systems (usually Zn/Al). All the metals have Lewis acidity.

For industrial purposes calcium and zinc/aluminum based catalysts are the most widespread (Table I).

The anionic coordination polymerization, initiated by Ca catalysts is one of the main subjects of our studies. Alkaline-earth amides and amide-alkoxides are the most active catalysts for the EO polymerization. They are active at temperatures $0°$ - $50°C$, which is remarkable considering that alkaline earth carbonates and oxides require much higher temperatures, usually above 70 °C.

Table I. X-ray microanalysis of metal content in the most popular PEO grades

Samples	Molecular mass	Integral intensity,imp
Union Carbide Polyox N-60K	2,000,000	Ca, 47
Seitetsu Kagaku PEO-1	150,000	Zn, 136
		Al, 631

The analysis was performed on a electron microscope model JEM-200 CX with EDS from TRACOR (USA).

The fact that calcium amide and amide-alkoxides are very active at temperatures below the melting point of PEO has a considerable industrial significance. To carry out the polymerization process either in bulk or solution would be technologically impractical due to problems of viscosity, heat transfer and rapid degradation of PEO in shear fields.The synthesis of PEO resins can be most effectively realized in a precipitant medium where the polymer is directly produced in the form of fine particles. The polymerization is a precipitation polymerization since the reaction medium (hydrocarbons) dissolves only the monomer. Below the melting point, the polymer produced remains in a granular form. The temperature is easily controlled by the rate of monomer feed. Some calcium based catalysts are very active thus the granular PEO contains only a small amount of the catalyst as impurity and no special purification of the polymer is required. The granular polymer can therefore be recovered by filtration and used without further processing. The details of the techniques used to manufacture PEO resins have not been disclosed.

Various calcium based catalysts of coordinate anionic polymerization of EO have been discovered. Systematic studies on catalyst activity and polymerization mechanism are scanty. The synthesis of an active catalyst is a manufacturer's knowhow. Most of the effective catalyst systems are published in patent literature. There are only two russian publications (12,13) that explain the synthesis of Ca catalysts in more details. Usually the catalyst systems represent poorly defined heterogeneous aggregates. The efficiency of these initiators is rather low: in most cases only from 10^{-2} to 10^{-3} of the introduced initiator is consumed. This mode of polymerization greatly differs from the olefin precipitation polymerization where extremely high catalyst efficiencies are achieved. Recent modifications of Ca catalysts are reported to be exceptionally effective for EO polymerization yielding 1800 g of polymer/ g Ca (*14*). Table II summarizes the catalyst efficiency of the published results.

We have performed a combined study of EO suspension polymerization using the catalyst system developed in our laboratory (*15*). This study comprises the kinetic measurements, the determination of the molecular characteristics of the polymer obtained, as well as the structure and morphology of the polymer particles.

Table II. Main characteristics of EO suspension polymerization in presence of different catalyst systems

Catalyst	Solvent	T,°C	Time, hrs	Polymer yield, moles EO / mol.l^{-1} cat.	M_v ×10^{-6}	Ref.
Ca(NH$_2$)$_2$	Benzene	30	12	40	8	12
Ca amide-alkoxide 1/3	Benzene	17	10	70	3	12
Ca amide-alkoxide modified	Heptane	40	8	170	4.5	15
(Ph)$_2$Ca	Heptane	10	9.5	270	6	13
Union Carbide	Heptane	31	4	1800	4	14

Kinetic Investigations. The polymerization kinetics were carried out in a glass reactor equipped with a stirrer and thermostated jackets. The catalyst suspension in heptane was introduced from a sealed ampoule under argon atmosphere. Gaseous EO was connected to the reactor and a constant pressure in the reactor was established. At the end of the process the polymer was filtered, washed with heptane and vacuum dried. Table III presents the results characterizing the MW, MWD and the polymer yield dependence on the reaction time.

The chemical mechanism of EO polymerization is relatively simple in view of the well known stability of alkoxide growing chains towards termination and transfer reactions. The so called "living-type" process is observed. According to Kazanskii et al (*13*) the rate of EO polymerization initiated by calcium catalysts in hydrocarbon media is a linear function of the monomer and catalyst concentrations so that we can write:

$$R = k_{eff} \cdot [Cat]_o \cdot [M]_o$$

The plot of the polymer yield (G_p/G_{Ca}) vs. time (Figure 1) shows a linear function with no decrease of the polymerization rate as usually observed (*12,13*), which means that the process is kinetically controlled and the growing centers are not being blocked by the formation of partially crystalline polymer.

Table III. The MW, the polydispersity and PEO yield as functions of the reaction time. Catalyst synthesis as described in (15); [Cat] = 0.09 mol.l^{-1}; 40°C; constant monomer concentration [EO] = 0.127 mol.l^{-1}.

Time, hrs	Yield: EO units in the polymer/ mol.cat.	M_v × 10^{-6}	M_n × 10^{-6}	M_w/M_n
0.25	5.0	0.2	0.35	8.3
0.5	11.5	0.69	0.84	6.0
1.0	20	0.89	1.57	4.2
2.0	39	1.14	1.80	4.0
4.0	74	2.89	-	-
8.0	160	4.2	-	-

M_n determined by GPC, Waters, column with Ultrahydrogel 2000, eluent water/acetonitrile (80:20) ; M_v from viscometric measurements at 35°C in water according to [η] = 6.4.10^{-5}M$^{0.82}$ (16).

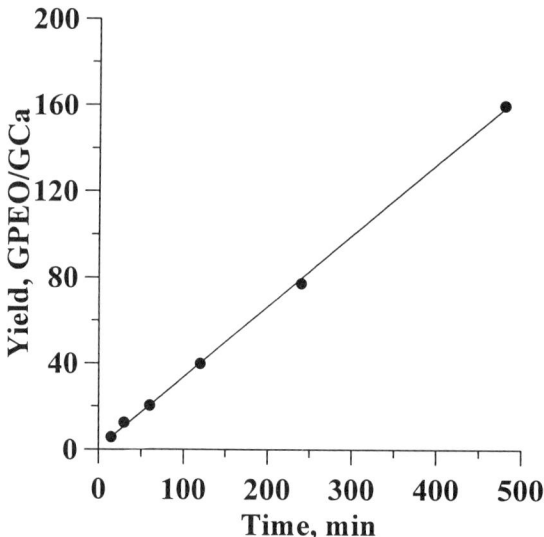

Figure 1. Polymer yield vs. time by constant monomer concentration: [EO]$_o$ = 0.127 mol.l^{-1}; heptane, [Cat]$_o$ = 0.09 mol.l^{-1}; reaction temperature 40 °C. G$_{PEO}$/G$_{Ca}$:EO units in the polymer / mol.cat.

The molecular weight of the polymer rises constantly during the first stage of the process (Table III, Figure 2). The polymers possess a relatively broad polydispersity. It decreases, along with the monomer consumption which indicates a nonsimultaneous appearance of active centers as well as some difference in their activity, obviously due to the different sites on the catalyst surface.

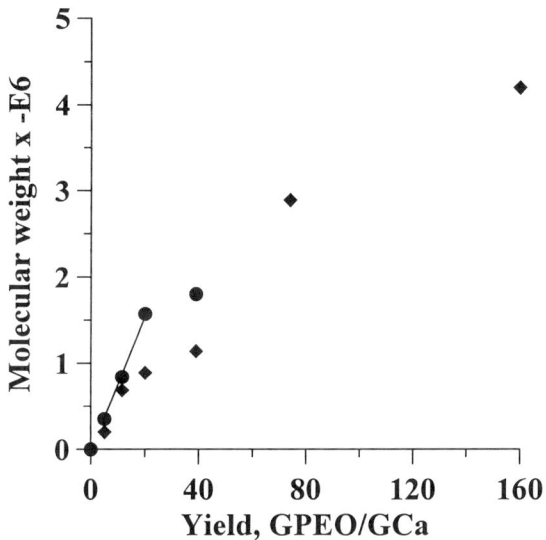

Figure 2. Molecular weight of PEO vs. yield. ●- M_n determined by GPC; ◆- M_v
Reaction conditions, see legend of Figure1.

Since the EO concentration ($[M]_o$) was fixed, we were able to measure the initiator efficiency **f**:

$$f = W_p/M_n : W_{Ca}/M_{Ca}$$

where W_P and W_{ca} are weights of PEO and calcium in the reaction mixture.

In our case a successful conversion into a high MW polymer requires from 0.057 to 0.16 % of catalyst.

The value of k_{eff} based on the data presented in Fig.1 is $4.37 \cdot 10^{-2}$ $l.mol^{-1}.sec^{-1}$
The polymerization rate constant k_p is expressed according to the equation:

$$k_{eff} = k_p \cdot f$$

Since the number of growing macromolecules participating in the early stages remains constant we can find the propagation rate constant using the value for **f** ($5.7 \cdot 10^{-4}$) and for k_{eff}; we have $k_p = 77 \pm 10$ $l.mol^{-1}.sec^{-1}$ which means that the growing species under study possess a marked activity as compared with the k_p^- value of 1.67 $l.mol^{-1}.sec^{-1}$ from experiments performed with $K^+ +[222]$ as the counterion and the $k_p^{+/-}$ value $1.2 \cdot 10^{-1}$ $l.mol^{-1}.sec^{-1}$ for Cs^+ as the counterion in THF solution at 20°C (*17,18*). The propagation rate constants for EO polymerization in solution and in suspension are presented in Table IV.

Table IV. Propagation rate constants for anionic polymerization of EO.

Solvent Diluent	T, °C	k_p +/- ($l.mol^{-1}.s^{-1}$) K^+	Cs^+	Ca^{++}	k_p^-	Ref.
THF[222]	20	0.025			1.67	18
THF	20	0.05	0.12			18
HMPA	40		0.2		22.0	19
heptane	10			2.8		13
heptane	40			77±10		this work

The rate constant in suspension polymerization is higher than the propagation rate constant for free ions k_p^-. A reasonable explanation for this enormous difference in the activity was given by Kazanskii (*13*). According to him the acceleration is due to the effective coordination between the EO and the active centers in the absence of the competing electron pair donor, THF, as well as to the localization of the growing centers on the catalyst surface in contrast to the associated and lower activity growing species in solution polymerization.

Morphological Investigation of the Forming Polymer Particles. The physico-chemical mechanism of the anionic suspension polymerization is far from clear (*20,21*). In contrast to analogous radical processes, where the active species are distributed in the volume of the polymeric phase, the active centers in this case are localized on the surface of the catalyst particle, placed under the layer of the polymer formed. The monomer addition to the growing chain takes place on the surface of the catalyst particle. Since the polymerization process is highly exothermal, a surface space of enhanced temperature ("hot zone") is created around the particle. In the "hot zone", the polymeric chain cannot start to crystallize immediately. Krusteva et al (*21*) assumed that the initial crystallization takes place at a distance of 30 μm from the catalyst surface. These authors present a model for the formation of the nascent PEO structure in different polymerization stages.

EO diffuses to the catalyst surface through the cavities between the crystal blocks. The monomer diffusion is one of the most serious technological problems.The diffusion rate decreases with the accumulation of polymer, which limits the accessibility to the growing sites. The diffusion limitation inside the microparticles is mostly related to the density of the particle and the size of the penetrating molecules. The morphology of the growing polymer particles becomes obviously an important factor, particularly in view of the marked crystallization tendency of PEO.

The anionic dispersion polymerization of EO produces PEO which is insoluble in the diluent.As a result of this, the catalyst particles transform into polymer particles rapidly, within a few minutes after the start of the reaction. Figure 3a and 3b represent electron micrographs of initial catalyst particles (3a) and of the same particles during the first stage of the polymerization process (3b). The size of the initial catalyst particles is in the range of 5-50 μm. Their shape is not well defined. The relatively broad distribution is due to the formation of agglomerates. The start of EO polymerization leads to fragmentation of the former catalyst particles (Figure 3b).

243

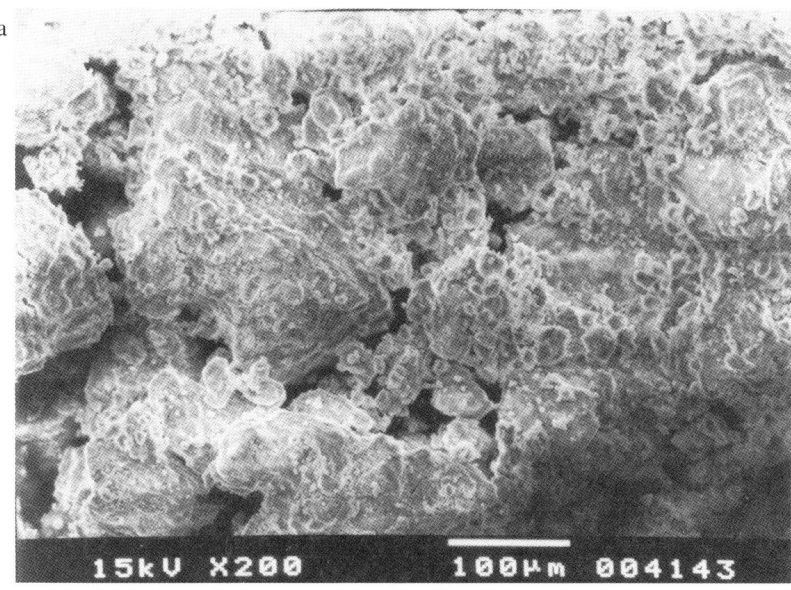

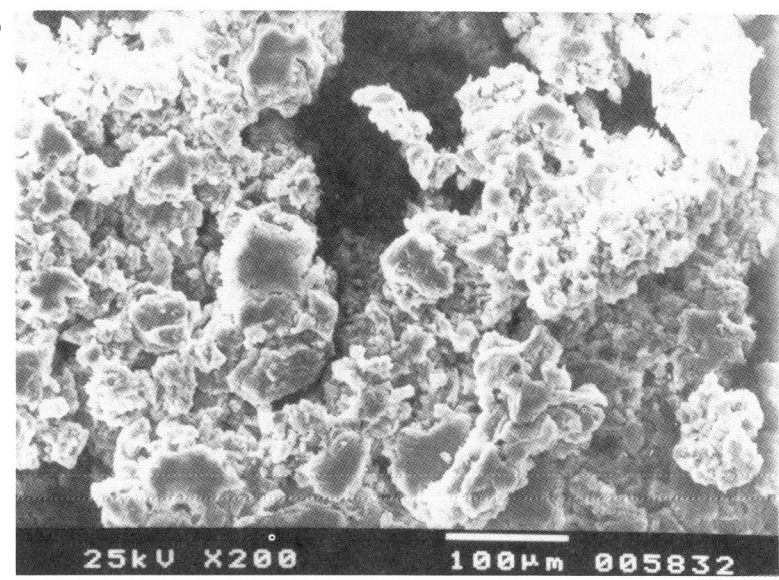

Figure 3. a) Electron micrograph (SEM) of catalyst surface. Catalyst - modified Ca-amide-alkoxide, obtained according to (4); b) Electron micrograph (SEM) of PEO particles formed after 15 min. of reaction time. Reaction conditions: see Table II and Figure 1 (legend).

Our primary aim was to trace the change of the particle size and number during the polymerization process in order to distinguish between processes of particle formation and agglomeration . The particle size distribution is determined essentially by the number of new particles which appear continously in the course of the polymerization, as well as by their subsequent growth and possible aggregation. It was found that the morphology of the nascent PEO strongly depends on the monomer conversion. A wide and bimodal distribution of the particle size (Figure 4) indicates that the particle formation has continued simultaneously with their agglomeration. The shape of the particles (Figure 3b) is irregular which can be explained with the agglomeration processes. Given the high PEO polarity and the liophobic character of the particles, a rapid aggregation is taking place.

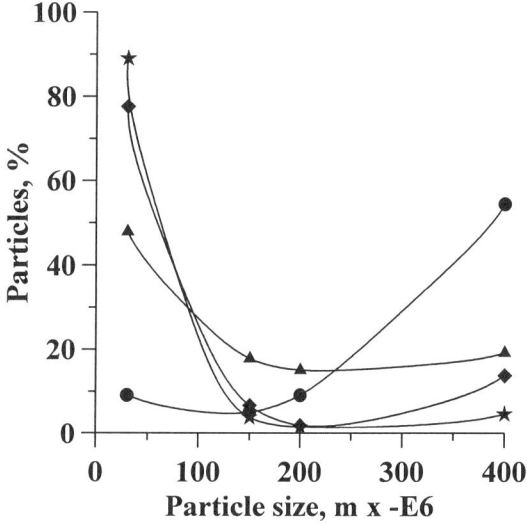

Figure 4. Particle size distribution in the course of the polymerization process: ★ - 15 min.; ◆ - 30 min.; ▲ - 1hr.; ● - 2hrs. Reaction conditions: see legend in Figure 1.

Microscopic examination of the texture and size of PEO grains shows that they are macroporous aggregates of smaller particles (Figure 5,a,b,c)

The PEO particles formed in the first 15 min. are agglomerates of the order of 10-100 μm in diameter (Figure 3b). They grow rapidly (Figure 5 a,b,c) increasing in size and reducing in number. The formation of void spaces between agglomerates creates penetrating channels, thus feeding the growing centers with monomer. The relatively small EO molecules rapidly diffuse into the pores and pass through the smallest aggregate particles. The polymerization rate remains constant (Figure 1) for a long period of reaction time, thus becoming independent on the particle size and number.

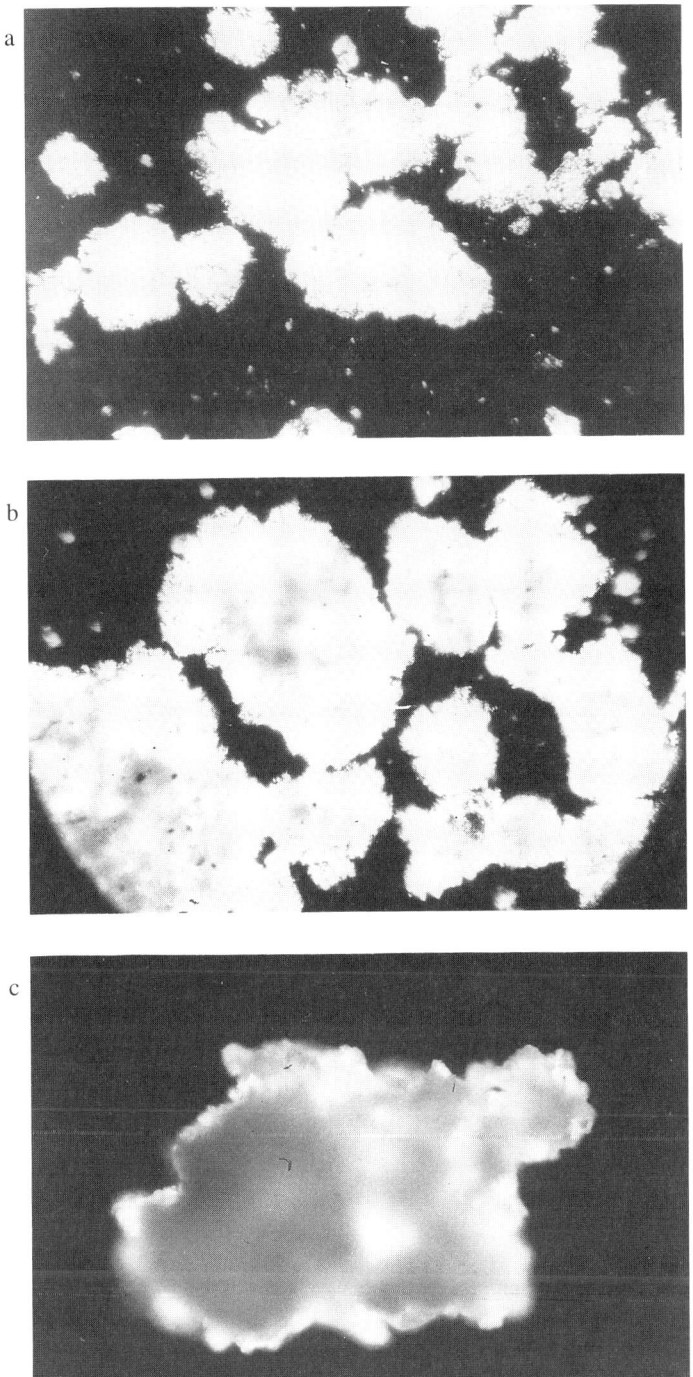

Figure 5. Optical micrographs of PEO particles growth during the polymerization process: a - 15 min.; b - 30 min.; c - 2 hrs.

The degree of crystallinity of the polymer formed on the catalyst surface can also influence the activity of the growing species. Dubrovskii and Kazanskii (20) suggest that a mechanical chain termination is likely to take place, due to the diffusion blocking of the active centers located within the large crystallites.

DSC was used to observe the change of the degree of crystallinity during the course of the polymerization (Table V). The molar fraction of the crystalline phase was estimated as the ratio between the experimentally determined enthalpy of melting and the enthalpy of melting of 100% crystalline PEO (ΔH = 8650 J/mole (21)).

Table V. Thermal properties of PEO obtained at various reaction times

Reaction time, hrs	T_m, °C	ΔH_m, J/mole	Crystallinity index
0.25	68.4	3351.5	0.39
0.5	69.3	4825	0.56
1.0	69.6	5347.3	0.62
2.0	70.1	5558.5	0.64

The crystallinity of nascent PEO particles rises with the increase in polymer yield. In the initial polymerization stages the polymer is characterized by a substantially low crystallinity index, most probably due to the presence of a great portion of catalyst residue that hinders the crystallization process.

It is very important to note that the particle size and shape strongly depend on the monomer concentration and the polymerization temperature. EO is a good solvent for its own polymer. It is our belief that in the case of low monomer concentration two particles with EO-swollen surfaces approach one another thus increasing the concentration of EO in the surface layer between the particles, which generates an osmotic pressure. To counteract this effect, the hydrocarbon diffuses into the region of higher EO concentration forcing the particles apart. Obviously, this type of disaggregation (renucleation) causes the formation of whiskers of PEO crystals that connect the individual particles, as it is seen in Figure 6.

Relatively large particles are formed at high EO concentrations. With the increase in particle size it becomes difficult to control the reaction temperature within the active centers area. At temperatures higher or close to the melting point of PEO the structure of the particle changes from porous to dense, leading to a considerable decrease of the polymerization rate.

The requirements for a successful polymerization process comprise the use of a catalyst system of high productivity as well as the formation of small polymer particles of narrow polydispersity.

The catalyst system developed by a Union Carbide group (*14*) is reported to be exceptionally effective (see Table II).The process of catalyst preparation includes aging at temperatures from 150 °C to about 225 °C, Most probably the thermal treatment of the catalyst causes the removal of ammonia molecules from the Ca^{+2} solvation shell, which facilitates the monomer coordination. Consequently the polymerization rate increases considerably.

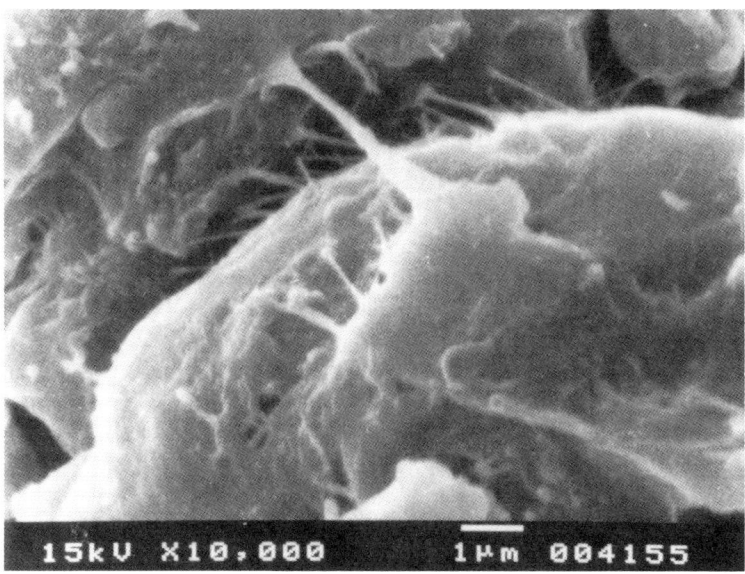

Figure 6. SEM micrograph of PEO particles obtained after 8 hrs. of reaction time.

The main requirement for a successful dispersion polymerization is the presence of nonionic surfactant or amphipatic polymer to stabilize the polymer particles by forming a dissolved protective layer around them. The most efficient type of dispersant is based on a block or graft copolymer which consists of two essential polymeric components - one soluble in the liquid medium interspersed by short segments, usually called "anchors" which are strongly adsorbed at the particle surface. Because of the industrial interest in this subject no detailed publications are available in the scientific literature. One of the future goals is the development of techniques for the preparation of fine uniform PEO particles of controlled size.

Ultraviolet Induced Crosslinking of Solid PEO. Poly(ethylene oxide) Networks (PEON) *(22).*

PEON are an important class of new materials with a number of applications, i.e., wound dressings (*23*), controlled release drug systems (*24*), phase transfer catalysts

(25), semipermeable membranes (26), solid electrolytes for batteries (27) and many others. Crosslinked PEO was first obtained as a product of γ-irradiation of dilute aqueous solutions of PEO (28). Previously we reported the thermal behavior of freeze-dried PEON obtained by γ-irradiation of 2% aqueous solutions at different doses (29). It was found that by increasing the dose up to 5 Mrad the crosslink density increased and then levelled off. (High-energy electrons)-radiated aqueous solutions also result in a crosslinked gel and this process is currently commercially practised. However, the removal of water from the PEON is an expensive and inconvenient process.

It has been recently shown that PEO can be successfully crosslinked by irradiation with UV light with or without benzophenone (BPh) as photoinitiator (30). In the latter paper, very little, irreproducible data on UV-induced crosslinking of high-molecular-weight PEO ($M_w = 5 \times 10^6$) in the presence of 0.5 mol % benzophenone have been reported. The assumption of these authors that the photocrosslinking is the result of the photolysis of the polyether seems unconvincing.

The ease, relative safety, and low cost of UV-induced crosslinking should provide significant advantages for many applications over other methods. It was considered worthwhile to study the photocrosslinking process of PEO resins in more detail in order to provide a practical method for producing well defined networks.

The samples for UV irradiation were cast in the form of films from methylene chloride solutions of PEO, or by pressing a homogenized mixture of PEO and photoinitiator at 120°C for 5 min under a pressure of 150 kG/cm^2. Usually, the duration of the irradiation was 40 min., the initial 10 minutes being necessary for reaching the standard emission of the UV lamp (TQ 150 ORIGINAL HANAU high pressure 150 W mercury lamp with a quartz tube and a cooling jacket). The thickness of the film samples varied between 200 and 250 μm.

UV irradiation experiments were conducted in an argon atmosphere. Most of the crosslinking experiments were carried out in the presence of benzophenoneas a photoinitiator. In order to remove BPh and the photoproducts arising from it, as well as to remove the soluble (sol) fraction after irradiation, the samples were subjected to extraction for 24 hrs with CH_2Cl_2.

Photochemical crosslinking results from hydrogen atom abstraction from PEO. BPh was chosen because of its high hydrogen abstraction efficiency i.e. a) high intersystem crossing efficiency ($k_{T1-S1} = 10^{10}$ sec^{-1}); b) high triplet energy (69 kcal/mol); and c) long triplet lifetime ($k_{T1-S0} = 10^4$ sec^{-1}) (31).The photochemical processes which it undergoes are presently well understood. On photon absorption, BPh undergoes several photophysical processes, affording an n, π* triplet state which then proceeds to perform the reduction of BPh and the PEO radical formation (32):

$$\text{PhCOPh} \xrightarrow{h\nu} {}^3[\text{PhCOPh}] \xrightarrow{-CH_2-CH_2-O-} \text{Ph}_2\overset{(\cdot)}{C}-\text{OH} + -CH_2-\overset{(\cdot)}{C}H-O-$$

The efficiency of benzophenone in initiating the crosslinking of PEO was compared with the following available photoinitiators: 4,4-bis (dimethylamino) benzophenone (Michler's ketone) (MK); acetophenone (APh); benzoin methyl ether (BME), and benzoin (Bn). Some commercial photoinitiators for radiation curing of polymers were also examined (Table VI). The UV crosslinking experiment with Michler's ketone failed, although it was well purified prior to use. Benzoin methyl ether and acetophenone provide an appreciable degree of crosslinking but less than benzophenone. With exception of Darocur 1173 the commercial photocuring agents used showed much lower hydrogen-abstracting capacity as compared to BPh, Bn and APh. The results confirm that benzophenone and benzoin are the best choice as photoinitiators for the UV-induced crosslinking of PEO.

Table VI. Effect of the type of photoinitiator on the crosslinking efficiency.

Photoinitiator	Thickness, μm	Gel fraction, %	Equil. swelling in H_2O	Equil. swelling in $CHCl_3$	M_c in $CHCl_3$
BPh	190	92.1	4.0	5.4	8,000
	200	91.7	4.4	5.3	8,000
Bn	210	90.0	4.7	7.2	17,000
	250	90.3	4.9	6.9	15,000
APh	240	75.3	11.8	21.5	110,000
	250	74.8	12.7	22.2	113,000
MK	230	3.7	-	-	-
	250	5.8	-	-	-
BME	250	77.0	10.3	13.6	60,500
	250	80.4	10.2	12.7	54,000
Darocur 1173	210	88.0	7.0	11.8	48,000
Darocur 953	290	68.3	13.8	36.1	157,000
Darocur 1116	250	65.0	25.0	27.5	135,000
Darocur 1664	230	74.5	12.7	13.8	60,500

PhI - Photoinitiator; BPh - Benzophenone; Bn - Benzoin; MK - Michler's ketone; APh - Acetophenone; BME - Benzoin methyl ether. M_c - number-average molecular weight was estimated from the Flory-Rehner model (33) for equilibrium swelling of PEON in chloroform.

In order to investigate whether the crosslink density varies throughout the thickness of the samples, two identical films were crosslinked as follows: the first was irradiated twice with the same side exposed to the UV light source; the second was also irradiated twice - first one surface of the film and then the opposite side. The data shown in Table VII indicate that both methods of irradiation produce a similar crosslinking efficiency. However, a small difference in structure was indicated by film behavior during the swelling measurement. When the disk prepared by single side irradiation was soaked in water, it rolled up rapidly; while the disk obtained by irradiation on both sides retained its form in water and remained flat. This observation indicates that PEO films are not completely homogeneous through their thickness in crosslink density.

Table VII. Crosslinking of PEO film samples exposed to UV irradiation only on one side or on both sides. UV irradiation under a flow of argon at [BPh]/[-EO-] = 5×10^{-3}. Samples from Polyox N-12 K, MW 1,000,000. Irradiation time = 40 min; Thickness - 250 μm.

Irradiation Conditions	Gel Fraction, %	Equilibrium Swelling in H_2O	Equilibrium Swelling in $CHCl_3$
I and II irradiation - one side exposed to UV	92.0	5.0	6.9
I irradiation - one side and II irradiation - opposite side exposed to UV	93.9	4.9	8.2

The effect of different factors as BPh concentration, time and temperature of irradiation on the crosslinking efficiency was followed on the samples prepared from PEO of $M_v = 1 \times 10^6$.

Benzophenone concentration was varied in the range of 1×10^{-3} to 1×10^{-2} moles BPh per mole of ethylene oxide (-EO-) units, which corresponds to 0.4 - 4.0 wt %. Table VIII shows the change of gel fraction (GF) and the crosslinking density as a function of benzophenone concentration. GF levels off at a maximum value of 91-95 % at BPh concentration of $4 - 5 \times 10^{-3}$ mol/mol (1.5 - 2 wt %). The same trend is observed in the degree of crosslinking. All subsequent crosslinking experiments were therefore conducted with a BPh concentration of 5×10^{-3} mol/mol.

Irradiation times above 20 min. are sufficient to achieve more than 90 % GF yield. Maximum crosslink density is observed in the interval of 30 to 60 min irradiation. At an irradiation time of 80 min, an increase of equilibrium swelling and a decrease in crosslinking density are observed at constant GF yield. The optimal duration of the irradiation in the employed experimental conditions is thus between 30 and 40 min.

Table VIII. Effect of [BPh]/[-EO-] mole ratio on the crosslinking efficiency.
PEO MW 1,000,000, Polyox N-12K; T = 25 °C; Irradiation time - 40 min.

[BPh]/[-EO-] x10^3	Thickness μm	Gel fraction,%	M_c in $CHCl_3$
1	180	78.9	88,000
3	150	79.9	30,000
4	210	92.1	26,500
5	190	90.3	8,000
10	190	92.0	8,000

The temperature dependence of UV initiated crosslinking was examined in the range from 25 to 90°C (Table IX). Above 50°C the PEO samples changed their appearance due to the melting processes but kept their form and size. An increase of gel fraction yield and particularly of equilibrium degree of swelling occurred with the rise in temperature. The gel fraction increases over the range of 30 °C to 60°C and then remains almost constant at 97 - 98 %. There is a linear dependence of the equilibrium swelling in water on the irradiation temperature up to 80°C.

Table IX. Effect of the irradiation temperature on the crosslinking efficiency.
Irradiation time 40 min.;[BPh]/[-EO-] = 5.10^{-3}, PEO MW 1,000,000 (Polyox N-12K)

T, °C	Thickness μm	Gel fraction, %	M_c in $CHCl_3$	Equilibrium swelling in H_2O	Degree of crystallinity DSC
25	190	90.3	8000	4.1	0.55
30	200	94.3	8500	5.0	0.67
60	210	96.4	11300	5.3	0.74
70	200	97.1	24200	5.5	0.74
80	220	97.2	26100	5.9	0.71
90	200	97.8	42000	7.9	0.70

The effect of the molecular weight of PEO on the UV crosslinking efficiency is shown in Table X. Varying M_v from 200,000 to 8,000,000 exerts an appreciable influence on the gel fraction formation: the best yield is obtained for PEO with M_v from 900,000 to 2,000,000. The good agreement between the two sets of results shows that samples prepared by dry mixing and pressing under heat can be effectively crosslinked by UV irradiation.

Clearly, PEO films prepared from an organic solvent solution or by dry blending and press molding can be efficiently crosslinked by UV irradiation in the presence of a hydrogen-abstracting photoinitiator. Benzophenone was found to be the most efficient photoinitiator, among several examined, and its optimal concentration is 4 - 5 x 10^{-3} mol/mol -EO- units. Increasing the irradiation temperature resulted in an increased gel fraction yield and reduced crosslink density. It is therefore possible to control the network density by varying the irradiation temperature.

Table X. Effect of the molecular weight of PEO on the crosslinking efficiency.
[BPh]/[-EO-] = 5×10^{-3}; Irradiation time - 40 min., T = 25°C, Film thickness 200-400 µm.

M_v	Cast samples			Press molded samples		
	GF, %	Equilibrium swelling in H_2O	in $CHCl_3$	GF, %	Equilibrium swelling in H_2O	in $CHCl_3$
200,000	54.1	-	-			
300,000	74.0	6.4	9.4			
400,000	82.0	4.9	7.2			
600,000	92.7	4.0	5.6	92.1	5.4	6.9
900,000	94.7	4.1	5.8	95.1	6.2	9.2
1,000,000	92.0	4.0	5.7	92.7	5.4	6.0
2,000,000	95.7	3.8	5.5	94.6	5.3	5.9

Table XI. UV crosslinking of blends of poly(ethylene oxide) with poly(vinyl acetate).
UV irradiation under a flow of argon. Irradiation time = 40 min. PEO, MW 1,000,000; PVAc, M_n = 68,600 (Union Carbide, AYAT)

[PEO] : [PVAc]	Irradiation temperature °C	GF, %	ES, in H_2O
	[BPh]/[-EO-+-VAc-] = 5×10^{-3}		
10 : 1	25	95.3	5.8
	70	96.8	7.3
5 : 1	25	86.8	6.5
	70	89.6	7.4
3 : 1	25	94.3	5.2
	70	91.7	6.1
2 : 1	25	72.4	7.0
	70	81.3	6.6
	without benzophenone		
10 : 1	25	56.5	9.6
	70	77.6	7.0
5 : 1	25	80.4	6.0
	70	86.5	4.7
3 : 1	25	41.2	-
	70	62.1	-
2 : 1	25	40.7	-
	70	59.9	-

Formation of PEO/Polyvinylacetate Networks

PVAc belongs to the so-called crosslinking-type polymers under the influence of UV radiation. Poly(ethylene oxide)-poly(vinyl acetate) (PEO-PVAc) blends were successfully crosslinked by UV irradiation. Composition dependence of the efficiency of UV induced crosslinking of the blends is shown in Table XI. Generally, irradiation carried out at a higher temperature (70°C) showed better results with respect to gel fraction yield and crosslink density. PEO-PVAc blends were crosslinked without using photoinitiator as well. The carbonyl group of the vinyl acetate polymer has enough photoactivity for hydrogen abstraction.

Conclusions

High molecular weight PEO can be effectively manufactured via anionic suspension polymerization by using different types of calcium catalyst systems. The mechanism of polymerization is closely associated with the catalyst structure, the mode of polymer particle formation and growth, the particle size distribution and the tendency of agglomeration as well as the use of effective steric stabilizing additives.

High molecular weight PEO can be effectively crosslinked via simple UV-irradiation technique. Different types of crosslinked PEO networks networks can be easily obtained. The UV-irradiation method provides possibilities for the formation of new materials based on PEO.

Acknowledgments

Financial support from the National Foundation for Scientific Research in Sofia (project X-401) and Union Carbide Corporation is gratefully acknowledged.

The authors wish to thank Prof. I.M. Panayotov for helpful discussions.

References

1. McClelland, C.P., Bateman, R.L., *Chem.Eng.News*, **1945**, *23*, 247.
2. Kazanskii, K.S., Ptitsyna, N.V., Kazakevich, V.K., Dubrovskii, S.A., Berlin, P.A., Entelis, S.G., *Dokl.Akad.Nauk SSSR*, **1979**, *234 (4)*, 858.
3. Ptitsyna, N.V., Ovszannikova, S.V., Gel'ger, Ts. M., Kazanskii, K.S., *Polymer Science U.S.S.R.*, **1980**, *22*, 2779.
4. Kazanskii, K.S., *Pure & Appl. Chem.*, **1981**, *53*, 1645.
5. Sigwalt, P., Boileau, S., *J.Polym.Sci., Polymer Symp.*, **1978**, *62*, 51.
6. Boileau, S. in *Anionic Polymerization: Kinetics, Mechanisms and Synthesis*, Mc Grath J.E. Ed., *ACS Symposium Series* **1981**, *166*, 283.
7. Szwarc, M., *Carbanions, Living Polymers and Electron Transfer Processes;* Interscience: New York,1968, p.647
8. Berlinova, I.V., Panayotov, I.M., Tsvetanov, Ch.B., *Eur. Polym.J.*, **1977**, *13*, 757.
9. Berlinova, I.V., Panayotov, I.M., Tsvetanov, Ch.B., *Vysokomol.Soed.*, **1978**, *B20*, 839.

10. Tsvetanov, Ch.B., Petrova, E.B., Panayotov, I.M., *J.Macromol.Sci.*, Chem., **1985,** *A22,* 1309.
11. Chang, C.J., Kiesel, R.F., Hogen-Esch, T.E., *J.Am.Chem.Soc.,* **1973,** *95,* 8446.
12. Tarnorutskii, M.M., Artamonova, S.G., Grebenshchikova, V.A., Filatov I.S., *Vysokomol. Soed.,* **1972,** *A14,(11),* 2371.
13. Kazanskii, K.S., Tarasov, A.N., Paleyeva, I.Ye., Dubrovskii, S.A., *Vysokomol. Soed.,* **1978,** *A20,(2),* 391.
14. Goeke, G.L., Karol,F.J., *U.S. Pat.* 4,193,892, **1980.**
15. Panayotov, I. M., Berlinova, I.V., Bojilova, M., Boyadjiev, A., Rashkov, I.B., Tsvetanov, Ch.B., *Bulg. Pat.* 25,142 , **1978.**
16. Faucher, J.A., *J.Appl.Phys.,* **1966,** *37,* 3962.
17. Kazanskii, K.S., Solovyanov, A.A., Entelis, S.G., *Eur. Polym.J.*, **1971,** *7,* 1421.
18. Deffieux, A., Boileau, S., *Polymer*, **1977,** *18,* 1047; Boileau, S., *ACS Symp. Ser.*, **1981,** *166,* 283.
19. Nenna, S., Figueruelo, J.F., *Eur. Polym.J.*, **1975,** *11,* 511.
20. Dubrovskii, S.A., Kazanskii, K.S., *Khim.Fyz..*, **1982,** *12,* 1681.
21. Yang, F., Dejardin, Ph., Frere,Y., Gramain, Ph., *Makromol. Chem.*, **1990,** *191,* 1209.
22. Doytcheva, M., Dotcheva, D., Stamenova, R., Orahovats, A., Tsvetanov, Ch., Leder, J., *J. Appl. Polym. Sci.*, in print.
23. Lang, S.L., Webster, D.F., *U.K. Pat.Appl.* 2,093,702 and 2,093,703, **1982**.
24. Lambov, N., Stanchev, D., Peikov, P., Belcheva, N., Stamenova, R., Tsvetanov, Ch., *Pharmazie*, **1995,** *50, H2,* 125.
25. Tsanov, T., Stamenova, R., Tsvetanov, Ch.B., *Polymer*, **1993,** *34,* 617; Tsanov T., Stamenova, R., Tsvetanov, Ch.B., *Polym.J.*, **1993,** *25,* 853; Tsanov, T., Vassilev, K.G., Stamenova, R., Tsvetanov, Ch., *J.Polym.Sci. Polym. Chem.Ed.*, **1995,** *33,* 2623.
26. Dennison, K.A., *Ph.D.Thesis*, MIT, Cambridge, MA, **1988.**
27. Hooper, A., North, J.M., *Solid State Ionics*, **1983,** *9&10,* 1161.
28. King, P.A., Warwick, N.Y., *US Pat.* 3,264,202, **1966.**
29. Minkova, L., Stamenova, R., Tsvetanov, Ch., Nedkov, E., *J.Polym.Sci., Part B: Polym. Phys.*, **1989,** *27,* 621.
30. Sloop, S.E., Lerner, M.M., Stephens, T.S., Tripton, A.L., Paull, D.G., Stenger-Smith, J.D., *J. Appl. Polym. Sci.*, **1994,** *53,* 1563.
31. Turro, N. J., in *Modern Molecular Photochemistry*, The Benjamin/ Cumming Co.,Inc., Menlo Park, California, **1978.**
32. Rabek, J. F., in *Mechanisms of Photophysical processes and Photochemical Reactions in Polymers*, J. Wiley&Sons, New York, **1987.**
33. Flory, P. J., Rehner, R., Jr. *J.Chem.Phys.*, **1943,** *11,* 521.

Chapter 19

Anionic Polymerization of Lactams: Some Industrial Applications

K. Udipi[1], R. S. Dave[2], R. L. Kruse[3], and L. R. Stebbins[3]

[1]Monsanto Company, 800 North Lindbergh Boulevard, St. Luois, MO 63167
[2]Morrison and Foerster, 2000 Pennsylvania Avenue, Washington, DC 20006
[3]Bayer Polymers Division, 800 Worcester Street, Springfield, MA 01151

> Anionic ring opening polymerization of lactams to generate polyamides has been studied quite extensively both in academia and in industry. Caprolactam is by far the most studied lactam and the nylon 6 prepared by this route compares favorably in properties with that prepared by conventional hydrolytic polymerization. Fast reaction kinetics, absence of byproducts, and the crystalline nature of the nylon so produced makes anionic polymerization a compelling choice for several industrial applications. This paper will review a few such industrial applications as in reactive extrusion, reactive thermoplastic pultrusion, and reaction injection molding.

Anionic ring opening polymerization of lactams to generate polyamides has been studied quite extensively by Sebenda[1], Wichterle[2], and Sekiguchi[3] among others in academia and Gabbert and Hedrick[4] in industry. Caprolactam is by far the most studied lactam and the nylon 6 prepared by this route compares favorably in properties with that prepared by conventional hydrolytic polymerization. Fast reaction kinetics, absence of by-products, and the crystalline nature of the nylon so produced makes anionic ring opening polymerization a compelling choice for several industrial applications such as reactive extrusion, reactive thermoplastic pultrusion, and reaction injection moldings. This paper will review reactive processing of caprolactam in the above three areas.

Anionic ring opening polymerization of lactams follows an activated monomer mechanism as against conventional activated chain end mechanism. That is, the chain growth reaction proceeds by the interaction of an activated monomer (lactam anion) with the growing chain end. A typical reaction path for the polymerization of caprolactam initiated by a bisimide (isophthaloyl bis caprolactam) and catalyzed by a Grignard species (caprolactam magnesium bromide) is shown in Figure 1.

A large number of initiators and catalysts are mentioned in literature but the most commonly employed initiators are N-acyl lactams while lactamate anions generated by the reaction of a Grignard or sodium with lactams are the preferred catalysts.

REACTIVE EXTRUSION

Reactive extrusion as it applies to polymers involves conducting a chemical reaction in an extruder on a preformed polymer to generate functionalized polymers such as compatibilizers for polymer blends or carrying out continuous conversion of low viscosity oligomers or monomers to polymers. Although single screw extruders were employed in the early stages of development,[5,6] in recent years, twinscrew extruders are preferred because of their ability to mix, devolatalize, and pump low viscosity liquids and to add or remove ingredients from the melt at various stages. Several different twinscrew extruders of varying configurations are available, but among those, counter rotating non-intermeshing, counter rotating intermeshing and corotating intermeshing extruders have gained the broadest acceptance.

Currently almost all of the ~ 1.2 billion lbs of worldwide supply of nylon 6 is manufactured by the hydrolytic, continuous or batchwise polymerization of caprolactam. It takes a few hours at temperatures in the 250-270^{0}C range to produce this polymer. Anionic ring opening polymerization of caprolactam on the other hand is fast, takes only minutes at < 250^{0}C to complete and as suggested earlier with no byproducts formed, it lends itself to a continuous polymerization process in an extruder. It is also less capital intensive. Evidence in literature dates back to 1969 of some early attempts by Illig[7] and others[8,9] to adapt the above chemistry to reactive extrusion. Bartilla[10-11] et al explored the sodium salt of caprolactam as a catalyst and acetyl caprolactam as the initiator in their study carried out in a 30mm corotating intermeshing twinscrew extruder. They also fed the ingredients to the extruder as solids at room temperature expecting to get a uniform mixing when melted. Nichols et al[12-14], employed caprolactam magnesium bromide as the catalyst and acetyl caprolactam as the initiator in their study in a 20mm counter rotating non-intermeshing twinscrew extruder. In all the cases cited above, they were able to successfully polymerize caprolactam to nylon 6 comparable in most respects to commercial hydrolytic nylon 6. In our own study[15] (Figure 2) carried out in a counter rotating nonintermeshing twinscrew extruder (L/D=48), we employed caprolactam magnesium bromide as the catalyst and isophthaloyl biscaprolactam (BAIT) as the initiator. It is imperative that the polymerization be carried out under anhydrous conditions and the twinscrew extrusion is quite conducive in this respect. All the materials employed in the polymerization are thoroughly dried and pumped through heat traced transfer lines under moisture-free conditions. The variables examined included initiator concentration (3 mmoles to 5 mmoles/100g monomer at a catalyst/initiator ratio of ~ 2.5), feed rates (6 lb/hr to 16 lb/hr), residence time in the extruder(50 secs to 170 secs), and the extruder temperature

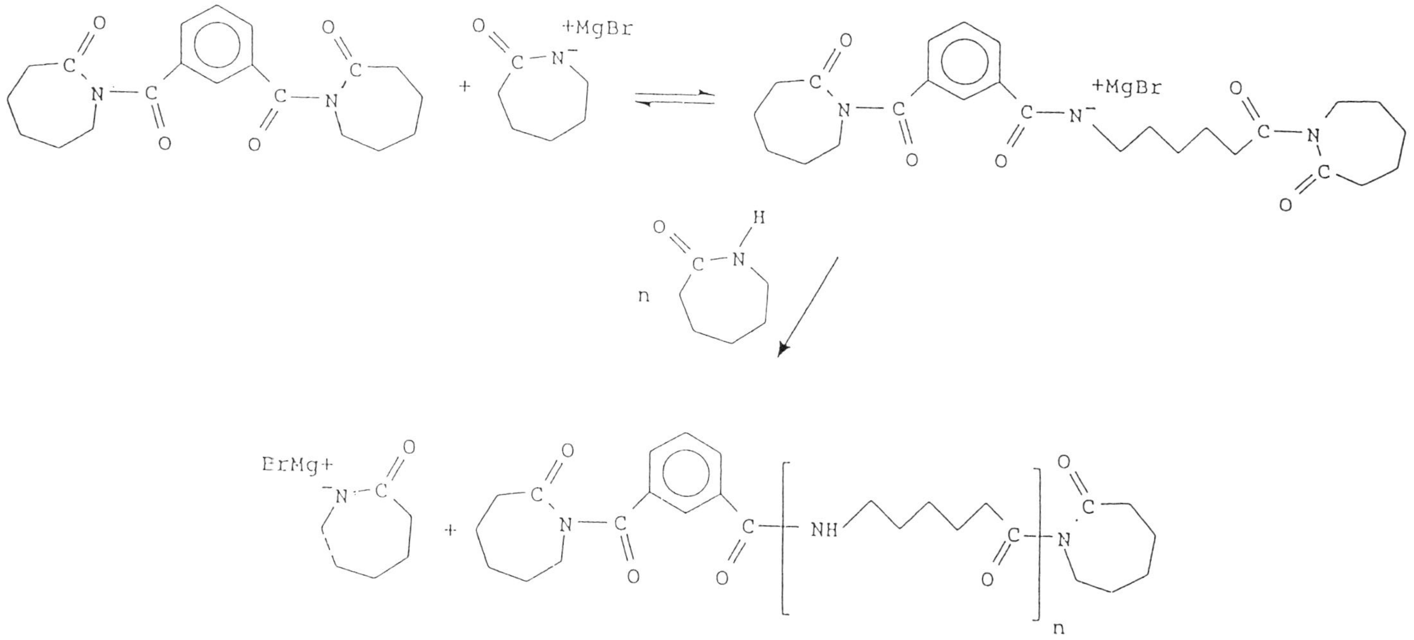

Figure 1. Bisimide initiated anionic polymerization of caprolactam

(100 to 230°C). In all cases the rates of monomer conversion were fast and the molecular weights attained were considerably higher than for hydrolytic nylon 6. A major distinguishing factor in favor of nylon 6 produced by reactive extrusion route as against conventional hydrolytic nylon 6 is the ease with which molecular weight is controlled. Thus in hydrolytic nylons, the commonly attained weight average molecular weight M_w is ~30Kg/mole. Although such molecular weights are adequate for most applications, they are not high enough for such applications as blow molding. Conventional nylons with higher molecular weights are obtained by solid state polymerization, wherein nylon pellets are heated below the melting point with either vacuum or a nitrogen sweep. Our studies as well as earlier studies have demonstrated that anionically polymerized nylons can be produced in the molecular weight range of M_w~ 80-120 Kg/mole without going through the solid state polymerization step.

All the nylon 6 polymers generated by the reactive extrusion route exhibit reasonably high levels of residual caprolactam monomer. This is typical of the ring chain equilibrium associated with ring opening polymerization. Even the nylon 6 produced by the hydrolytic polymerization contains residual monomer which is removed by either washing or application of high vacuum. The level of residual monomer in reactively extruded nylon 6 can be brought down by optimizing the extrusion variables to the levels in commercial hydrolytic nylon 6. Further reduction is feasible by injecting water into the extruder barrel towards the end of polymerization and subsequent devolatalization before exiting from the extruder die. In our work we have been able to lower the residual monomer to <2% by injecting water and devolatalization .This steam washing step also deactivates the catalyst to arrest subsequent monomer generation. Active residual catalyst can attack the terminal units to regenerate free caprolactam monomer.

Reactive extrusion is also an effective process tool to prepare polymer blends and block and graft copolymers. Indeed, several reports in literature and our own work bear this out. McGrath et al[16] and also Van Buskirk and Akkapeddi[17] have shown that polysulfones participate in the anionic polymerization of caprolactam to form graft copolymers. Similarly, polyetherimides[17,18] (Figure 3) and styrene-maleic anhydride and styrene- maleimides[18] react during the polymerization of caprolactam to form graft copolymers with improved performance at elevated temperatures under load. Figures 4 and 5 compare the dynamic mechanical properties of nylon 6 control and a 85/15 copolymer of nylon 6 and polyetherimide. It is apparent from the tanδ peaks that the glass transition temperature of the copolymer is about 30°C higher than the control nylon 6 and the modulus of the copolymer is also considerably higher at elevated temperatures.

REACTIVE THERMOPLASTIC PULTRUSION

Pultrusion is one of the most cost-effective means of manufacturing composites having unidirectional fibers and constant cross-sectional areas. It is a continuous process unlike most other composite manufacturing processes.

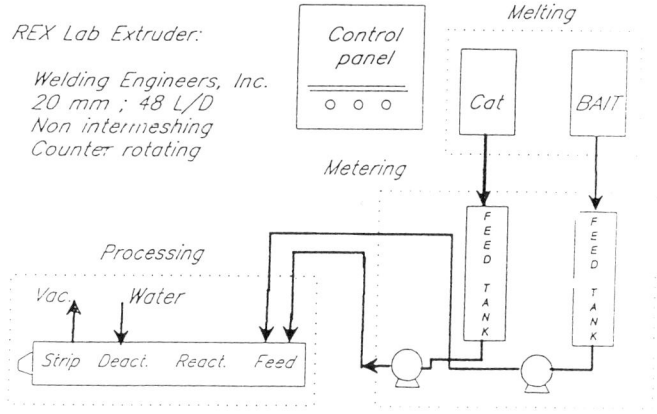

Figure 2. Reactive extrusion set-up for polymerization of caprolactam

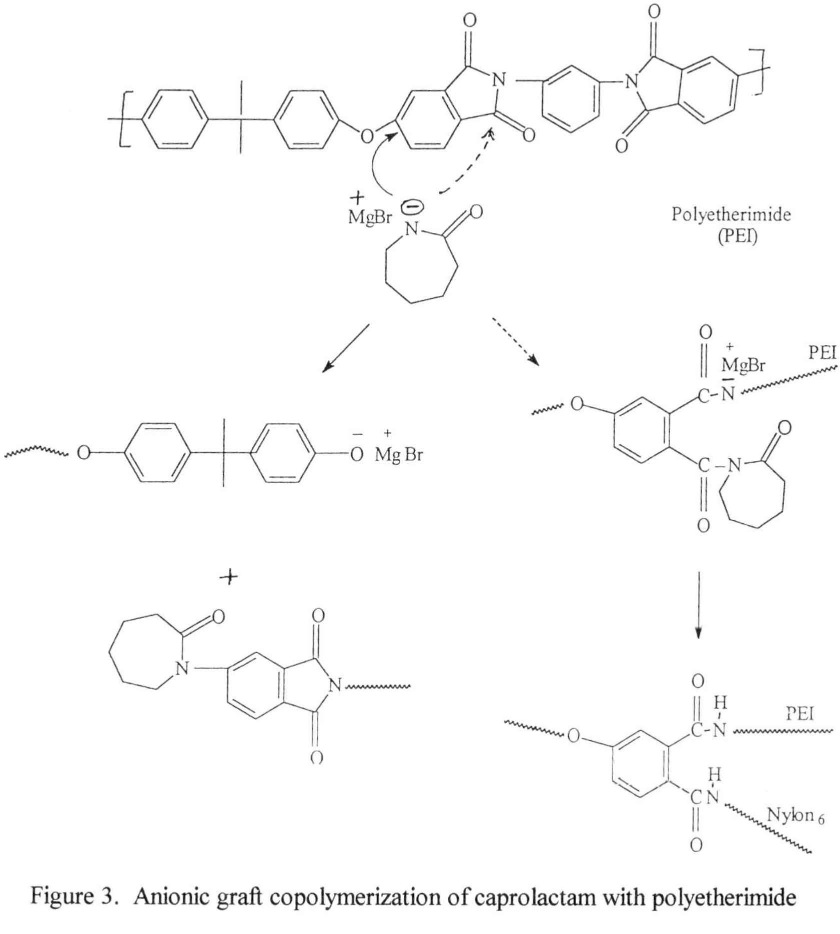

Figure 3. Anionic graft copolymerization of caprolactam with polyetherimide

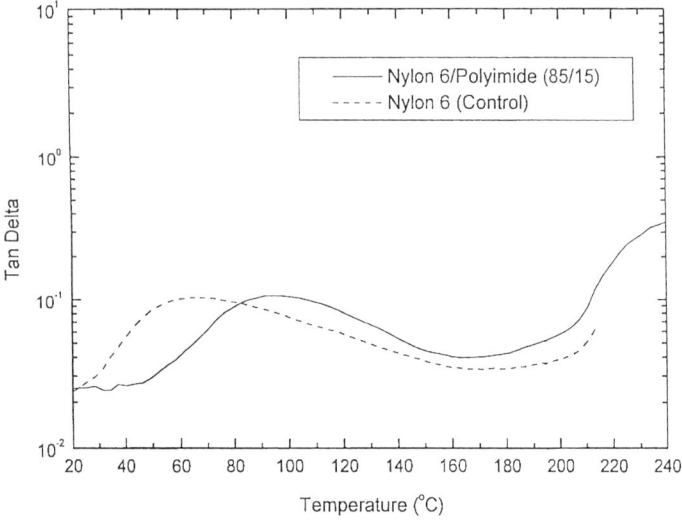

Figure 4. Dynamic mechanical analysis (tanδ vs temperature) of graft copolymer of caprolactam and polyetherimide (85/15)

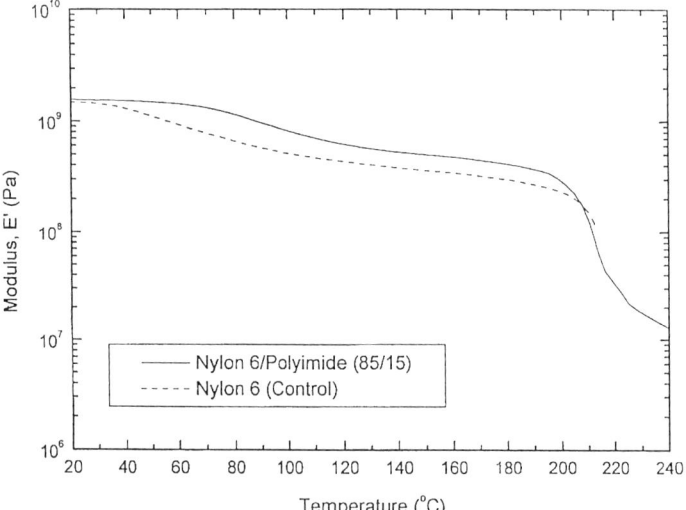

Figure 5. Dynamic mechanical analysis (modulus vs temperature) of graft copolymer of caprolactam and polyetherimide (85/15)

Composites offer high stiffness and strength, low weight, and excellent chemical and corrosion resistance. Matrices employed to hold the fibers together can be thermoplastic or thermosetting in nature. Although most of the composites currently made use thermosetting matrix, thermoplastic composites have the advantages over thermoset matrices with respect to impact strength, formability, weldability and recyclability. Major end markets for pultruded composites include electrical equipment, consumer goods, chemical and petrochemical processing and related industries, as well as construction and transportation industries.

Although the composite market is growing at a slower rate than common expectations, pultrusion is a rapidly growing segment[19] and the thermoplastic pultrusion is forecasted to grow at ~ 35-40% a year in the coming years.

A typical thermoplastic matrix pultrusion process involves impregnation of the molten polymer into fiber bundles or mat. This is a very difficult process, since thermoplastic matrices typically have high melting points and high melt viscosities, requiring high temperature and high pressure processes. As a result, fiber wetting and impregnation by the molten polymer can be difficult and preparation of high quality thermoplastic matrix composites is both difficult and expensive. In reactive thermoplastic pultrusion, low viscosity monomers are mixed with reinforcing fibers and as they are pulled through a heated zone and exit die, the monomer polymerizes around the fibers in the die.

In nylon 6 pultrusion, the process essentially involves rovings of glass fiber reinforcement pulled through a "feed zone" where they mix with caprolactam monomer, catalyst and initiator[20] (Figure 6). It is not unusual to use about 75 weight percent glass fibers. The fast polymerization kinetics help the monomer to fully polymerize before the shaped composite exits the die. There are many variables that need to be controlled to produce a reproducible product of high quality at competitive costs. Input variables include pressure, temperature, equipment position and rates, catalyst and initiator concentration, fiber orientation (degree of twist during pulling) and operator ability. Output variables include fiber volume fraction, interfacial characteristics, part thickness, void content, and the degree of byproduct generation.

For Nylon 6 pultrusion, the mixing temperature is a few degrees above the melting point of caprolactam while the reaction temperature in the die is of the order of 160^0C. The major variable of concern is the amount of moisture in the reaction mixture. Above 100ppm of water severely inhibits polymerization. Other variables include the amount of catalyst and initiator in the reaction system , which will determine the conversion rate during polymerization. The pultruded composite is generally pulled at about 2 ft /min although it is believed that the same can be increased to about 5 ft/min. The total cycle time for the continuous process is less than 10 minutes. A typical 75% glass reinforced pultruded nylon 6 composite exhibits a flexural strength of 200Kpsi and flexural modulus of 5.5Mpsi.

REACTION INJECTION MOLDING

Reaction injection molding (RIM) is a polymer process operation whereby reactive liquid components are mixed by impingement, injected into a mold, and polymerized therein to form a plastic part.[21] Polyurethanes, by virtue of fast reaction kinetics and absence of byproducts are the most commonly encountered materials in this field and account for about 95% of the total market volume. Nylons, epoxies, polyesters, and polycyclodienes constitute the rest. All the above RIM materials are thermosets except nylon 6. RIM offers several advantages over other polymer processes particularly in the manufacture of large molded articles in that it is more practical, economical, and energy efficient.

Although reaction injection molded nylon 6 is conceptually feasible through polymerization of caprolactam in the mold, the first to emerge into commercial reality were the nylon block copolymers[22]. Although nylons are considered as tough thermoplastics, they tend to be notch sensitive; that is they are not resistant to crack propagation and this often results in brittle or catastrophic failures. Incorporation of a suitable elastomer either as a block copolymer with nylon or as a dispersed phase helps overcome this deficiency. Nylon block copolymers prepared via reaction injection molding consist of alternating elastomer and nylon 6 segments with the latter forming terminal blocks. The elastomeric component provides toughness and ductility whereas the crystalline nylon 6 blocks offer strength, rigidity, chemical resistance, and high melting temperature. Incorporation of elastomer in nylon 6 RIM is commonly achieved by in-situ copolymerizing caprolactam with terminally functional low molecular weight polyethers or polybutadienes (Figure 7). Alternately caprolactam is polymerized in the presence of a polymeric diol using bisacyllactam as coupler/initiator.

Properties of the block copolymers are controlled over a broad range by selecting the type of elastomer and the elastomer content.(Table I)

In another method, nylon 6 RIM is toughened through particulate rubber modification. Core-shell rubbers with soft, crosslinked butadiene based rubbery core and grafted styrene-acrylonitrile shell with anhydride functionality have been found quite useful in toughening preformed nylons[23]. This approach however is not applicable to nylon 6 RIM system since the polymer is formed in situ in the mold. Attempts to toughen nylon 6 RIM by dispersing dried graft rubber crumb in caprolactam monomer and then polymerizing the same in a mold have not been very successful and this has been ascribed to the poor quality of rubber dispersion. However, a dispersion obtained by mixing the graft rubber latex with caprolactam monomer and stripping off the water, upon polymerization produced toughened nylon 6 with moderate notched Izod impact. Incorporation of a reactive monomer like hydroxypropyl methacrylate (HPMA) in the shell of the graft polymer produced polymers with considerably improved impact properties (Table II). It is believed that this reactive monomer participates in the caprolactam polymerization to form a covalent bond between the grafted rubber particle and nylon matrix[24,25]. (Figure 8)

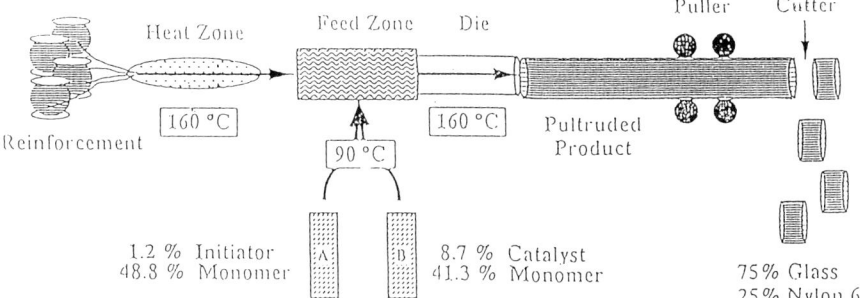

Figure 6. Fiber glass-nylon 6 reactive pultrusion set up

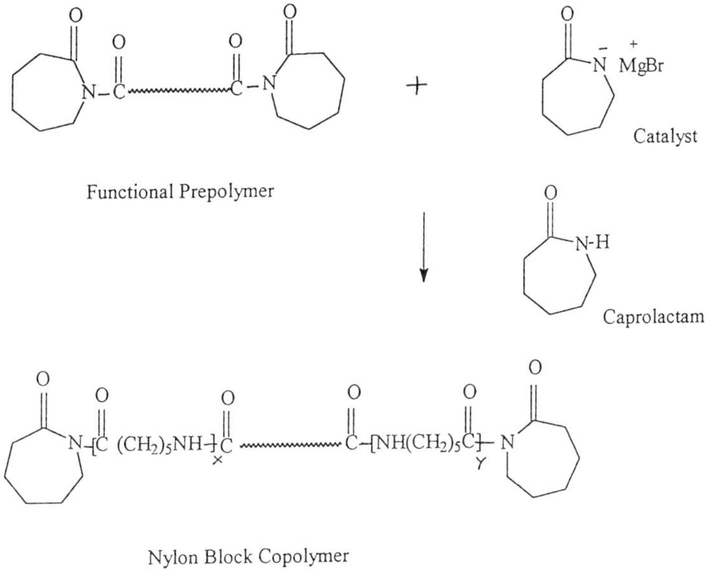

Figure 7. Anionic block copolymerization of caprolactam with terminally functional elastomeric prepolymer

TABLE I

PROPERTIES OF NYLON 6 -POLYETHER BLOCK COPOLYMERS

PROPERTY	% POLYETHER			
	0	10	20	40
Shore D Hardness	84	83	78	62
Tensile Strength, MPa	74.4	53.7	44.1	36.5
Tensile Elongation, %	30	35	285	490
Flexural Modulus, GPa	2.69	1.93	1.52	0.21
Notched Izod Impact, J/m	32	85	998	NoBreak

TABLE II

PROPERTIES OF PARTICULATE RUBBER MODIFIED NYLON 6 RIM
(%Rubbery Core:12, Core:Shell =100:25)

Shell Composition	Tensile Strength MPa	Tensile Elongation %	Flexural Modulus MPa	Notched Izod J/m
Nylon 6 RIM (Control)	64.1	28	2828	37.3
S/AN (70/30)	51.7	27	2345	127.9
S/AN/HPMA (60/30/10)	54.8	51	2103	287.8

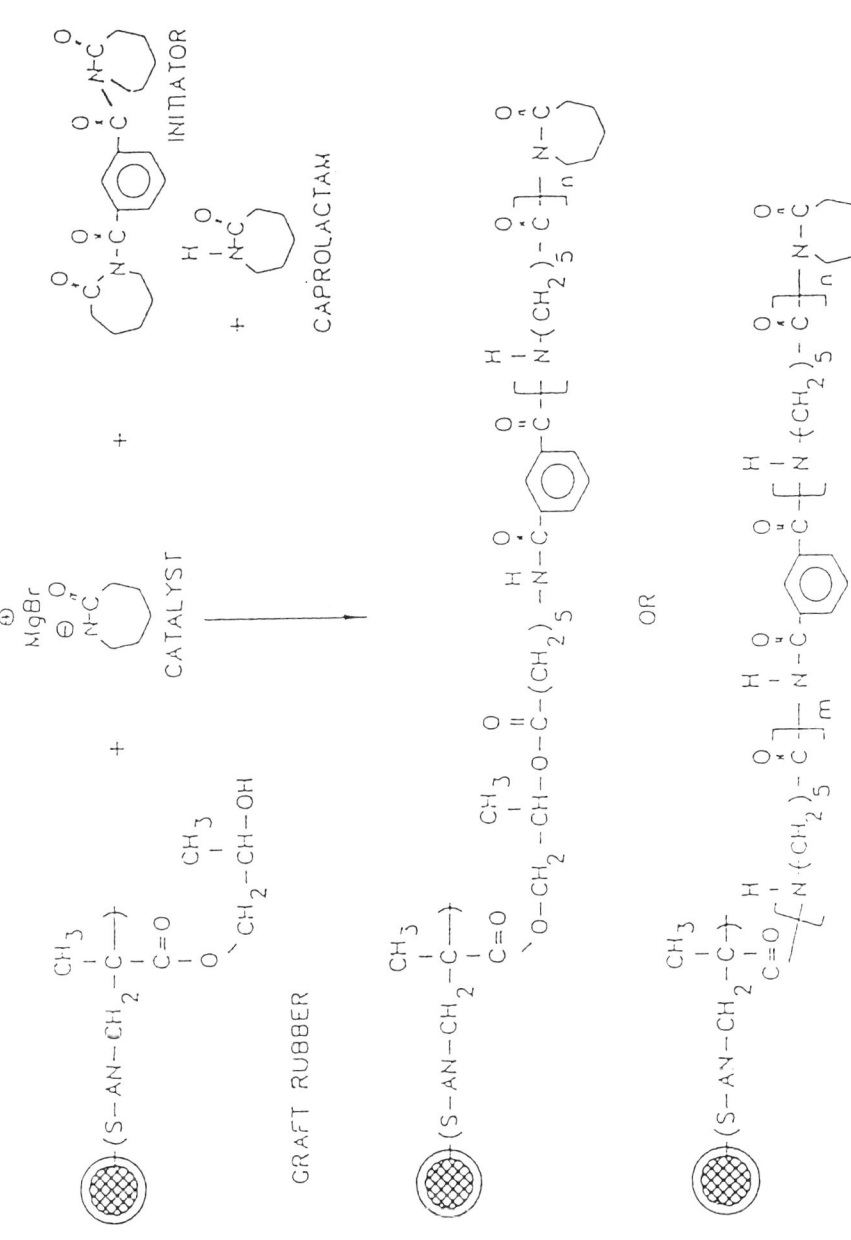

Figure 8. Anionic graft copolymerization of caprolactam with hydroxy functional core shell rubber

ACKNOWLEDGEMENT

The authors would like acknowledge the help of Drs. G. A. Gutierrez and C. G. Hagberg in conducting the reactive extrusion experiments, J. Hurlbut for his help in dynamic mechanical analysis and B. Moran for his help in particulate rubber modification of nylon 6 RIM.

REFERENCES

1. Sebenda, J., *Pure and Applied Chemistry*, **1976**, 48, 329
2. Wichterle, O., Sebenda, J., and Kralicek, J., *Fortsch.Hochpolymer Forsch*, **1961**, 2, 578
3. Sekiguchi, H., *Ring Opening Polymerization,* Vol 2, (Eds. K.J.Ivin and T.Saegusa), Elsevier, New York, pp 809-918
4. Hedrick, R.M., and Gabbert, J.D., AIChE National Meeting, Detroit, Michigan, 8/17/81
5. Reinking, K.;.Voge, H.; and Hechelhammer, W U.S.Patent 3,634,574 (Bayer, Germany), 1972
6. Blazen, M.; and Potin, P.; U.S.Patent 4,067,861 (Ato Chimie, France), 1978
7. Illig, G., Modern Plastics, **1969**, 46, 8, 70
8. Walker, A.C. and Kinsey,I.H. British Patent 1,289,349 (Polymer Corporation, USA), 1972
9. Joris, G.G. and Miller, R.H., U.S.Patent 3,484,414 (1969)
10. Bartilla, T., 13th IKV-Kolliguium, Aachen, 1986
11. Menges, G., and Bartilla, T., 2nd Annual Meeting, Polymer Processing Society, Montreal, 1986
12. Nichols, R.J.; Golba, J.C.Jr and Shete, P.K., AIChE Diamond Jubilee Meeting, Washington, D.C., 1983
13. Nichols, R.J., Golba, J.C., and Johnson, B.C., Polymer Processing Society, Montreal, 1986
14. Tucker, C.S. and Nichols, R.J., Proceedings of ANTEC '87, 1987 pp117
15. Udipi, K., Gutierrez, G., and Hagberg, C.G., unpublished data
16. McGrath, J.E., Robeson, L.M.,and Matzner, M., *Poly.Sci. and Technol.*,**1974**, 4, 195
17. Van Buskirk, B. and Akkapeddi, M.K., Polymer Preprints, American Chemical Society, 29 (1), 1988
18. Udipi, K. and Stebbins, L.R., unpublished data
19. Astrom, B.T., Proceedings of the 3rd International Symposium on Polymers for Advanced Technologies, Pisa, Italy June 1995
20. Dave', R. S., Udipi, K., Kruse, R.L., Stebbins, R.L., and Williams, D.E., Conference of the Center for Composite Materials, Virginia Tech, April, 1995
21. Macosko, C.W., *RIM: Fundamentals of Reaction Injection Molding*, Hanser Publishers, NY, 1989
22. Dupre, C.R., Gabbert, J.D., and Hedrick, R.M., Polymer Preprints, American Chemical Society, 25, 2, 296 (1984)
23. Baer, M., U.S.Patent 4.306,040 (1981)
24. Udipi, K., U.S.Patents 4,882,382; 4,994,524; 5,051,469; 5,189,098
25. Udipi, K., *J.Appl.Polym.Sci.*, **1988,** 36, 117

Chapter 20

Preparation of Poly(2,6-dimethyl-1,4-phenylene ether) (PPE) and Polyamide-6 Blends via Activated Anionic Polymerization of PPE and ε-Caprolactam Solutions

I. Chorvath[1], M. D. M. Mertens[2], A. A. van Geenen[2], R. J. G. van Schijndel[1,3], P. J. Lemstra[1], and H. E. H. Meijer[1]

[1]Center for Polymers and Composites, Eindhoven University of Technology, 5600 MB Eindhoven, Netherlands
[2]DSM Research, 6160 MD Geleen, Netherlands

Activated anionic polymerization of homogeneous poly (2,6-dimethyl-1,4-phenylene ether)(PPE)/ε-caprolactam solutions at temperatures below the melting point of polyamide-6 (PA-6) using the catalyst, lactam magnesium bromide, and the co-catalyst, ε-caprolactam blocked hexamethylene diisocyanate, provides a route for *in situ* preparation of PPE/PA-6 blends.
The polymerization of the solutions using the catalyst and the co-catalyst results in PPE/PA-6 blends with either the morphology of a continuous PPE and dispersed PA-6 phase or vice versa, depending on the PPE concentration. The particle size of the dispersed PA-6 phase in the continuos PPE matrix is determined by the viscosity i.e. the PPE concentration. High ε-caprolactam conversions could be achieved.
Before polymerization was performed, the phase behavior of PPE/ε-caprolactam solutions in the absence of a catalyst and co-catalyst of polymerization was investigated.

In the past, extensive studies have been conducted on the polymerization of initially homogeneous systems based on a polymer dissolved in a monomer (*1-19*). Special attention was paid to the morphology control and the mechanical properties of polymer blends prepared in this way. Upon the polymerization of homogeneous polymer/monomer solutions, a new polymer is synthesized from the monomer. Due to the immiscibility of the majority of polymers, the homogeneous solution will undergo phase separation. A number of polymer blends were prepared in this manner applying for example epoxy resins, styrene, methacrylates, etc. as solvents for polymers.
Especially, a remarkable number of polymers and rubbers were reported to be soluble in epoxy resins. Some of the resulting solutions were used as precursors for

[3]Current address: ATO-DLO, 6700 AA Wageningen, Netherlands.

the preparation of toughened epoxies. Via polymerization of for example CTBN rubber/epoxy solutions, rubber toughened epoxies were prepared (*1,2*). The preparation of thermoplastic toughened highly crosslinked epoxies was based on the polymerization of, for instance, solutions of tetrafunctional epoxies with high T_g polymers, such as poly(ether imide) (PEI), poly(ether sulphone) (PES) (*3-5*).

With respect to the morphology of these blends, the previously dissolved polymer was intended to later form the dispersed phase, while the polymer (crosslinked epoxy), originating from "solvent" resulted in the continuous matrix.

This is also the case with high impact polystyrene (HIPS) and acrylonitrile-butadiene-styrene (ABS) polymer, where initial solutions of butadiene in styrene and butadiene in a styrene-acrylonitrile mixture, respectively, are used. For HIPS and ABS there is an additional phase seen inside the rubber domains which contain occluded polystyrene, and occluded copolymer styrene and acrylonitrile respectively.

Another example of polymerization of monomers in polymer solutions or gels is found in the preparation of sequential-IPN's (*6-8*). Two subdivisions are often made in sequential IPN's; thermoplastic IPN's (*9,10*), which contain physical crosslinks rather than chemical crosslinks, and semi-IPN's (*6,11,12*), where one of the polymers is linear.

Recently, a new polymer processing route was developed in our laboratories employing monomers as reactive solvents for thermoplastics. The research originated from the antagonism between on the one hand requirements placed upon the ultimate properties of polymeric materials, often determined by the molecular characteristics, and on the other hand requirements on processability (*13*). In general, the application of solvents lowers the processing viscosity and/or processing temperature. This is also the case with reactive solvents. By using concentrated solutions of thermoplastic/reactive solvent, upon polymerization, phase separation is induced and in contrast with the examples mentioned above, this process is followed by phase inversion. After polymerization, the polymerized reactive solvent becomes the dispersed phase in the thermoplastic polymer matrix.

This concept was explored for the processing of intractable polymers, for example PPE combined with epoxy resin as reactive solvent (*14-17*) as well as of tractable polymers, for example PE combined with styrene and butyl-methacrylate (*18*) as reactive solvents. Besides an extension of processing characteristics of polymers, the concept can also result in materials with very fine morphologies, difficult to prepare by conventional blending processes.

In the literature, another example of a polymer blend where the originally dissolved polymer forms the continuous phase can be seen in the case of high impact PMMA/EVA blend. This blend was prepared from solutions of poly(ethylene-co-vinylacetate) in methyl methacrylate (MMA) by "cast polymerization". Also in this case, the polymerized monomer PMMA forms the dispersed phase, while the continuous phase is EVA (*19*).

An ever present need for faster and more economic processes created an interest in monomers (reactive solvents), which can be converted into polymers by fast polymerization. It is known that activated anionic polymerization of ε–caprolactam proceeds at high polymerization rates, reaching temperature dependent equilibrium conversion. Polymerization can be performed either below or above the melting point of PA-6 (220°C). In both of the industrial applications, monomer casting (20,21,22) and reaction injection molding (RIM) (23-26) polymerizations proceed below the melting point of PA-6 and result in products with conversions higher than 98%. Activated anionic polymerization of ε–caprolactam in an extruder, which proceeds above the melting point of PA-6, has been reported by several authors (27-33). Generally, compared to the cast PA-6 products, a lower conversion was found, resulting in the plastification of PA-6 by the residual monomer.

Even though ε–caprolactam was used as a solvent for oligomers during the development of RIM Nylon (NYRIM), containing an elastomeric phase (26), it was seldomly used as a solvent for polymers. In the literature PPE, poly(ether imide) and polysulfone (34) were reported to be soluble in ε–caprolactam, but no phase diagrams were published. The concentrated solutions were polymerized in an extruder via activated anionic polymerization of ε–caprolactam at temperatures above the melting point of PA-6. The blends of PA-6 and a high T_g polymer prepared in this way, possessed a morphology of a PA-6 matrix and a high T_g polymer dispersed phase. The presence of high T_g polymer and/or high polymerization temperature resulted in a decreased monomer conversion (34).

This paper discusses a detailed study of the phase behavior of PPE/ε–caprolactam solutions and the morphology of the PA-6/PPE blends obtained via the activated anionic polymerization of these homogeneous PPE/ε–caprolactam solutions at temperatures below the melting point of PA-6.

Experimental

Materials. Poly (2,6-dimethyl-1,4-phenylene ether) (PPE) with a viscosity-average molecular weight of 30 kg·mol^{-1}, supplied by General Electric Plastics (Bergen op Zoom, The Netherlands) and AP grade ε-caprolactam, supplied by DSM Research (Geleen, The Netherlands) were used. Activated anionic polymerization of ε-caprolactam was performed with catalyst NYRIM C1 (lactam magnesium bromide) and co-catalyst C 20 (ε-caprolactam blocked hexamethylene diisocyanate in ε-caprolactam), supplied by DSM Research (Geleen, The Netherlands). This combination of catalyst/co-catalyst was chosen in order to obtain a system with a long inhibition period, which allows sufficiently long mixing times in order to obtain the initial reactive mixtures as homogeneous solutions.

PPE/ε–caprolactam Solutions. Solutions of PPE/ε–caprolactam with a polymer concentration $\varphi \leq 50$ wt% (φ is the weight percentage of polymer) were prepared by mixing and heating the polymer powder and flakes of ε-caprolactam in a glass tube or flask, placed in an oil bath at 150°C. Solutions with higher polymer concentration $\varphi > 50$ wt% were prepared from a mixture of PPE powder and powdered ε-caprolactam in a co-rotating twin screw mini-extruder in a temperature range of 150-200°C, depending on the PPE concentration. This co-rotating twin screw mini-extruder with a volume of 6 cm^3 was developed by DSM Research, Geleen, The Netherlands.

The phase behavior of PPE/ε–caprolactam solutions was investigated by cloud point measurements and Differential Scanning Calorimeter (DSC) (Perkin-Elmer DSC-7).

Cloud Point Measurements. The cloud point curve of the polymer solutions was measured using Laser Light Scattering (at 488 nm). Homogeneous solutions of PPE/ε-caprolactam were immediately, after preparation, placed between glass slides in a Linkam THMS 600 hot stage preheated to 150°C. A laser beam was sent through the sample. Subsequently, homogeneous samples were then cooled from 150°C to 30°C at a rate of 2°C/min. Dilute PPE solutions turned turbid upon cooling, which resulted in the scattering of light. A cloud point temperature was recorded as the temperature at which first turbidity, thus phase separation occurs.

Calorimetric Investigations. Different thermal transitions of PPE/ε-caprolactam solutions were studied by means of DSC. Solutions, φ ≤ 50 wt%, were treated in two different ways before being subjected to DSC scans, in order to favor or suppress the possible crystallization of PPE. To promote crystallization, homogeneous solutions were slowly cooled in an oil bath at a rate of 20°C/hour to room temperature. The samples were then heated in a DSC from -50°C to 200°C at a rate of 10°C/min. After 5 min at 200°C the samples were cooled at a rate of 5°C/min to -50°C. Subsequently, the samples were heated to 200°C at a rate of 10°C/min, followed by a final cooling to -50°C at a rate of 10°C/min. This allowed the study of the melting and crystallization of ε-caprolactam and, possibly, PPE.

To suppress crystallization, homogeneous solutions were poured into liquid nitrogen (quenched). The samples were allowed to reach room temperature in an exsiccator in the presence of a water absorbing agent, to avoid water uptake. The samples were then heated in a DSC from -50°C to 200°C at a rate of 10°C/min to study the glass transition temperature (T_g).

The more concentrated solutions, φ > 50 wt%, were given the same thermal history as the dilute solutions and were then heated in a DSC from -50°C to 200°C at a rate of 10°C/min, followed by cooling to -50°C at a rate of 50°C/min. The samples were then heated to 200°C at a rate of 10°C/min. The glass transition measured in the last run was used for the phase diagram.

Preparation of Polyamide-6 (PA-6). Initially, two solutions were prepared in test tubes (ext. diameter 25 mm, length 200 mm) in a Lauda oil bath at a temperature of 165°C. One was a solution of catalyst in ε–caprolactam and the other was a solution of co-catalyst in ε–caprolactam. The catalyst NYRIM C1 in concentration 0.6 mol% and the co-catalyst C20 in concentration 0.9 mol% were used. A reactive mixture was formed when the two solutions were intensively mixed. Polymerizations were conducted under static conditions (no stirring) at a temperature of approximately 165°C, i.e. below the melting point of PA-6 (220°C).

PPE/PA-6 Blends. PA-6/PPE blends were prepared via the activated anionic polymerization of PPE/ε–caprolactam solutions. In order to do so, catalyst NYRIM C1 (0.6 mol%) and co-catalyst C20 (0.9 mol%), (unless specified otherwise), were mixed in PPE/ε–caprolactam homogeneous solutions to form the reactive mixtures. Polymerizations were conducted at a temperature of approximately 165 °C if not specified otherwise.

The viscosity of the starting solutions was the limiting factor for fast and sufficient mixing. Therefore, the reactive mixtures were prepared in three different ways, depending on the PPE concentration.

Reactive mixtures containing PPE concentrations of $\varphi < 20$ wt% were prepared in the test tubes in a Lauda oil bath at a temperature of approximately 165°C. Initially, two solutions were prepared, one solution of catalyst and PPE in ε-caprolactam and one solution of co-catalyst and PPE in ε-caprolactam. Upon the mixing of these solutions a reactive mixture was formed and polymerized.

More concentrated reactive mixtures containing 20 to 50 wt% PPE were prepared in a glass flask placed in an oil bath using an electric stirrer, to provide sufficient mixing. First a solution of PPE in ε-caprolactam was prepared. After homogenization, co-catalyst was added and dissolved while stirring. Then, the catalyst was mixed in. After approximately 30 seconds, the stirrer was removed and polymerization was performed under static conditions.

Reactive mixtures containing PPE concentrations of $\varphi > 50$ wt% were prepared as follows. PPE and powdered ε-caprolactam were premixed at room temperature and added to the mini-extruder (conical corotatory twin screw extruder). After homogenization via recirculatory extrusion at 170°C under a nitrogen atmosphere the co-catalyst was mixed in. Finally, the catalyst was added and the reactive mixture was mixed for 5 minutes. Here, the catalyst and the co-catalyst were used in concentrations of 3.6 mol% and 5.5 mol% respectively. The polymerization was continued under static conditions in a glass tube under a nitrogen atmosphere at a temperature of approximately 170°C.

Morphology of PPE/PA-6 Blends. To visualize the morphology of the resulting PPE/PA-6 blends via scanning electron microscopy (SEM), (Cambridge Stereoscan 200), the samples were fractured in liquid nitrogen, treated in chloroform, in order to extract PPE, and then coated with gold/palladium.

The morphology of blends containing $\varphi > 50$ wt% PPE was visualised via Transmission Electron Microscopy (Philips TEM CM 200). The samples were stained with a mixture OsO_4/ formaldehyde and ultramicrotomed using a diamond knife (Reichert Ultracut E Microtom).

Monomer Conversion. As an indication of monomer conversion in the blends and PA-6, the water extractable content (WEC) was determined. A sample was powdered and dried for 3 hours in a vacuum oven at a vacuum of 27 mbar and a temperature of 45°C. The samples were extracted in a round-bottom flask with boiling water, in 6 cycles of 11 minutes extraction and 1 minute cooling, according to the standards (35). Phosphoric acid was added to remove carbonate. Subsequently, the carbon content of the solution was determined, according to DSM 0522-E (36).

The effect of residual ε-caprolactam and oligomers on the T_g's and T_m of the final polymer blends was investigated by Dynamic Mechanical Thermal Analysis (DMTA) (Polymer Laboratories Mk III) after the samples were dried in a vacuum oven at 40°C for 60 hours to remove water. Measurements were performed in a tensile mode with a frequency of 1 Hz, heating from room temperature to 250°C at a heating rate of 2°C/min.

Results and Discussion

Preparation of polymer blends from a polymer/monomer solution via the polymerization of a monomer is often reported to start from a homogeneous solution (*14,18,37,38*). Therefore, prior to the preparation of PPE/PA-6 blends from PPE/ε-caprolactam solutions via activated anionic polymerization of ε-caprolactam, the phase behavior of PPE/ε-caprolactam solutions in the absence of the catalyst and co-catalyst was first investigated.

PPE is reported to be a crystallizable polymer which has difficulty in crystallizing from melt (*39*), because the melting and glass transition temperatures are very close together. PPE can, however, crystallize from solution (*40-43*). Generally, for semi-crystalline polymers, good and poor solvents can be considered. When a good solvent is applied, a characteristic melting point depression with an increasing solvent concentration occurs, and is theoretically described by the Flory-Huggins equation (*44*).

However, when a poor solvent is used, the situation becomes more complex, since liquid-liquid (L-L) demixing can take place upon cooling. During cooling, phase separation will start locally (binodal or spinodal) but this process is arrested when L-L demixing interferes with crystallization (*45,51*).

The same distinction between good and poor solvents is recognized for amorphous polymers. In the case of a good solvent, the solvent acts as a plasticizer and thus T_g decreases with an increasing solvent concentration as described by the Fox equation (*46*).

Application of a poor solvent for amorphous polymer results in a more complex phase behavior. L-L demixing, induced by cooling, interferes with the glass transition. This kind of phase behavior was demonstrated in literature for solutions of PPE/epoxy (*14*). The intersection point of the demixing line and T_g line is referred to as the Berghmans point (*47*).

The phase behavior of PPE/ε-caprolactam solutions has not yet been reported. Therefore, it is initially studied by cloud point measurements and DSC.

Calorimetric Investigations.

Solutions, $\varphi \leq 50$ wt%, Thermal History of Slow Cooling. When solutions with a PPE concentration, $\varphi \leq 50$ wt%, were heated in a DSC at a rate of 10°C/min (after previous slow cooling in an oil bath to promote crystallization), two melting endotherms were recorded. This is illustrated for PPE/ε-caprolactam solution, $\varphi = 5$ wt%, in Figure 1A. The first melting endotherm corresponds with the melting of ε-caprolactam, while the second endotherm corresponds with the melting of PPE.

Subsequent cooling of the solutions in a DSC, at a cooling rate of 5°C/min, resulted in the occurrence of two exotherms for samples with a PPE concentration $\varphi \leq 35$ wt%. This is demonstrated for PPE/ε-caprolactam solution, $\varphi = 5$ wt% in Figure 1B. The large exotherm is ascribed to the crystallization of ε-caprolactam, while the small exotherm originates from the formation of a crystalline PPE phase. No crystallization of PPE was observed in a DSC for solutions with a PPE concentration $\varphi > 35$ wt%.

Solutions, $\varphi \leq 50$ wt%, Thermal History of Quenching. In order to measure the T_g of PPE/ε-caprolactam solutions with a PPE concentration $\varphi \leq 50$ wt%, the solutions were prior to a DSC scan, quenched in liquid nitrogen to suppress crystallization.

Heating of the solutions with a PPE concentration $\varphi \leq 35$ wt% at a rate of 10°C/min in a DSC showed a glass transition, followed by a crystallization exotherm of PPE. For a solution containing 35 wt% PPE, a typical example of a DSC heating scan is shown in Figure 1C, revealing a T_g of 45°C. However, the T_g calculated, assuming ε-caprolactam to be a good solvent for PPE, is -2.2°C. The calculation was made according to the Fox equation (46) using the T_g's (determined by DSC) of the pure components PPE (210°C) and ε-caprolactam (-54°C).

A difference between the measured and calculated T_g's was also observed for the solutions PPE/cyclohexanol as reported in literature (48). This difference and the fact that all of the dilute solutions have the same measured T_g, is the consequence of L-L demixing preceding vitrification (48).

With respect to the presently studied solution containing 35 wt% PPE, the occurrence of crystallization immediately after heating above the glass transition temperature, as seen in Figure 1C, demonstrates the high rate of crystallization of PPE in the presence of ε-caprolactam.

Only the glass transition of PPE and no crystallization was observed in a DSC for solutions with a PPE concentration range of 36-50 wt%.

Solutions, $\varphi > 50$ wt%. Cooling of the homogeneous solutions containing $\varphi > 50$ wt% at a cooling rate of 5°C/min in a DSC shows no crystallization exotherm of PPE or ε-caprolactam. Regardless of the previous thermal history of slow cooling or quenching of the homogeneous solutions, upon heating at a rate of 10°C/min in a DSC, only the glass transition was recorded.

Binary Phase Diagram of PPE/ε-caprolactam solutions. The binary phase diagram was constructed by plotting the melting points of PPE and ε-caprolactam, the crystallization and glass transition temperatures of PPE, and the temperature at the onset of turbidity (cloud point) as a function of the PPE concentration. The temperature at the end of the melting endotherm of polymer was taken as the melting point. The onset of the crystallization exotherm was taken as the crystallization temperature. No corrections to the experimental data were made for the dynamic character and non-equilibrium nature of the experiments. The phase diagram of PPE/ε-caprolactam solutions is represented in Figure 2.

As shown in Figure 2, the melting temperature of the polymer (measured at 10°C/min) decreases slightly with an increasing solvent content in the concentration range of $\varphi \leq 50$ wt%. The melting temperature of neat PPE, 250 °C, was taken from literature (48). No experimental data could be obtained in the concentration range $\varphi > 50$ wt%. The concentration dependence of the melting point is typical for the melting of a polymer in a rather poor solvent system (49).

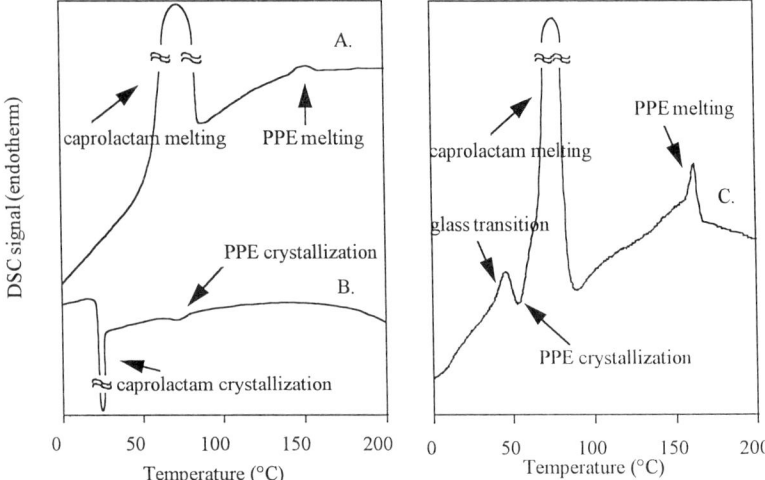

Figure 1. Typical DSC scans for PPE/ε-caprolactam solutions:
A. heating at a rate of 10°C/min, history of slow cooling, φ = 5 wt%
B. cooling at a rate of 5°C/min, φ = 5 wt%
C. heating at a rate of 10°C/min, quenched sample, φ = 35 wt%

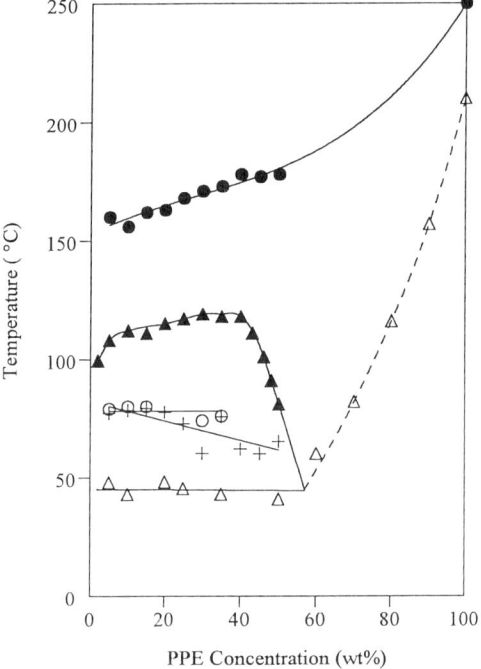

Figure 2. Phase diagram of PPE/ε-caprolactam solutions, (●) melting of PPE, (▲) cloud point, (+) melting of ε-caprolactam, (O) crystallization of PPE, (Δ) glass transition, (- - -) calculated T_g composition line.

The melting point of the solvent ε-caprolactam (measured at 10°C/min) slightly decreases with an increasing PPE concentration in a concentration range of $\varphi \leq 50$ wt%. At higher concentrations, viz. in the vitrified system, the melting endotherm of ε-caprolactam was not detected.

The cloud point curve (measured at 2°C/min) reveals an upper critical solution temperature (UCST) behavior and due to the polydispersity of PPE, it does not correspond to the binodal curve. Consequently, the cloud point curve does not provide information concerning the composition of the co-existing phases, i.e. the polymer poor phase and the polymer rich phase formed as a result of entering into the two phase region. Cooling of concentrated solutions, with a PPE content $\varphi > 57$ wt%, results in the formation of transparent glasses.

The crystallization temperature of PPE stays constant in a concentration range $\varphi \leq 35$ wt%. The constant character of crystallization temperature is considered to be indicative of the occurrence of L-L demixing preceding crystallization (45). Crystallization takes place in the concentrated domains which are formed in the process of L-L demixing (50). A similar proximity of the crystallization temperature of PPE and L-L demixing was reported to have a positive influence on the overall crystallinity of PPE in PPE/cyclohexanol solutions (48). Crystallization of PPE was not observed upon cooling of solutions with a concentration $\varphi > 35$ wt%.

Similarly, the T_g of PPE becomes constant for solutions with $\varphi \leq 57$ wt%. This constant of the T_g-concentration line was only measured for the quenched samples. In the concentration range $\varphi > 57$ wt% the T_g decreases with a decreasing PPE concentration and fits the Fox equation.

In summary, the phase diagram of PPE/ε-caprolactam differs from what is commonly observed in polymer/solvent systems since both crystallization and vitrification can interfere with L-L demixing depending on the thermal history and cooling rate.

Quenched PPE/ε-caprolactam solutions show an invariant T_g in the L-L demixing domain. This is characteristic for the interference of L-L demixing with the concentration dependent T_g (47). The cloud point curve and the glass transition line intersect at a PPE concentration of $\varphi = 57$ wt%. The intersection point, referred to in literature as Berghmans point (47), has several important consequences. Solutions with a polymer concentration lower than the concentration corresponding to the Berghmans point, will undergo phase separation upon fast cooling. However, complete phase separation is not reached, because the phase separation process will be arrested as soon as the PPE-rich domains vitrify at a PPE content of $\varphi = 57$ wt%. As a result, all phase separated solutions possess a T_g of approximately 45°C. Solutions with $\varphi \geq 57$ wt% will vitrify upon cooling and will be homogeneous over the entire temperature range from 57 to 100 wt%.

Upon slow cooling, L-L demixing is followed by crystallization of PPE in the concentration range $\varphi \leq 35$ wt%. Crystallization of PPE proceeds in the PPE-rich domains as indicated by the constant crystallization temperature.

From the above mentioned results, it can be concluded that ε-caprolactam can be classified as a poor solvent for PPE. In spite of this fact, homogeneous solutions PPE/ε-caprolactam can be prepared in a broad temperature-concentration region.

The lower border of this region is defined by the cloud point curve for polymer concentrations $\varphi \leq 57$ wt% and the T_g-φ curve for concentrations $\varphi > 57$ wt%. The upper border of the temperature-concentration region is governed by the thermal and oxidative stability of constituents.

Preparation of PPE/PA-6 blends from PPE/ε-caprolactam solutions via activated anionic polymerization of ε-caprolactam requires the incorporation of a catalyst and a co-catalyst which are compatible with the PPE/ε-caprolactam solution. Catalyst NYRIM C1 and co-catalyst C 20 were found suitable for this requirement. Due to their incorporation, the binary phase diagram of Figure 2 is not strictly valid. However, from extensive solubility tests, we concluded, that addition of the catalyst and co-catalyst only enhances miscibility of the system, most likely, because the catalyst and co-catalyst are derivatives of the solvent, ε-caprolactam.

Before the morphology development and morphology of PPE/PA-6 blends are discussed, some aspects of the polymerization will first be introduced.

Activated Anionic Polymerization of ε-caprolactam. The activated anionic polymerization of ε-caprolactam proceeds under inert and anhydrous atmosphere at fast rates until an equilibrium is reached between ε-caprolactam and PA-6. The position of this equilibrium is significantly influenced by the reaction temperature (52,53). It was reported in literature, that at high polymerization temperatures in an extruder above the melting point of PA-6, the conversion of ε-caprolactam was low (27,28). The residual ε-caprolactam needs to be removed because it drastically decreases the T_g of PA-6 (54,33,55) and introduces toxicity of the product.

One way to obtain a high conversion of ε-caprolactam and to avoid necessary extraction of residual ε-caprolactam is to carry out the anionic polymerization of ε-caprolactam at lower temperatures, below the melting point of PA-6 (120-170°C) (52).

The mechanism of activated anionic polymerization of ε-caprolactam in the presence of a catalyst and co-catalyst was discussed in literature by several authors (56-58). Often a metal lactamate is used as the catalyst. A co-catalyst can be either a carbamoyl or an acyl lactam (59,60). The metal lactamate prefers to react with the co-catalyst to form a resonance stabilized anion in the first step. When the metal lactamate forms an anion, the counter (cat)ion is released. The co-catalyst is able to form a complex with this cation and thus increases the concentration of free anions. The complex formation depends on the co-catalyst and cation used. Therefore, co-catalyst and catalyst form their preferred combinations (61). The total reaction time depends on the catalyst/co-catalyst system that is used, and on the concentration of catalyst and co-catalyst.

General Considerations on Morphology Development. Upon activated anionic polymerization of a homogeneous PPE/ε-caprolactam solution, a new component, PA-6, is formed. From this point on the system is considered to be a three component system containing PPE, PA-6 and ε-caprolactam.

The process of polymer blend formation from polymer/monomer solutions via chain growth polymerization of monomer can be in general illustrated by a ternary phase diagram. The schematic ternary phase diagram PPE/ε-caprolactam/PA-6 is illustrated in Figure 3.

With respect to the system studied, several comments have to be made concerning the ternary phase diagram. Generally, equilibrium can not be attained in viscous systems, where diffusion of matter is limited. Secondly, the ternary phase diagram reflects only an isothermal situation. Activated anionic polymerization of ε-caprolactam is known to be a highly exothermic reaction (57). The nonisothermal character of the system can affect the overall solubility as well as the viscosity of the phases. Thirdly, it is important to note, that a ternary phase diagram is valid for polymers with invariable molecular weight. The molecular weight of PA-6, however, increases gradually with polymerization. Finally, for the polymerization temperatures below the melting point of PA-6, it should be noted that liquid-solid transition (crystallization) can occur. Despite of the limitations imposed by the above mentioned comments, the schematic ternary phase diagram can still serve as a useful tool for the discussion of the morphology of PA-6/PPE blends prepared via activated anionic polymerization of PPE/ε–caprolactam solutions.

Two regions can be distinguished in Figure 3. The area above the binodal curve represents the single-phase region. The area below the curve represents a two phase region.

Starting from an initial homogeneous solution A, upon polymerization the composition of the polymerizing mixture follows the dotted line. After a small amount of ε-caprolactam is converted into PA-6, the binodal line, representing compositions of two liquid phases in equilibrium is crossed. The system becomes metastable which eventually leads to L-L phase separation in a two phase system. One phase is PA-6 rich, the other PPE rich, both containing monomer. The composition of the phases is determined by the intersection of the corresponding tie lines and the binodal curve. The ε-caprolactam conversion at this point is expected to be very low.

When the initial PPE content is low, phase separation results in a dominant phase rich in PPE and a lean phase rich in PA-6 (AX in Figure 3). The volume ratio of the PA-6 rich phase to the PPE rich phase is according to 'lever-rule' equal to the ratio of the segments of the tie line, for example 1-2/1-3. The volume fraction occupied by PA-6 is low and, therefore, at this point PA-6 forms the dispersed phase. As polymerization proceeds (AX), the volume of the PA-6 rich phase increases at the expense of the PPE rich phase. A point (A1) will come when that fraction of the volume occupied by the PA-6 rich phase is large enough for phase inversion to occur. Under assumption of equilibrium conditions, the PPE rich phase would become a dispersed phase.

The conversion at which phase inversion takes place depends on the initial concentration of PPE in the solution. For a polymer solution B with higher initial PPE concentration than A, a higher conversion (point B1) is required for phase inversion to take place in the reacting mixture. This can be seen from the intersection of the reaction line BY with the tie line.

Phase inversion can be limited by the viscosity of the system. The mobility in the system has to be high enough for phase inversion to take place. Mobility decreases with increasing ε-caprolactam conversion, due to the increase in viscosity. For a certain concentration of PPE, the conversion required for the formation of a PA-6 continuous phase may be so high that the viscosity of the system is too high for phase inversion to take place. In that case, PPE forms the continuous phase in the reacting mixture from the time when phase separation sets in until the end of polymerization.

Despite the fact that the formation of a PA-6 continuous phase is favored, it will form at the end dispersed phase. In the literature (*13,14,18*), however, this kind of morphology, when the previously dissolved polymer results in the continuous matrix, is sometimes referred to as a phase inverted morphology.

Starting from even more concentrated polymer solutions (C), the volume fraction occupied by PPE rich phase is along the entire reaction line CZ higher than the volume fraction of the PA-6 rich phase. Formation of a PPE continuous phase is favored, and the blend will exhibit a morphology of a PPE continuous phase and a dispersed PA-6 phase.

Morphology. The effect of PPE concentration on the morphology of PPE/PA-6 blends, prepared *in situ* by activated anionic polymerization of a homogeneous PPE/ε-caprolactam solution, was investigated. The morphology strongly depends on the concentration of PPE, see SEM (Figures 4, 5, 6) and TEM (Figure 7) micrographs.

Blends with PA-6 as Continuous Phase. At low PPE concentrations, 1.1 and 2.2 wt% (Figure 4b,c) a morphology of a PA-6 matrix containing dispersed PPE particles is found. The size of the PPE particles increases with increasing PPE concentration.

At higher PPE concentration, 4.4 wt% (Figure 5), two coexisting morphologies were observed. In the first, dominant morphology, (Figure 5a), PA-6 forms the matrix, while the PPE phase forms a dispersed phase. On the contrary, in the second, lean morphology, (Figure 5b), PPE forms the continuous phase, while PA-6 forms a dispersed phase. With a further increase of PPE concentration to 5.5 wt%, the second morphology of PPE continuous, PA-6 dispersed phases (Figure 5d) becomes dominant, while the first morphology turns into the lean (Figure 5c). The occurrence of a dual phase morphology in the blends was observed up to 6 wt% of PPE.

Blends with PPE as continuous phase. At PPE concentrations over 6 wt% PPE morphologies, referred to in literature as the phase inverted morphologies, (*13,14,18*) were always found with PPE as the continuous phase. At relatively low PPE concentrations, 8.4 wt%, large dispersed PA-6 particles (30 μm) were found (see Figure 6a). Upon increasing the PPE concentration, the diameter of the dispersed PA-6 particles decreases significantly, down to 1 μm at PPE concentration 40 wt% (see Figure 6e).

In general, the final blend morphology resulting from phase separation of the polymerizing system polymer/monomer, assuming static conditions, is determined by the competition between polymerization and phase separation. With respect to the rather complex character of polymerization of the system studied, the effect of increasing PPE concentration on the polymerization rate is difficult to estimate. However, a higher PPE concentration, i.e. a higher viscosity of the reactive mixture results in the reduction of phase separation rate. Polymerization in the viscous environment slows down particle coalescence during phase separation and prevents the coarsening of the separated structure.

The morphology of the blend containing 62 wt% PPE is presented in Figure 7. The majority of PA-6 particles in the PPE matrix is of a diameter of approximately 35 nanometers, while a smaller portion is of a diameter of approximately 300 nanometers. This bimodal particle size distribution could be explained with respect to the sample

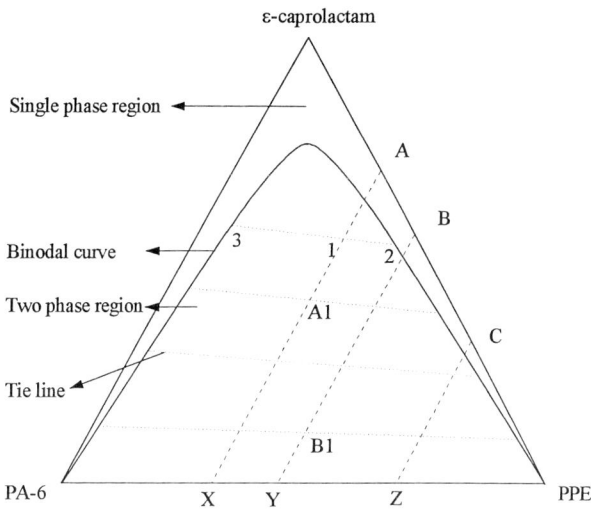

Figure 3. Schematic ternary phase diagram PPE/ε-caprolactam/PA-6. For details see text.

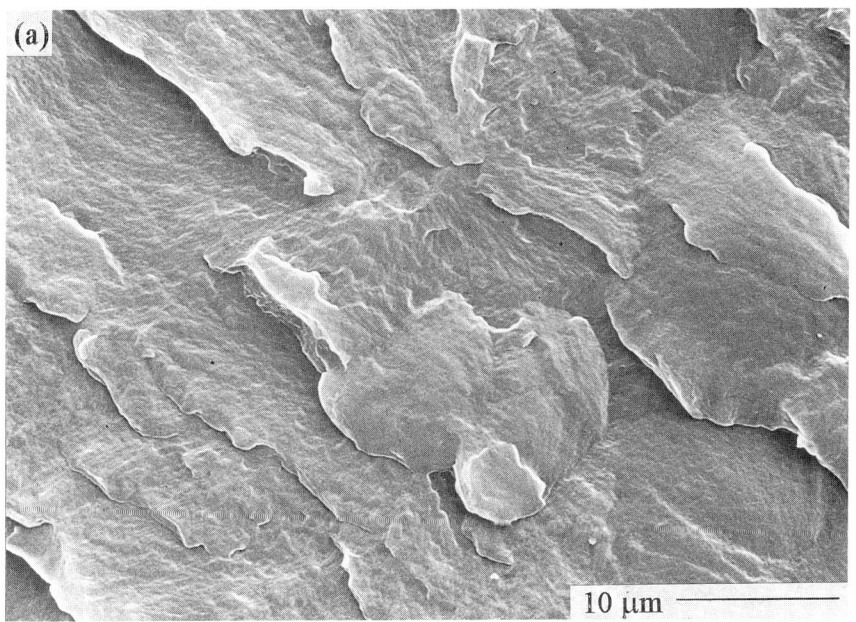

Figure 4. SEM micrographs of (a) PA-6 and PPE/PA-6 blends with a PPE content of (b) 1.1 wt%, (c) 2.2 wt%. PPE phase extracted by chloroform.

Continued on next page.

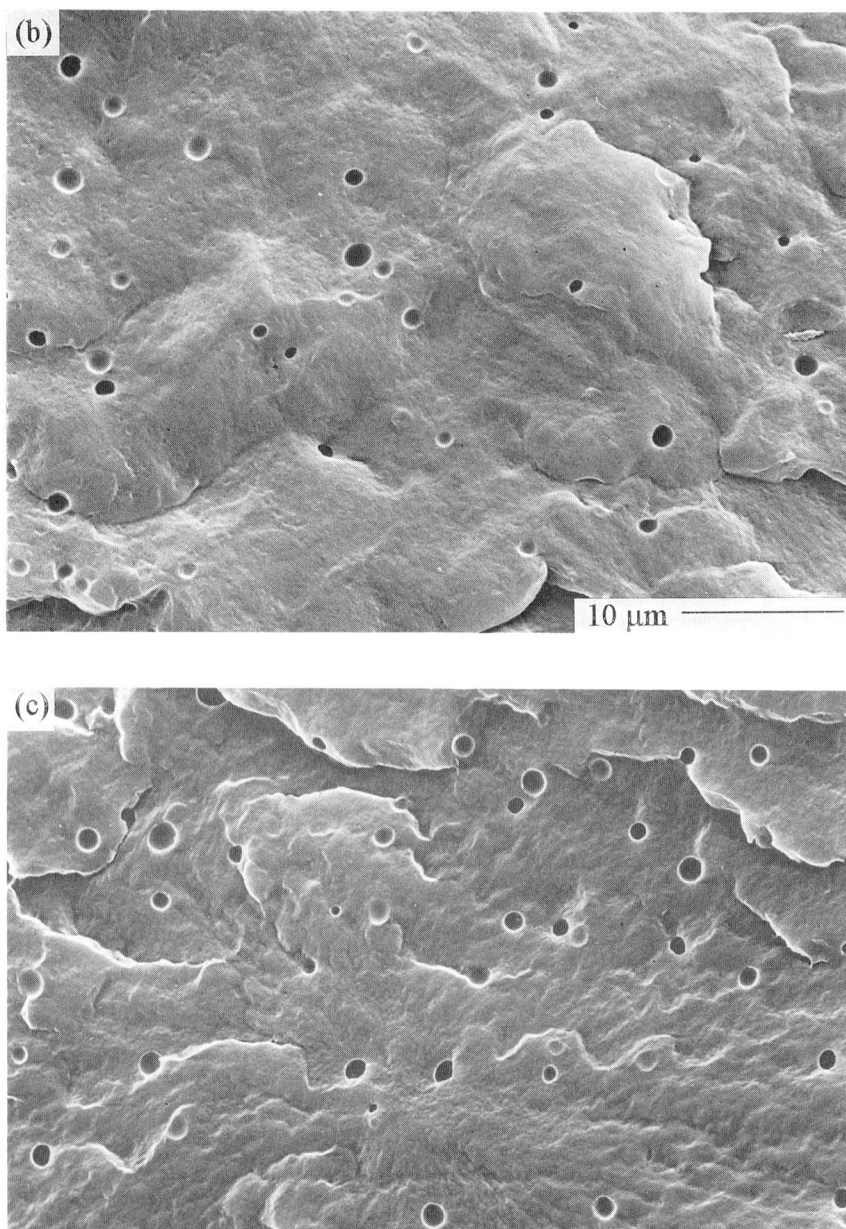

Figure 4. *Continued.*

281

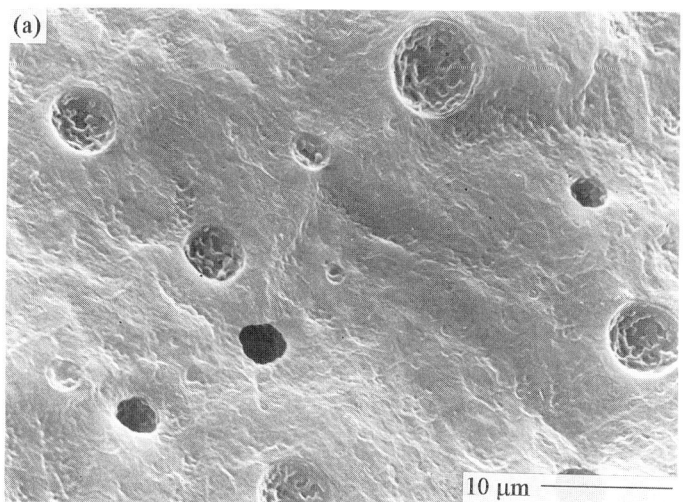

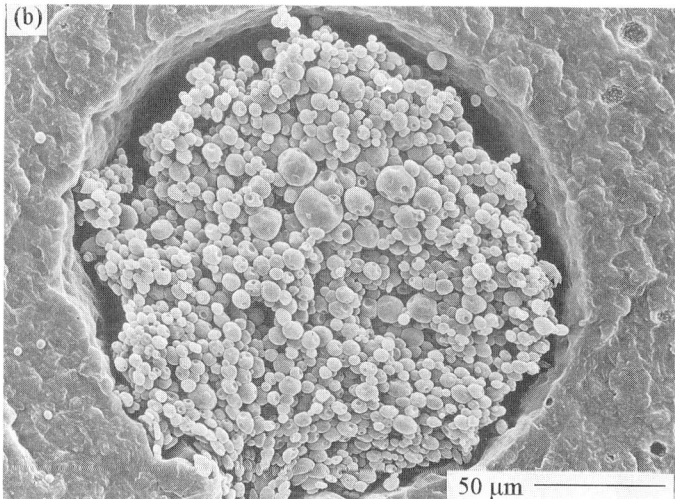

Figure 5. SEM micrographs of PPE/PA-6 blends with a PPE content of (a,b) 4.4 wt%, (c,d) 5.5 wt%. For details see text. PPE phase extracted by chloroform.

Continued on next page.

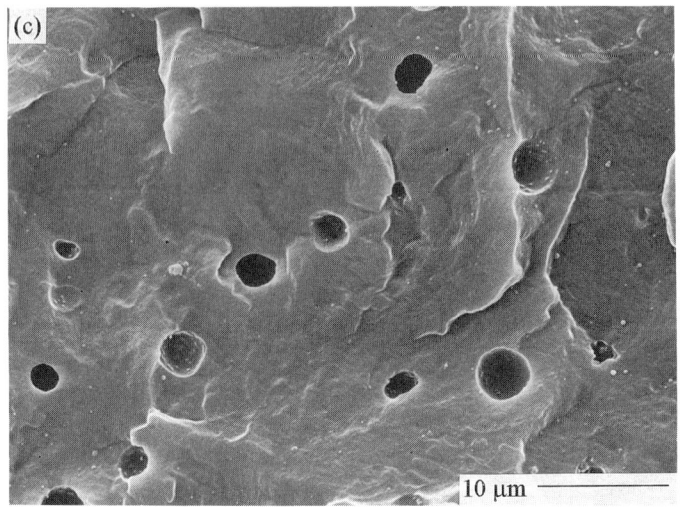

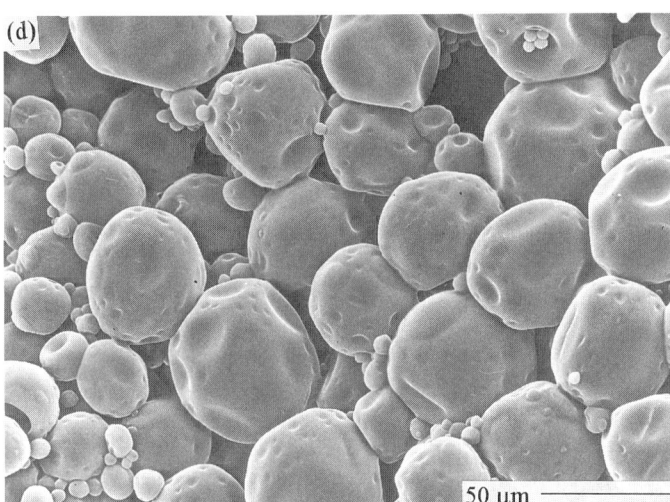

Figure 5. *Continued.*

283

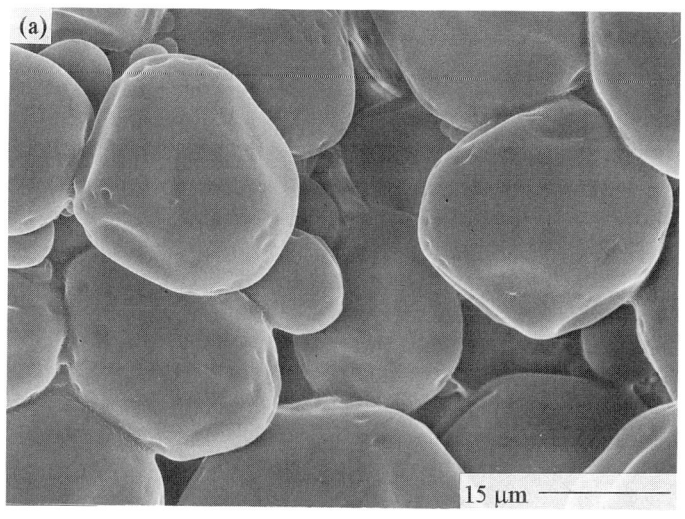

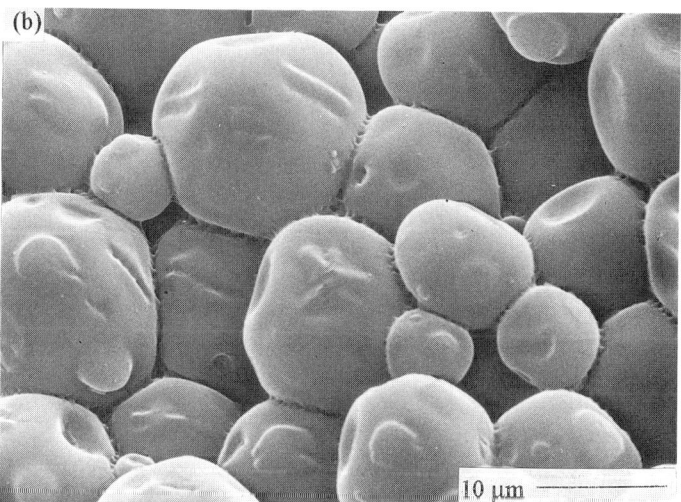

Figure 6. SEM micrographs of PPE/PA-6 blends with a PPE content of (a) 8.4 wt%, (b) 12.4 wt%, (c) 16.0 wt%, (d) 27.0 wt%, (e) 40.0 wt%. PPE phase extracted by chloroform.

Continued on next page.

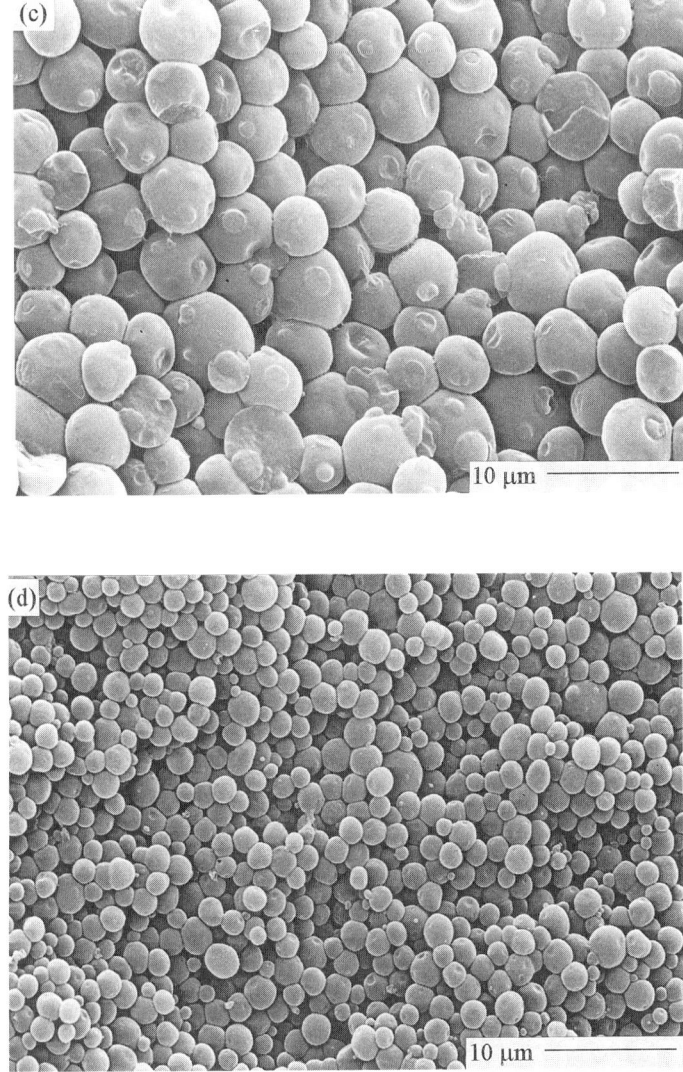

Figure 6. *Continued.*

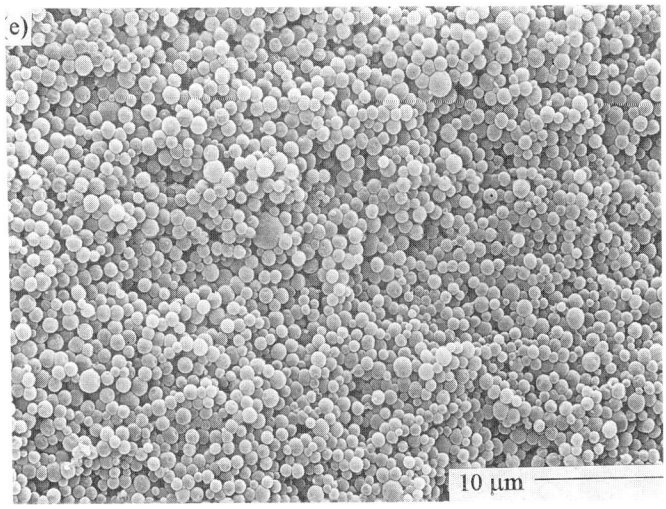

Figure 6. *Continued.*

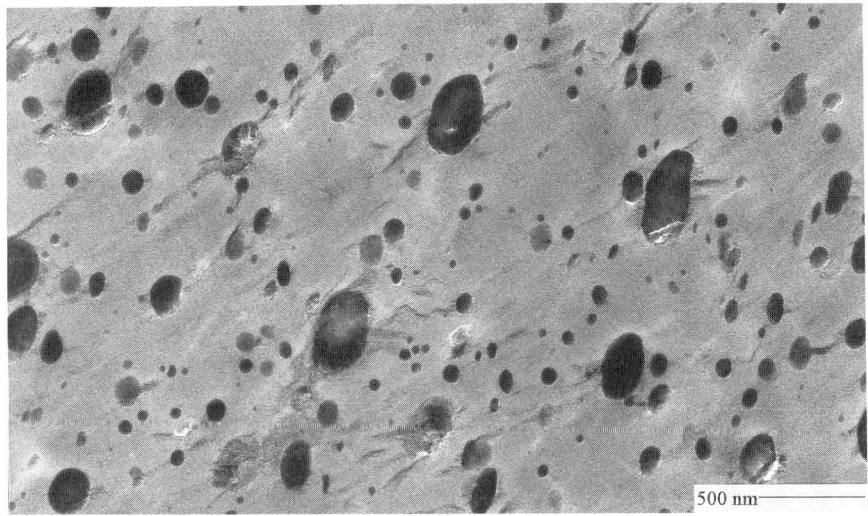

Figure 7. TEM micrograph of PPE/PA-6 blend with PPE content of 62.0 wt%.

preparation. Although polymerization of the reactive mixture PPE/ε-caprolactam/catalyst/co-catalyst was performed under static conditions (no stirring) in a glass tube, preceding homogenization of the reactive mixture was performed in the mini-extruder, while stirring. This homogenization could be accompanied in the later stage by premature polymerization resulting in partial coalescence of PA-6 particles. The effect of the high viscosity of reactive mixture on the final morphology discussed above holds also here. During the preparation of this blend, however, also a higher concentration of catalyst and co-catalyst was used (see experimental), which results in an increased polymerization rate. In spite of the chemical and physical complexity of the system, an increased polymerization rate would provide less time for coarsening of the morphology.

Monomer Conversion. Information about the conversion of ε-caprolactam in the PPE/PA-6 blends can be derived from WEC. WEC represents the weight portion of unreacted ε-caprolactam and oligomers present in the sample. The results are shown in Table I. The WEC in blends containing up to 40 wt% of PPE is rather low, indicating a high ε-caprolactam conversion. A very small influence of PPE content on WEC was seen in blends with φ < 20 wt%. Blends with a higher concentration of PPE showed slightly higher values of WEC. A possible explanation of a lower conversion could be the decreased activity of the catalyst. Catalyst NYRIM C1 is known to be sensitive to water. With respect to the longer preparation times of reactive mixtures containing higher concentration of PPE φ >20 wt%, water can be introduced into the system via air moisture.

Table I. WEC, T_g and T_m of PA-6, T_g of PPE dependent on PPE content in PA-6/PPE blends.

PPE (wt%)	WEC (%)	T_g PA-6 (°C)	T_m PA-6 (°C)	T_g PPE (°C)
0	1.9	76.9	225.9	-
1.1	1.8	76.0	225.7	-
2.2	1.9	76.3	225.0	-
4.4	2.2	75.0	225.0	-
8.4	2.9	72.0	-	225.1
12.4	2.2	71.3	-	225.1
28.0	5.1	50.0	-	225.0
40.0	4.0	49.0	-	223.0

The effect of residual ε-caprolactam and oligomers on the T_g of PA-6, on the T_g of PPE, and on the melting point of PA-6 in these PPE/PA-6 blends was characterized by means of DMTA. The tensile modulus and tan δ were measured as a function of temperature. Plots of tan δ versus temperature of PPE/PA-6 blends showed two peaks. The maximum of the first peak occurred in the temperature range 50-77°C and is attributed to the T_g of PA-6. The maximum of the second peak was observed at approximately 225°C.

With respect to this, it should be noted, that neat PPE and neat PA-6 possessed the T_g and the T_m respectively at approximately the same temperature, 225°C. However, the tan δ values at the maximum differ for neat PA-6 and neat PPE by a factor of 15. For PPE/PA-6 blends, the second peak could be a result of the overlap of two peaks corresponding to the melting of PA-6 and glass transition of PPE, respectively. Considering tan δ values at the maximum, the corresponding temperatures can be ascribed to either T_m of PA-6 or the T_g of PPE. Experimental DMTA data are summarized in Table I.

A more detailed analysis of DMTA plots yields that:
-for PPE/PA-6 blends with a PA-6 continuous phase (φ ≤ 4.4 wt%), the second peak could be ascribed to the melting of PA-6. T_m of about 225°C does not change with PPE content. No other peak which could be ascribed to the T_g of plastified PPE was observed.
-for PPE/PA-6 blends with a PPE continuous phase (φ ≥ 8.4 wt%), the second peak could be ascribed to the glass transition of PPE and this does not vary in the concentration range studied. No other peak which could be ascribed to the T_m of PA-6 was observed at lower temperatures.

Since there were no signs of plastification of PPE, or T_m depression of PA-6, it can be assumed that the residual monomer and oligomers are located in the PA-6 amorphous phase. From the literature, (54,55,33) it is known, that the depression of the T_g of PA-6 with increasing ε-caprolactam and oligomers content does not follow a typical T_g-solvent concentration relationship for the system polymer/good solvent described by Fox (46). The specific T_g depression of PA-6, with increasing ε-caprolactam and oligomers concentration, consists of only a minor influence up to 3 wt% ε-caprolactam and oligomers, followed by a drastic 26°C drop of T_g in the range 4-6 wt% (54). Our experimental data correspond well with these data (54) and confirm the assumption of the preferred location of the ε-caprolactam and oligomers in the amorphous PA-6 phase.

Conclusions

ε-caprolactam is used as a solvent for PPE. Upon cooling of the solutions, liquid-liquid phase separation can interfere with either crystallization or vitrification, depending on the cooling rate.

PPE/ε-caprolactam solutions can be processed at a temperature below the T_g of PPE (approximately 165-170 °C). Activated anionic polymerization of the solutions at these temperatures provides a route for *in situ* preparation of PPE/PA-6 blends.

PPE/PA-6 blends with a PPE matrix and a PA-6 dispersed phase or vice versa were prepared, depending on the initial PPE concentration. The morphology of the previously mentioned blends seems to be controlled by the viscosity of the reacting mixture, predetermined by the PPE concentration and can consist of submicron sized, nonadhering PA-6 particles in the PPE matrix.

The content of residual ε-caprolactam in blends in general is very low. Residual ε-caprolactam and oligomers seem to be located in the amorphous phase of PA-6.

Acknowledgments

This research is financially supported by DSM Research (Geleen, The Netherlands). The authors would like to thank Prof. H. Berghmans (K.U. Leuven, Belgium) for the stimulating discussions on the binary phase diagram.

Literature Cited

1. Garg, A.C., Mai, Y. *Composite Science Technology* **1988**, *31*, 179.
2. Levita, G. *Adv. Chem. Ser.* **1989**, *222*, 93.
3. Mulhaupt, R. *Chimia* **1990**, *44*, 43.
4. Pearson, R.A. *Adv. Chem. Ser.* **1993**, *223*, 405.
5. Hedrick, J.C., Patel, N.M., Mc Grath, J.E. *Adv. Chem. Ser.* **1993**, *223*, 293.
6. Sperling, L.H. *Interpenetrating Polymer Networks, Adv. in Chem. Ser.* **1994**, *239*, 3.
7. Kim, S.C., Klempner, D., Frisch, K.C., Ragidan, W., Frisch, H.L. *Macromolecules* **1976**, *9*, 258.
8. Devia, N., Manson, J.A., Sperling, L.H., Conde, A. *Macromolecules* **1979**, *12*, 360.
9. Gergen, W. P. *Kautschuk Gumm*, **1984**, *37*, 284.
10. Siegfield, D. L., Thomas, D. A., Sperling,, L. H. *J. Appl. Polym. Sci.* **1981**, *26*, 141.
11. Hourston, D. J., Zia, Y. *J. Appl. Polym. Sci*, **1983**, *28*, 2139.
12. Hourston, D. J., Zia, Y. *J. Appl. Polym. Sci*, **1984**, *29*, 629.
13. Meijer, H. E. H., Venderbosch, R. W., Goossens, J. G. P., Lemstra, P. J. *High Perform. Polym.* **1996**, *8*, 133.
14. Venderbosch, R.W., Meijer, H.E.H., Lemstra, P.J. *Polymer* **1994**, *35*, 4349.
15. Venderbosch, R.W. *Processing of Intractable Polymers using Reactive Solvents*, PhD. Thesis, TUE: Eindhoven, The Netherlands, 1995; 7-34.
16. Venderbosch, R.W., Meijer, H.E.H., Lemstra, P.J. *Polymer* **1995a**, *36*, 1167.
17. Venderbosch, R.W., Meijer, H.E.H., Lemstra, P.J. *Polymer* **1995b**, *36*, 2903.
18. Goossens, J. G. P., Rastogi, S., Meijer, H. E. H. *Polymer*, submitted.
19. Ohnaga, T., Sato, T., Kojima, T., Inoue, T. *Polym. Prepr. Jpn.* **1992**, *41*, 3825.
20. Bongers, J., Mooij, H. In *Nylon Plastics Handbook*, Kohan, M.I., Hanser/Gardner Publication 1995, 542.
21. Meyer, R.V., Fahnler, F., Dhein, R., Michael, D. *EP 167.907*, 1986.
22. Mengason, J. In *Nylon Plastics*, Kohan, M.I. Ed., Wiley-Interscience New York, 1973, 580.
23. Hendrick, R.M., Gabbert, J.D. *A New RIM System from Nylon 6 Block Copolymers: Chemistry and Structure*, Presented at the AIChE National Summer Meeting, Detroit MI, August 17, 1981.
24. Gabbert, J.D., Hendrick, R.M. *Properties of RIM Nylon 6 Block Copolymers*, Presented at the AIChE National Summer Meeting, Detroit MI, August 17, 1981.
25. Mooij, H. *Composites* **1992**, *3*, 384.

26. Bongers, J., Mooij, H. In *Nylon Plastics Handbook*, Kohan, M.I., Hanser/Gardner Publication 1995, 545.
27. Kye, H., White, J.L. *J. Appl. Polym. Sci.* **1994**, *52*, 1249.
28. Hornsby, P. R., Tung, J. F., Tarverdi, K. *J. Appl. Polym. Sci.* **1994**, *53*, 891.
29. Menges, G., Bartilla, T. *Polym. Eng. Sci.* **1987**, *27*, 1216.
30. Berghhaus, U., Michaeli, W. *Soc. Plast. Eng. ANTEC '90 Proc.*, 1990, 1929.
31. Wichterle, O., Sebenda, J., Kralicek, J. *U.S. Patent* (filed December 15, 1959) 1965, 3,200,095.
32. Illing, G., Zahradnik , F. *U.S. Patent* (filed July 21, 1964) 1968, 3,371,055.
33. Kolarik, J., Janacek, J. *J. Polym. Sci. Part. C.* **1967**, *16*, 441.
34. Van Buskirk, B., Akkapeddi, M. K. *Polym. Prep.* **1988**, *29*, 557.
35. DSM analytical method DSM 0522-E
36. DSM analytical method CRO 0256- 01-E
37. Inoue, T. *Prog. Polym. Sci.* **1995**, *20*, 119.
38. Nachlis, W. L., Kambour, R. P., MacKnight, W. J. *Polymer* **1994**, *35*, 3643.
39. Karasz, F. E., O'Reilly, J. M. *J. Poly. Sci. Part B, Polym. Lett.* **1965**, *3*, 561.
40. van Emmerik, P. T., Smolders, C. A. *J. Poly. Sci. Part C* **1972**, *38*, 73.
41. Koehen, D. M., Smolders, C. A. *J. Polym. Sci., Polym. Phys. Edn.* **1977**, *15*, 167.
42. Wenig, W., Hammel, R. M., MacKnight, W.J., Karasz, F. E. *Macromolecules* **1976**, *9*, 23.
43. Aerts, L., Berghmans, H. *Bull. Soc. Chim. Belg.* **1990**, *99*, 931.
44. Flory, P.J. *Principles of Polymer Chemistry*, Cornell University Press, Ithaca, 1956.
45. Stoks, W., Berghmans, H. *J. Pol. Sci.: Part B: Polym. Physics* **1991**, *29*, 609.
46. Fox, T.G. *Bull. Am. Phys. Soc.* **1956**, *2*, 123.
47. Arnauts J., Berghmans H., Koningsveld, R. *Makrom. Chemie* **1993**, *94*, 77.
48. Berghmans, S., Mewis, J., Berghmans, H., Meijer, H. E. H. *Polymer* **1995**, *36*, 3085.
49. Bakhuis Roozeboom, H. W. In *Die heterogene Gleichgerichte vom Standpunkte der Phasenlehre*, Friedrich Vieway und Sohn, Braunschweig, 1918, No 2, Part 2, 129.
50. Burghardt, W. R. *Macromolecules* **1989**, *22*, 2482.
51. Vandeweerdt, P., Berghmans, H., Tervoort, Y. *Macromolecules* **1991**, *24*, 3547.
52. Mark, H.F., Bikales, N.M., Overberger, C.G., Menges, G. In *Encyclopedia of Polymer Science and Engineering*, 2nd ed., 1988, Vol. 11, 338.
53. Bongers, J., Mooij, H. In *Nylon Plastics Handbook*, Kohan, M.I., Hanser/Gardner Publication 1995, 25.
54. Sauer, J.A., Lim, T. *J. Macromol. Sci. Phys.* **1977**, *B13*, 419.
55. Carbuglio, C., Ajroldi, G., Casiraghi, T., Vittadini, G. *J. Appl. Polym. Sci.* **1971**, *15*, 2487.
56. Sebenda, J. *J. Macromol. Sci.-Chem.* **1972**, *A6 (6)*, 1145.
57. Reimschuessel, H.K. In *Ring Opening Polymerization*, Chap.7, Frisch, K.C., Reegen, S.L., Eds., Dekker, M., New York, 1969, 313-323.
58. Sebenda, J. In *Comprehensive Polymer Science*, Allen, S.G., Eds., Pergamon Press, New York, 1989, 3, 511.

59. Sekiguchi, H. In *Ring-Opening Polymerization*, Chap. 12, Ivin, K., Saegusa, T., Eds., Elsevier, London, 1984, 2, 809.
60. Sebenda, J. In *Lactam-Based Polyamides*, Puffr, R., Kubanek, V., Eds., CRC, Boca Raton, 1991, 1, 42.
61. van Geenen, A.A. DSM Research/SPPS Research Communications 1990, 1, 14.

Chapter 21

Nitroaryl and Aminoaryl End-Functionalized Polymethylmethacrylate via Living Anionic Polymerization: Synthesis and Stability

R. J. Southward[1], C. I. Lindsay[2], Y. Didier[2], P. T. McGrail[2], and D. J. Hourston[1]

[1]IPTME, University of Loughborough, Leics. LE11 3TU, England
[2]ICI Plc, Wilton, Middlesbrough, Cleveland TS90 8JE, England

Polymethylmethacrylate incorporating terminal nitroaryl and aminoaryl functionality has been synthesised via living anionic polymerisation. It has been shown that nitroaryl-ended polymers of precisely controlled molecular weight and narrow polydispersity can be synthesised with about 80% end-functionality. Reductive treatment of these polymers yields aminoaryl-ended polymers with greatly reduced functionality regardless of the conditions employed. The nitroaryl end groups also exhibit a degree of thermal instability. The combined use of Nuclear Magnetic Resonance Spectroscopy and Matrix Assisted Laser Desorption Mass Spectrometry has been shown to be a powerful tool for the characterisation of polymer end groups.

End-functionalisation of polymethylmethacrylate (PMMA) with reactive groups is of interest because it offers the possibility of covalently linking PMMA with property-enhancing polymers and additives with which it cannot at present chemically bond. The resultant materials may be of value because of their intrinsically novel properties or may function as compatibilising agents allowing access to new types of polymer blend (e.g. PMMA/condensation polymers).
Living anionic polymerisation offers a potential route to the synthesis of well-defined end-functionalised polymers of controlled molecular weight and narrow polydispersity. Indeed syntheses of a variety of end-functionalised polystyrenes and polybutadienes by this route have been increasingly reported in the literature in recent years[1-6].
Reported herein is the synthesis of primary amino-ended PMMA via end capping of the living polymer anion. The incorporation of this protonic end-group offers specific challenges in terms of the protection chemistry required to afford the desired end-functionality in high yield. This is made more complex by the

synthetic requirements of a sterically hindered initiator, low temperature (-78° C) and ultra-pure reagents for living anionic polymerisation of methyl methacrylate in combination with a moisture-free, inert atmosphere to avoid termination reactions involving the ester group in the monomer[7-10].

Experimental.

Materials. Tetrahydrofuran (THF) was dried by refluxing over sodium/benzophenone prior to use. Methyl methacrylate was dried and purified by refluxing over calcium hydride, passing through alumina and molecular sieve columns and distilling from triethylaluminium immediately prior to use. Lithium chloride was dried at 150°C in vacuo overnight. 1,1-Diphenylethylene was dried over molecular sieves and vacuum distilled prior to use. All other reagents were used as supplied.

Synthesis of Nitroaryl-ended PMMA. A solution of n-butyllithium in hexanes (1.6M) was added dropwise to a solution of 1,1-diphenylethylene/lithium chloride (1/10 molar ratio) in THF (10cm^3) until a deep red colour just persisted. A further stoichiometric amount of butyllithium solution was added and the deep red solution stirred for 1 hour to form 1,1-diphenylhexyllithium. The solution was diluted with THF (250cm^3) at -78°C after which methyl methacrylate (10cm^3, 93.5mmol) was added dropwise over the course of ca. 30 minutes to give a colourless solution which was stirred for a further hour. A five-fold excess 4-nitrobenzylbromide or 4-nitrobenzoylchloride was added and the solution stirred overnight during which time the temperature gradually rose to ca. 0°C. The volume of solvent was reduced in vacuo and the resultant viscous solution added to a ten-fold excess of stirred acidified (5% HCl) methanol. The nitroaryl-functionalised PMMA was obtained as a white powder in >95% yield after drying. Syntheses of nitrobenzyl- and nitro-benzoyl ended poly(methyl methacrylate)s of a variety of molecular weights were achieved by varying the molar ratio of methyl methacrylate:1,1-diphenylhexyllitium.

Reduction of Nitroaryl-ended PMMA. Nitroaryl-ended PMMA was dissolved in the solvent of choice incorporating dispersed catalyst (tin or palladium/charcoal) under nitrogen at room temperature. The reducing agent was added dropwise and any changes in the temperature of the reaction mixture were closely monitored. The reaction mixture was stirred, with heating if necessary, and the course of the reaction followed by thin layer chromatography and ^{1}H nmr spectroscopy. Table II summarises the products obtained under different reduction conditions.

Product Characterisation. The nature and level of PMMA end functionality was determined in each case by ^{1}H nmr spectroscopy (400MHz, Cl2CDCDCl2, 50°C). Gel Permeation Chromatography (GPC) was used in conjunction with ^{1}H nmr to determine molecular weight parameters. The identities of selected polymers were confirmed by ^{13}C nmr (300MHz, CDCl$_3$, 23.7°C) and/or Matrix Assisted Laser Desorption/Ionisation (MALDI) mass spectrometry.

Results & Discussion.

Synthesis of Nitroaryl-Ended PMMA. The strategy for the synthesis of nitroaryl-ended PMMA is illustrated in Figure 1. In this approach living anionic PMMA is generated by initiating the polymerisation of methyl methacrylate in THF with an appropriate amount of 1,1-diphenylhexyllithium. This technique has been shown to yield living polymer with narrow polydispersity and high initiator efficiency providing the following conditions are met[7-11]:

1. The methyl methacrylate monomer is distilled from calcium hydride and triethylalumimium in order to remove trace moisture and impurities, and all other reagents are rigorously dried and purified.

2. The temperature is maintained below -60°C to avoid self termination of the PMMA anion via a cyclisation reaction involving the three terminal monomer units[11].

3. A 10/1 lithium chloride/intiator ratio is used in order to protect the living polymer chain end via reversible complexation

4. High dilution conditions are employed.

The living PMMA anion produced under the above conditions was quenched with either 4-nitrobenzoylchloride or 4-nitrobenzylbromide at -78°C to give nitroaryl-ended PMMA. Nitroaryl-ended PMMAs of a variety of molecular weights were synthesised via this route with ca. 80% end-functionality, as shown in Table I. The molecular weights of the polymers were readily controlled by appropriate choice of the methyl methacrylate/initiator molar ratio.

Table I. Molecular Weight & Functionality Data for Nitroaryl-ended PMMA.

Polymer	$Mn/10^3$ (calc)	$Mn/10^3$ (^{1}H nmr)	$Mn/10^3$ (GPC)	$Mw/10^3$ (GPC)	Mw/Mn (GPC)	X	% NO_2 end groups
(1)	2	1.96	3.07	3.29	1.07	CH_2	82
(2)	5	5.26	6.46	6.8	1.05	CH_2	79
(3)	5.2	5.65	5.07	5.53	1.09	CO	84
(4)	10	8.6	9.9	11.2	1.13	CO	82
(5)	20	20	17	21.8	1.28	CO	80

Characterisation of Nitroaryl-Ended PMMA. The degree of end-functionality was determined via ^{1}H nmr spectroscopy by comparison of the integral of the aryl protons associated with the end group with that of the aryl protons of the initiator group (Figures 2 & 3). Molecular weight data was obtained by Gel Permeation Chromatography and by comparison of the ^{1}H nmr integrals of the aromatic initiator protons with those of the methyl ester protons in the PMMA chain. The excellent agreement between target and achieved molecular weights in conjunction with the narrow polydispersities illustrated in Table I demonstrates the living nature of the polymerisation and high initiator efficiency.

MALDI mass spectrometry was performed on selected samples to reinforce the characterisation of the nitroaryl-ended PMMA. Over recent years this technique has been increasingly used to characterise molecular weights and end groups of synthetic polymers as a consequence of its ability to characterise very precisely the masses of molecular polymer ions[12]. Resolution of the technique is critically dependent upon polymer molecular weight and polydispersity. In the work described here it was found that resolution of ca. 4 mass units was attainable for the near-monodisperse nitroaryl-ended polymers of molecular weights ca. 5,000.

MALDI mass spectra of nitrobenzyl- and nitrobenzoyl-ended PMMA are depicted in Figures 4 & 5 respectively. In each case a Gaussian envelop of major peaks corresponding to the sodium-coordinated molecular ions $[M+Na]^+$ is apparent with the peak separation corresponding to one monomer unit. Comparison of the experimental and calculated molecular masses $[M+Na]^+$ for each peak (taking into account the isotopic abundances of the elements) confirms the identities of the polymers as nitrobenzoyl- and nitrobenzyl-ended PMMA in the respective cases. In addition to the major peaks in Figures 4 & 5 three distinct series of minor peaks are evident for both polymers, corresponding to the end-cyclised and end-protonated polymers A, B, F & G. These observations are consistent with the ^{1}H nmr data for the polymers which indicate that about 20% of the polymer chains do not have nitroaryl end groups.

The chemoselectivity of the termination reaction was determined by ^{13}C nmr spectroscopy. The living PMMA anion exists in a resonance stabilised form (Figure 6) which can in principle be trapped by the nitroarylhalide terminating agent at the carbanion to yield the desired adduct **(I)** (Figure 1) or at the enolate oxyanion to yield polymer **(II)** (Figure 7). The two structures were not distinguishable by ^{1}H nmr spectroscopy or MALDI mass spectrometry. However, analysis of the carbonyl resonances in the ^{13}C nmr spectrum of nitrobenzoyl-ended PMMA indicated the presence of a distinctive ketone resonance (195ppm) which could only be accounted for via the end group in structure **(I)** (the carbonyl resonance arising from the nitrobenzoyl end group in structure **(II)** is in an enolate ester environment). Comparison of calculated and measured ^{13}C nmr chemical shifts associated with the end groups in the nitroaryl-terminated polymers confirmed the structure **(I)**.

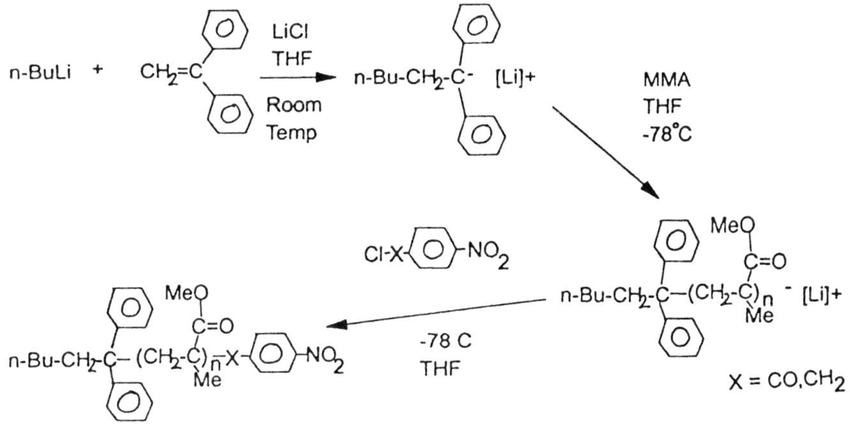

295

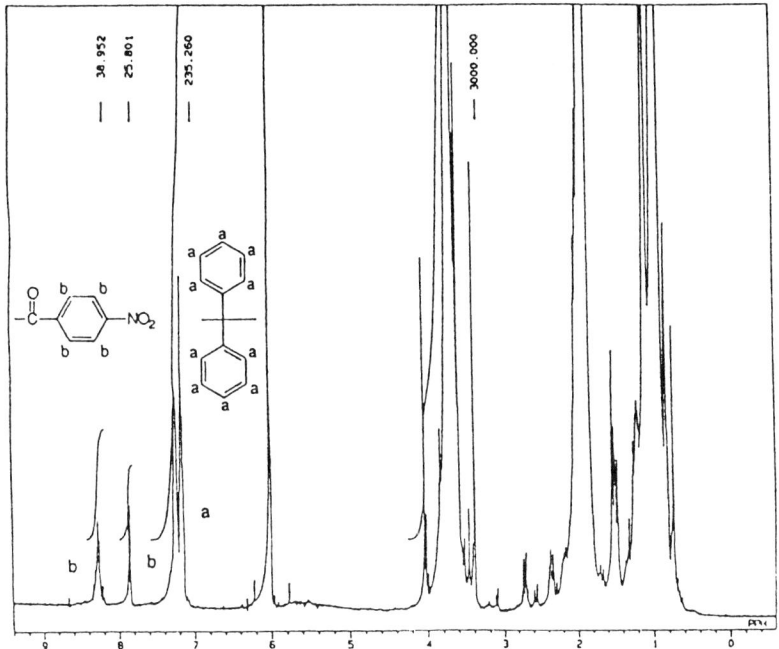

Figure 1. Synthesis of Nitroaryl-Ended PMMA.

Figure 2. ^{1}H nmr Spectrum of Nitrobenzoyl-Ended PMMA

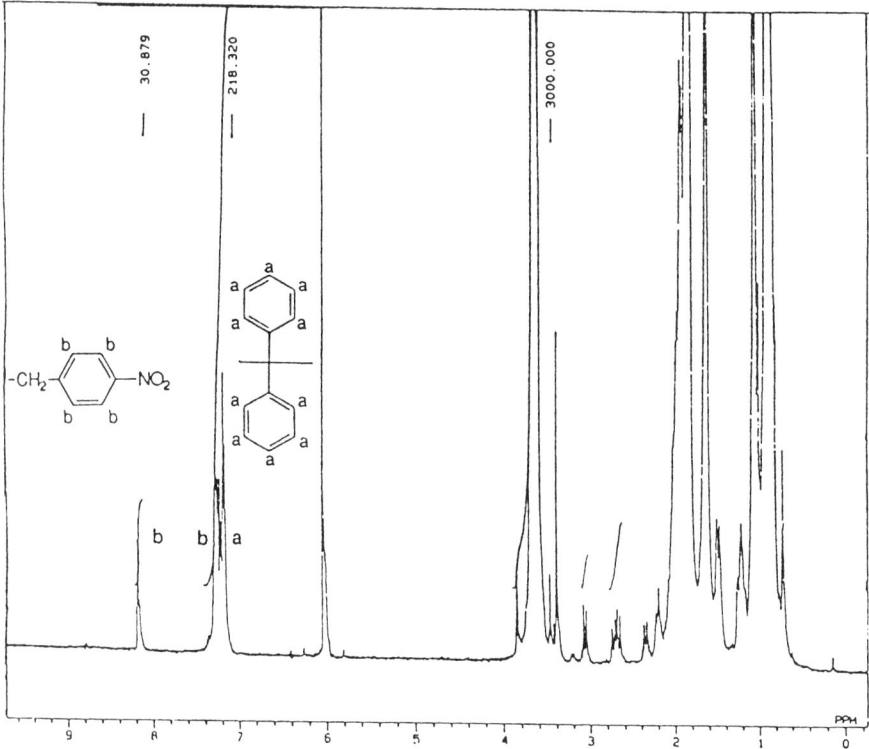

Figure 3. ^{1}H nmr Spectrum of Nitrobenzyl-Ended PMMA

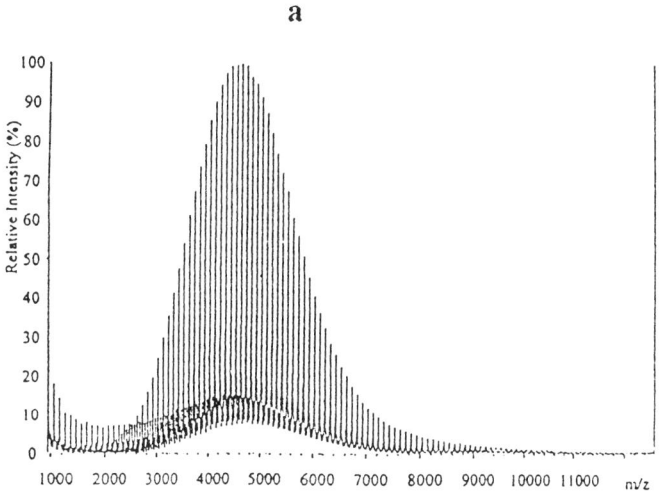

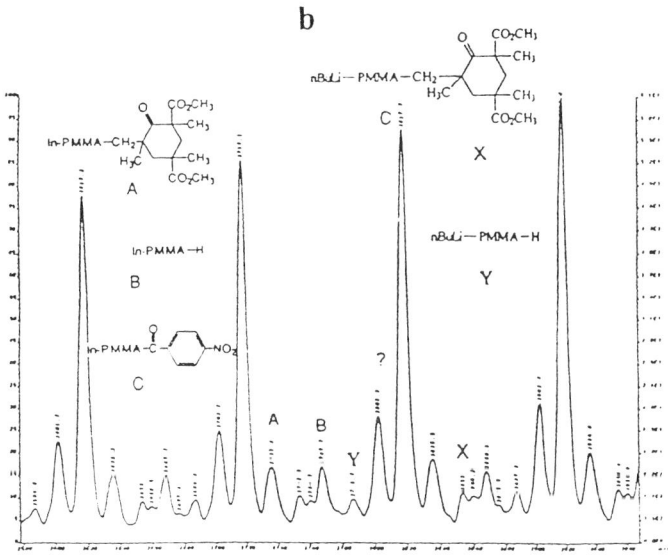

Figure 4. MALDI Mass Spectrum of Nitrobenzoyl-Ended PMMA

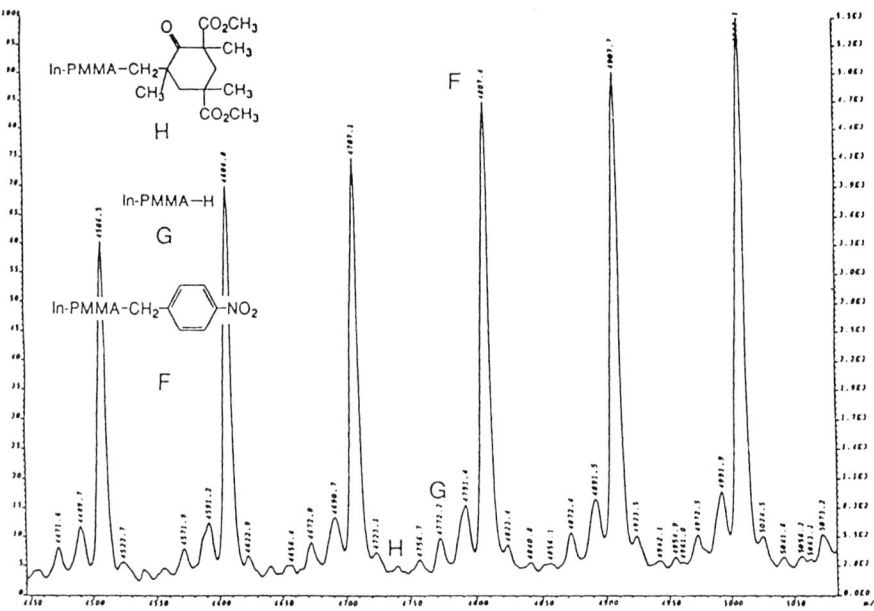

Figure 5. MALDI Mass Spectrum of Nitrobenzyl-Ended PMMA

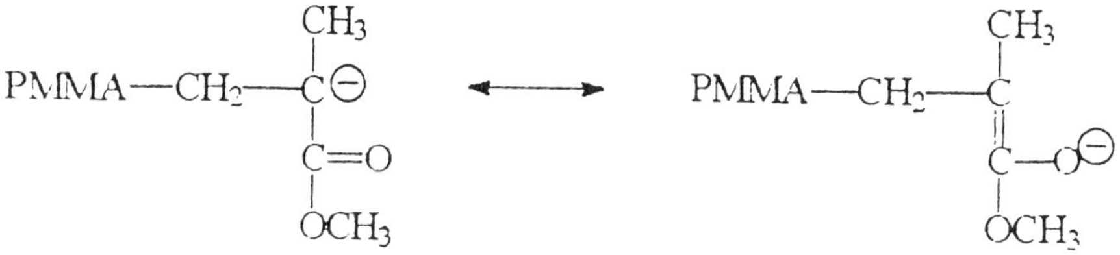

Figure 6. Resonance Stabilisation of PMMA Anion

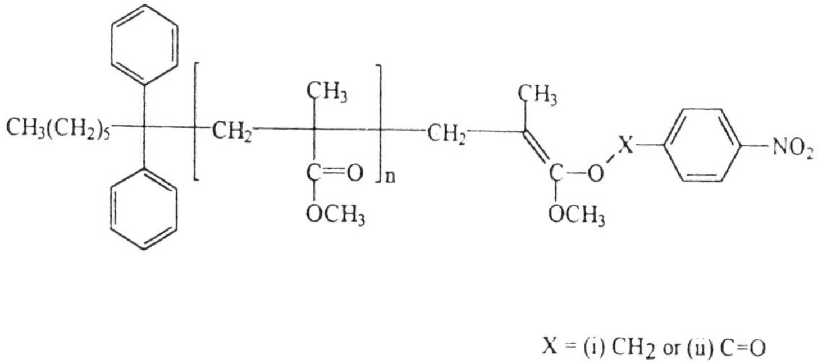

X = (i) CH_2 or (ii) C=O

Figure 7. Alternative End-Structure of Nitroaryl-Ended PMMA
(not observed as a product)

Thermal Stability of Nitroaryl-Ended PMMA. Some thermal instability of the nitroaryl-ended polymers was observed. This was significantly more pronounced for the nitrobenzoyl ended polymer, in which a loss of 30% functionality was observed upon heating at 67° C (refluxing THF) for 14 hours, than for the nitrobenzyl-ended analogue in which only 10% functionality was lost after similar treatment for 2 days. It is important to note that no loss of the initiator-derived diphenylhexyl moiety or change in molecular weight was observed under thermal treatment.

Synthesis of Primary Aminoaryl-Ended PMMA. Reduction of nitroaryl-terminated PMMA to yield primary aminoaryl-ended PMMA (Figure 8) was attempted using a variety of conditions (Table II). Each set of conditions was shown to be suitable for the specific reduction of the nitro-group in the presence of carbonyl functionality via model reactions in which 4-nitrobenzophenone was quantitatively reduced to 4-aminobenzophenone.

Table II. Conditions & Product Functionalities for Reductions of Nitroaryl-Ended PMMA.

Parent Polymer	Reducing Agent	Catalyst	Solvent	Temp (°C)	Reaction Time (Hours)	%NO_2 end groups
(2)	H_3PO_2(aq)	Pd/C	THF	23	16	0
					31	0
(3)	H_3PO_3(aq)	Pd/C	CH_2Cl_2	23	18	0
					31	0
(2)	H_3PO_3(aq)	Pd/C	THF	23	24	0
(4)	c-C_6H_{12}	Pd/C	$HCONMe_2$	130	2	0
					4	0
					6	0
(5)	HCl(aq)	Sn	THF	23	1	46
					5	0
					21	0

It is clear from the data in Table II that all attempts to reduce nitroaryl-ended PMMA to its primary amino-terminated analogue resulted in loss of end functionality (quantified by ^{1}H nmr spectroscopy) regardless of temperature, solvent, reducing agent or catalyst. Treatment of polymer (5) with a tin/hydrochloric acid mixture for 5 hours appeared to yield primary amino-ended PMMA with no loss of functionality but attempts to reproduce this result proved unsuccessful. ^{1}H nmr spectroscopy demonstrated that in all cases the loss of functionality occurred as a consequence of the removal of the functionalised aromatic end groups and that no change in molecular weight was associated with this process.

Despite the relatively low level of aminoaryl end functionality in the reduced polymers evident from the ^{1}H nmr spectra (Figure 9), the aminoaryl-ended PMMA appeared as the major series of peaks in the MALDI mass spectra (Figure 10) with only minor peaks apparent for other species. This paradox highlights the qualitative nature of MALDI mass spectrometry which arises from its varying sensitivity to different species.

The instability of the nitroaryl- and aminoaryl- end groups under thermal and reductive treatment is somewhat surprising as their removal requires carbon-carbon sigma bond cleavage. A more detailed study of the nature of the end-group in terminally functionalised PMMA and its regiochemistry is now required to determine the factors which affect the thermal and chemical stability of such species.

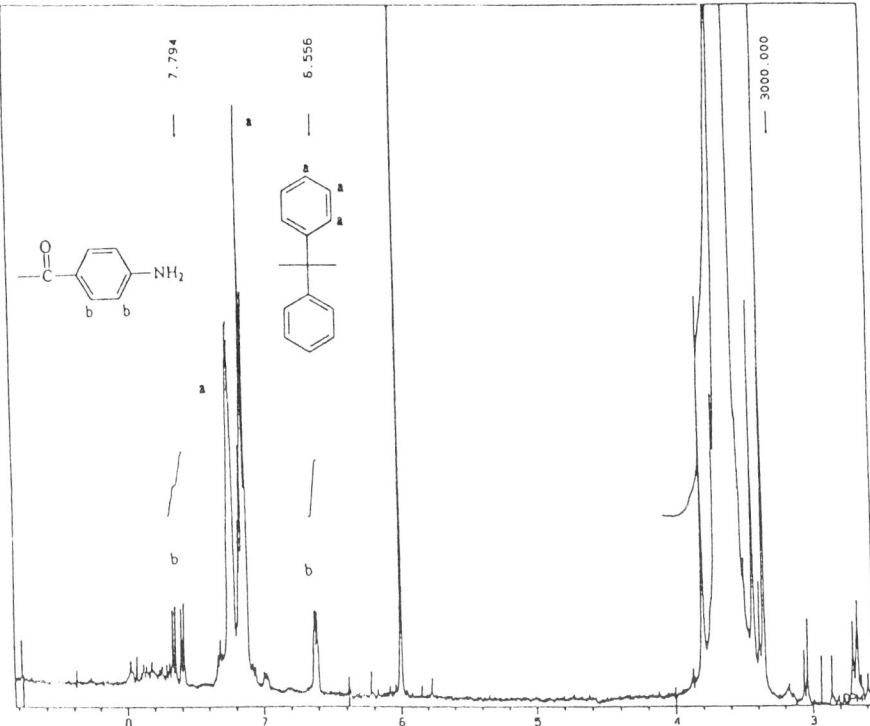

Figure 8. Reduction of Nitroaryl-Ended PMMA.

Figure 9. ^{1}H nmr Spectrum of Aminobenzoyl-Ended PMMA

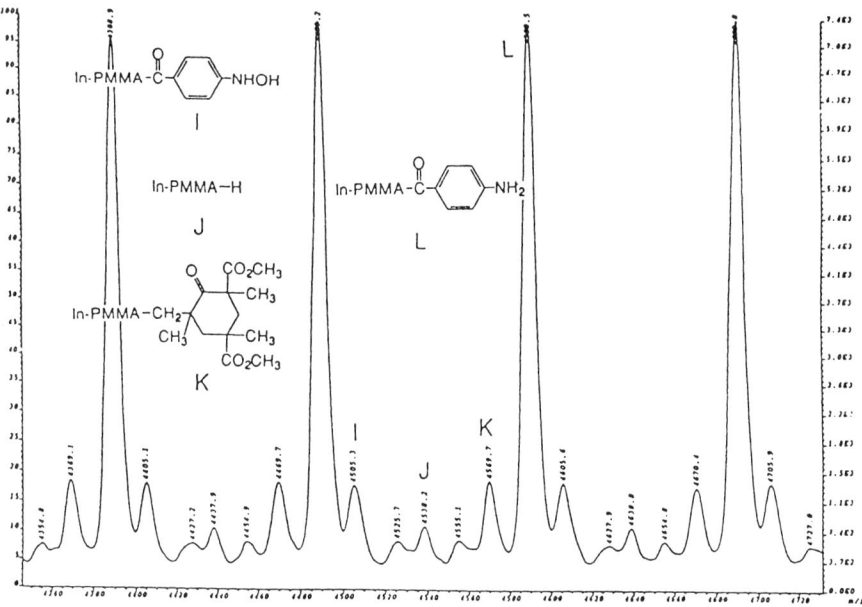

Figure 10. MALDI Mass Spectrum of Aminobenzoyl-Ended PMMA

Conclusions.

Addition polymers incorporating unusual end groups are attracting increasing interest in view of their potential chemical versatility. The studies described herein illustrate the importance of achieving a fundamental understanding of the factors affecting the synthesis, characterisation and properties of such polymers.

Acknowledgements.
We are grateful to the following people for their contributions to the characterisation of the polymers described in this paper: J. Heron, A. Bunn, H. T. Yates, A. T. Jackson at ICI Wilton.

Literature Cited.

1. Quirk, R. P.; & Cheng P-L. *Macromolecules*, **1986**, *19*, 1291
2. Quirk, R. P.; Zhu, L-F. *Makromol. Chem.*, **1989**, *190*, 487
3. Quirk, R. P.; Yin, J; Guo, S-H; Hu, X-W; Summers, G; Kim, J; Zhu, L-F.; Schock, L. E; *Makromol. Chem. Macromol. Symp.*, **1990**, *32*, 47
4. Bresentsky, D. M.; Hove, T. R.; Macosko, C. W.
   J. Polym. Sci. Part A: Polymer Chemistry, **1995**, *33*, 1957
5. Takenka, K.; Shiomi, T.; Hiraishi, Y.; Hirao, A.; Nakahama, S.
   ACS Polym. Prepr., **1996**, *37*, 658
6. Morita, K.; Nakayama, A.; Ozawa, Y.; Oshima, N.; Fujio, R.; Fujimaki, T.; Lawson, D. F.; Scheffler, J. R.; Antkowiak, T. A.
   ACS Polym. Prepr., **1996**, *37*, 700
7. Allen, R. D.; Long, T. E.; McGrath, J. E. *Polymer Bulletin*, **1986**, *15*, 127
8. Teyssie, Ph.; Fayt, R.; Hautekeer, J. P.; Jacobs, C.; Jerome, R.; Leemans, L.;

   Varshney, S. K. *Makromol. Chem. Macromol. Symp.*, **1990**, *32*, 61
9. D. Hunkel, D.; Muller, A. H. E.; Janata, M.; Lochmann, L.
   ACS Polym. Prepr., **1991**, *32*, 301.
10. Varshney, S. K.; Hautekeer, J. P.; Fayt, R.; Jerome, R.; Teyssie, Ph.
    Macromolecules, **1990**, *23*, 2618
11. Webster, O. W.; Sogah, D. Y.; Farnham, W. H.; Hartler, W. R.; Rajanbabu, T. V.
    J. Am. Chem. Soc., **1983**, *105*, 5706
12. Hillenkamp, F, In *Advances in Mass Spectrometry*, Cornides, I; Horvath, Gy.; Vekey, K., Eds.; John Wiley & Sons, Chichester, England, 1995, p95.

Chapter 22

Anionic Copolymerization of Dimethyl- and Diphenylcyclosiloxanes: Changes of Microstructure in the Course of Copolymerization

B. G. Zavin, A. Yu. Rabkina, I. A. Ronova, G. F. Sablina, and T. V. Strelkova

Institute of Organoelement Compounds RAS, Moscow, 117813, Russia

The changes in the macromolecules composition and their microstructures in the course of the equilibrium anionic ring opening copolymerization of unstrained cyclic organosilicon monomers- cyclotetrasiloxanes $(Me_2SiO)_4$ and $(Ph_2SiO)_4$ have been studied. It is found, that the process is accompanied by intensive chain transfer reactions from the very beginning. That manifests itself in (i) continuous interconversion of the active centers occuring from the each of co-monomers, (ii) appearance of the mixed cyclosiloxanes and (iii) changes in microstructure of the copolymer chains formed. For these reasons common methods for determination of relative activity of the comonomers, for example Mayo-Lewis relations, are unapplicable in this case. The evaluation of the microstructure of copolymers resulted has been fulfilled by the computing tecnicque, based on a comparison of intensities of the corresponding triads at copolymer chains in the NMR ^{29}Si- spectra. It is found that copolymers of identical composition obtained in various conditions differs by microstructure of macromolecules and show a different tendency to crystallization..

Polymethylphenylsiloxanes containing Ph_2SiO- or $MePhSiO$- units along with Me_2SiO- in macromolecular chains are of great interest from the point of view of polymer materials with improved properties (e.g. heat resistance, reduced tendency to crystallization, stability to UV radiation etc.) *(1, 2)*. The most convenient method of synthesis of linear siloxane polymers is ionic ring opening polymerization (ROP) of cyclosiloxanes *(3 - 5)*. Unfortunately, high sensitivity of the Si-Ph bonds to electrofilic reagents make it impossible to use cationic initiators. That is why anionic ROP is practically the only possible method for preparation of high molecular weight polymethylphenylsiloxanes *(6-9)*. Owing to high availability of unstrained cyclotetrasiloxanes they are most advantageous objects for preparation of copolymers at commercial scales. Processes of ROP of unstrained cyclosiloxanes (as compared both other heterocyclic or vinyl monomers) are known to be entropy controlled *(5, 10, 11)*. This determines its main peculiarities, in the first of all its equilibrium character and occurence of inter- and intramolecular chain transfer reactions (CTR) *(5, 11, 12)*. These reactions lead to redistribution of monomer SiO-units along of siloxane backbone, resulting both in appearance of a set of cyclic oligomers and in changes of MWD of macromolecules formed *(12)*. In the case of copolymerization both the changes of polymer composition and microstructure of growing chains are possible.

In this paper we report the results of the investigation of anionic ring opening copolymerization of cyclic organosilicon monomers: octamethylcyclotetrasiloxane [Me_2SiO]$_4$ (A_4) and octaphenylcyclotetrasiloxane [Ph_2SiO]$_4$ (B_4). In the case of copolymerization of two or more cyclosiloxanes the rates, yields and also composition of copolymers formed depend on a nature of cyclic monomers used. Furthermore, the physical and mechanical properties of polyorganosiloxane copolymers formed are influenced not only by their composition, but by microstructure of copolymer chains *(8, 9, 13)*. Reaction conditions may affect in a large extent on the mechanical properties and performance of the siloxane elastomers formed. So, to control properties of the copolymers obtained it is necessary to elucidate effects of reaction conditions upon composition – structure – properties relationships in these materials.

For these reasons the goals of the investigation are study of the changes in composition and microstructure of linear polydimethyl(diphenyl)siloxanes during anionic polymerization of the unstrained cyclotetrasiloxanes (A_4 and B_4) and estimation of the influence of copolymer composition (contents of A- and B-units) and of the copolymer microstructure (i.e. distribution of the units along chain) upon its physical properties.

We have studied the changes composition and microstructure of methylphenylsiloxane copolymers during anionic copolymerization of unstrained cyclotetrasiloxanes A_4 and B_4 in a wide range of monomers ratio.

Experimental section.

Materials. Commercial octamethylcyclotetrasiloxane (A_4), octaphenylcyclotetrasiloxane (B_4) and diphenylsilanediol (DP) were purified before use. A_4 was redistillated to give >99,8 purity, B_4 and DP were recrystallized from $CHCl_3$ -benzene and acetone, correspondingly.

Copolymerization was carried out both in bulk and in solution.

A. Bulk copolymerization. Weighed amounts of A_4 and B_4 were charged at a three neck vessel fitted with mechanical stirrer, thermometer, argon inlet and reflux condenser. The mixture was heated to 160 0C. Polymerization was initiated by [$KOSiPh_2$]$_2$O (0,2 w.% on the basis of K). Heterogeneous reaction mixture became homogenous after 1 - 5 hours heating, depending on the current concentration of B_4 at reaction system.

B. Solution polymerization. Weighed amounts of A_4, DP, toluene and KOH were disposed at above mentioned system, connected with trap to remove water-toluene azeotrope. After complete removal of water temperature was elevated to 160 ^{0}C during 2 h. and heating was continued during 2 h. to distillate off residual toluene.

C. Control of the copolymerization. In all cases a samples of reaction mixtures and obtained copolymers were treated with 2 fold excess of (CH_3)$_3$SiCl to neutralize the initiator. The course of the reactions was controlled by GLC-method. Process was stopped once a constant composition of low molecular cyclosiloxanes was attained and further change of the composition was not observed.

A samples taken at course of the process were separated by gel filtration method. The yields of high- and low molecular weight fractions were determined and the composition of every fraction analyzed by GLC and NMR -^{1}H and ^{29}Si- methods. For determination of yields of copolymer and oligomers the samples were separated on column 1300x10 mm filled with co-polystyrene-divinylbenzene gel (4 mol.% DVB) at flow rate 0,5 ml/cm^2.sec (benzene as eluent). The limit of separation ability under conditions employed of gel-filltration is 1700; it is approximately doubled compared to MW of monomer B_4. Thus, all particles with MW more than 1700 fall in high molecular weight fraction. On the example of A_4 polymerization it has been shown that the polymer yields data determined both by this technicque and by usual GLC or weight methods practically coinside.

Physical measurements. NMR ^{1}H- and ^{29}Si- spectra were obtained on Bruker-20054 model NMR spectrometer at 200 MHz (^{1}H-) and at 39,76 MHz (^{29}Si-). The measurements carried on concentrated solutions in CDCl$_3$. A relaxation agent - Cr(acac)3 was added to each of the samples before the ^{29}Si- NMR spectra was obtained in order to eliminate the nuclear Overhauser enhancement and short the relaxation times (T_1). Quantitative ^{29}Si- NMR spectra were obtained by using pulse supression and a 4 0 s recycle time to maximise the signal/noise ratio. According to results of measurements the ratios of A:B units determinated by ^{1}H- and ^{29}Si- NMR methods coinside for each analyzed sample.

Determination of transition temperatures of the copolymers was conducted by DSC-method under heating at interval from -150⁰ C to 250⁰ C (20⁰ C/min).

Results and Discussion

The anionic ring opening copolymerization in case of cyclosiloxanes includes the treatment of: (i) time dependent changes of yields and composition of the copolymers influenced by reaction conditions, (ii) the mechanistic peculiarities of the process, (iii) relationship of reaction conditions - macromolecular structure of the copolymers formed and (iv) dependence of the low temperature behavior of the copolymers from their microstructure.

Experiments shows (Fig.1,2) that an increase in B_4 content in the system led both to a decrease in equilibrium yield and in viscosity, and also in the polymerization rate. In the case when the content of B_4 is more than 80 mol.%, polymerization didn't occured even under heating for 2-3 days. This indicate that B_4 trends to formation of polymer chains insignificantly.

At the course of the ROP of cyclosiloxanes a propagation of polymer chains proceeds by addition of cyclic monomer to a silanolate end group. As a result of this step the regeneration of the active centers occur:

$$\sim SiO^-,K^+ + Si\frown O \longrightarrow \sim SiOSi\frown O^-,K^+ \qquad (1)$$

In the case of copolymeryzation A_4 and B_4 active centers of two kinds corresponding to each of the comonomers are formed. If chain transfer reactions (CTR) do not take place fraction of each kind of active centers is determined by the ratio of B_4/A_4 and by the values of copolymerization constants $r_1 = k_{11}/k_{12}$ and $r_2 = k_{22}/k_{21}$ (where k_{11}, k_{12}, k_{22} and k_{21} are partial rate constants for corresponding reactions involving active centers (1) or (2), each reacting with the same or with other monomers). In CTR case, the copolymer chain provides the additional active centers (according to scheme 2 a-d), what complicates analysis of kinetics of the process:

$$\sim A^-, K^+ + \sim A-B\sim \longrightarrow \sim A-A\sim + \sim B^-,K^+ \qquad (2a)$$

$$\sim A^-, K^+ + \sim B-B\sim \longrightarrow \sim A-B\sim + \sim B^-,K^+ \qquad (2b)$$

$$\sim B^-, K^+ + \sim B-A\sim \longrightarrow \sim B-B\sim + \sim A^-,K^+ \qquad (2c)$$

$$\sim B^-, K^+ + \sim A-A\sim \longrightarrow \sim B-A\sim + \sim A^-,K^+ \qquad (2d)$$

Since the contributions of chain propagation and chain transfer reactions are changing continiously, it leads not only to redistribution of active centers but also to

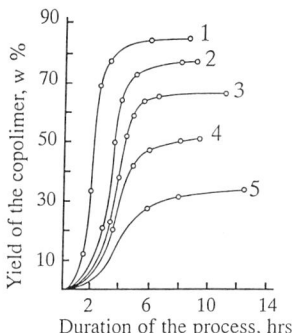

Fig. 1:
Time dependence of the yield of the copolymers.
(Polymerization in bulk, 160° C, [KOH] = 0,2 w.% quits on K), contents of B_4, molar %: 1 - 10, 2 - 30, 3 - 50, 4 - 60, 5 - 70

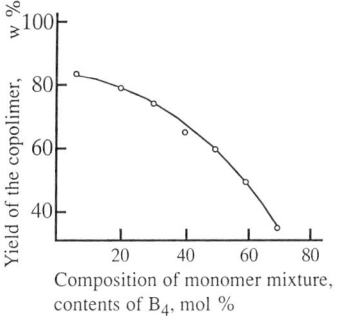

Fig. 2:
Dependence of equilibrium yield of copolymers from the starting composition of monomers mixture. (Bulk polymerization, 160° C, [KOH] = 0,2 w.% quits on K)

changes in copolymer composition. For this reason, it is interesting to elucidate changes in composition of copolymers formed at the various stages of the process. The composition of the copolymers was determined from NMR ^{1}H- and ^{29}Si- data based upon intensities of corresponding sygnals in the regions 0 - 2,0 and -19,4 - -22,3 ppm for A- units and 7,2 - 7,8 and -45,5 - 48,0 ppm for B- units (Fig. 3a). Besides, two groups of sygnals at the NMR ^{29}Si- spectra under -7,6 – -9,9 ppm were found. These sygnals corresponds to Me$_3$SiO-groups attached to Me$_2$SiO- (A) or Ph$_2$SiO- (B) terminal units of copolymer chains (Fig. 3b).

The data on the copolymer composition for early stages of copolymerization (when yield of polymer does not exceed 20-25 w.%) are listed at table 1. The results indicate to a significant enrichment of copolymer macromolecules by B-units as compare to starting ratio A_4:B_4 at the polymer system.

Table 1. Copolymer composition for various systems at early stages of the process.

	Content of B- units, molar % , at:					
reaction system:	20	30	40	50	60	70
copolymer formed:	49,9	61,3	68,6	76,7	84,0	90,7
yield of copolymer: (w.%)	13,0	17,2	16,0	15,4	13,0	11,6

This is due to higher reactivity of B_4 (as compared with A_4) in the reaction with nucleofilic species at anionic processes. On other side, the data on copolymer composition point to lower content of Ph$_2$SiO-units at equilibrium compared to total content of those in the system. Besides a mixed cyclosiloxanes of general formula A_mB_n appear at low molecular fraction as a product of CTR. This is found for all systems, even at low conversions of monomers (<10 w.%). It means that the rates of two competitive reactions- chain transfer and chain propagations- are comparable. In other words, in the course of chain propagation Ph$_2$SiO-units are being removed out copolymer chains. NMR^{29}Si- data characterizing the composition of terminal silanolate groups demonstrate a possible mechanism of this phenomena. According to spectral data not only end but penultimate groups of the macromolecules formed also are enriched by Ph$_2$SiO- groups. The results of the end group enrichment by B-units for various system at equilibrium state are quoted below:

Table 2. A comparison of B-units contents (mol %) at polymer chains and at ends of macromolecules.

Total (at the polymer system):	10,0	20,0	30,0	50,0
Particular (at the ends of chains):	60±10	73±10	86±10	100

Experiments show that under conditions studied usual methods of comparison of comonomers reactivity (for example, by means of copolymerization constant in the case of vinyl polymerization) are inapplicable. For all cases investigated it was found that experimental values of r_1 and r_2 vary in a wide range for the same system, depending on monomer conversion (Fig. 4). It is the main reason of the inapplicability of Mayo-Lewis (14) relationship to analysis of equilibrium anionic ROP of unstrained cyclosiloxanes. (15, 16).

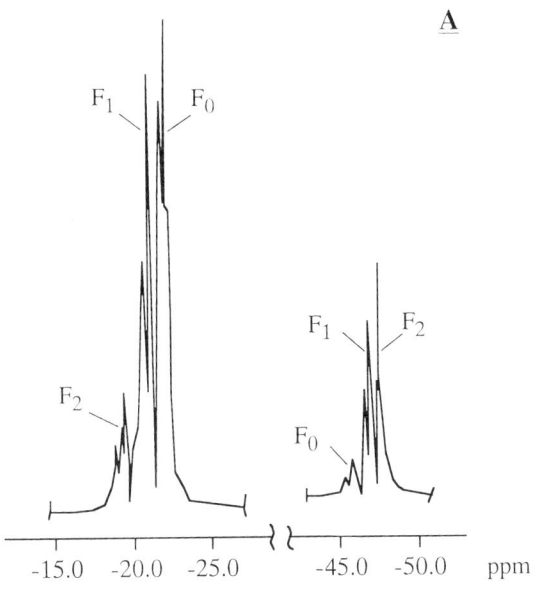
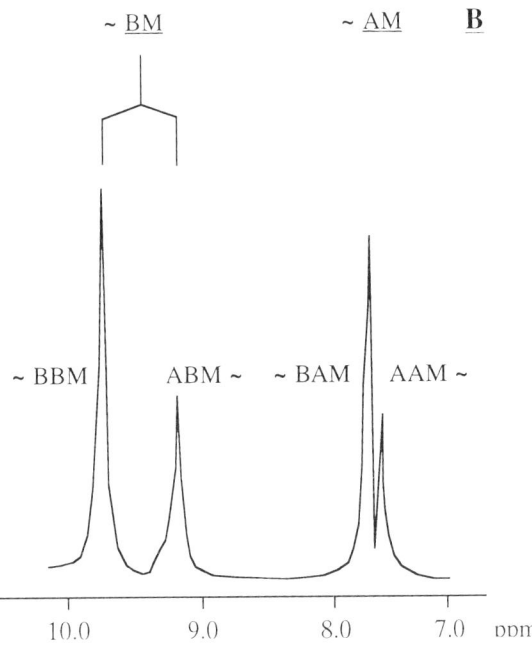

Fig.3: NMR ^{29}Si- spectra of the copolymers.
A: A- and B- centered triads at polymer chains;
B: A- and B- terminated end groups.

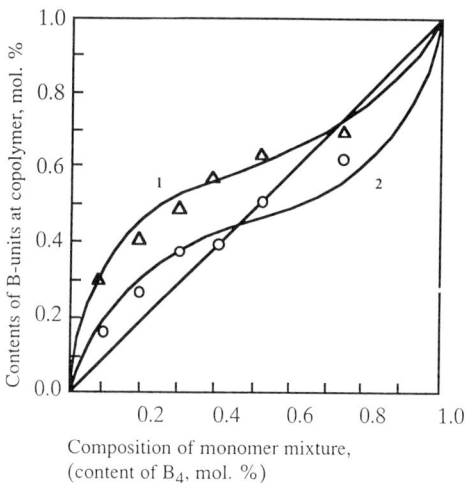

Fig. 4:
Relative reactivity of comonomers according to Mayo - Lewis correlations.
(yield of copolymers: 1 - <12 %,
                      2 - < 27%)

According to GLC-data the yields of cyclic trimers and overall yield of cyclic oligomers increase significantly with content of B_4. At initial contents of B_4 more than 60 mol.% the total equilibrium yield of mixed cyclotrisiloxanes $A_{3-n}B_n$ (n= 2,3) is comparable with those of cyclotetrasiloxane $A_{4-n}B_n$ (n=2,3,4). Taking into account that the concentration of each of cyclic oligomers may be considered as a measure of its thermodynamic stability, one can assume that cyclic tri- and tetramers in diphenylsiloxanes series does not differ so substantially as in the case of dimethylsiloxanes ones.

As mentioned above, the CTR leads to continious changes of structure of polymer chains. It is known that physico-mechanical properties are dependent upon block length as well as on composition of copolymers. Changes of structural parameter is important for practical applications. We tried to search these changes in order to elucidate relationship "reaction condition - structure" at different stages of copolymerization.

For that goal the NMR-spectra information reflecting the distribution of both A- and B- sequences along polymer chains was used *(17-20)*.

Analysis of the integral intensity values for corresponding A- and B-centered triads in NMR ^{29}Si- spectra allows to reveal a distribution both A- and B- units along polymer chains and to determine an average lengths of corresponding $[-A_X-]$ and $[-B_X-]$ microblocks in the chains of the copolymers.

The next computing technique has been suggested.

Let the frequency of occurence for any triad at copolymer chain is determined by value of probability p, which means the probability of addition of the next unit to preceding unit. Assuming that this probability is unaffected by length of chain it is possible to determine the probabilities that a few A- (or B-) units will be add one by one to form a microblocks of various length $[-A_X-]$ or $[-B_X-]$.

Let introduce the next symbols:

R and R_A - is a total quantity and amount of A- type blocks, respectively;

x_a - the share for A- units with respect to the total quantity of the units;

p_{AA} - the probability of attachment for A- units to A- sites at polymer chains (regardless of the number of preceding units);

p_{AB} - the probability of attachment for the B- units to A- sites at polymer chains;

The similar designations x_b, p_{BB}, p_{BA} may be introduced for B- type units.

The appearance of each alternate unit along polymer chain signifies the termination of the preceding and origin of the the next block. Considering that number of A-B and B-A contacts along chain is equal, $R_A = R_B = R/2$.

The values of x_a, x_b, p_{AA}, p_{AB}, p_{BB}, p_{BA} were calculated on the base of relations:

$$x_a = k_a / (k_a + k_b) = A_m / 100; \qquad (3a)$$

$$x_b = k_b / (k_a + k_b) = 1 - A_m / 100 \qquad (3b)$$

$$p_{AA} = (A_m - R/2) / A_m \qquad (4)$$

$$p_{AB} = R / 2 \cdot A_m \qquad (5)$$

$$R_A = k_A \cdot A_m = k_B \cdot (1 - A_m) = R_B \qquad (6)$$

$$R_B = K_B \cdot (1 - A_m) \qquad (7)$$

The average values of k_a and k_b for A- and B- centered triads are calculated from integral intensity data of NMR ^{29}Si- spectra according to *(17)*. The calculated values of x_a, x_b, p_{AA}, p_{AB}, p_{BB}, p_{BA} may be applied further for the evaluation of block dimensions.

It is reasonable, that the probability of the emergence of single A- units p_{1A}, i.e. the probability of -BAB- type consequences at polymer chains is determined by relation:

$$p_{1A} = p_{AB}^2 \cdot x_a \qquad (8)$$

Then sites involving two A- units appears with probability p_{2A}:

$$p_{2A} = p_{AA} \cdot p_{AB}^2 \cdot x_a \qquad (9)$$

The probability of the emergence of triplet of A- units, p_{3A} at chains is determinated as follow:

$$p_{3A} = p_{AA}^2 \cdot p_{AB}^2 \cdot x_a \qquad (10)$$

Correspondingly, the probability of the emergence of multi- A-units microblocks, p_{nA}, is determined by total relation:

$$p_{nA} = (p_{AA})^{n-1} \cdot p_{AB}^2 \cdot x_a \qquad (11)$$

Analogous relation may be derived also for the case of multi- B-units microblocks:

$$p_{nB} = (p_{BB})^{n-1} \cdot p_{BA}^2 \cdot x_b \qquad (12)$$

The number of very extended microblocks are too small since corresponding values of p_{nA} or p_{nB} quickly decrease with increasing of the value n (length of the microblock). From relations (11,12) it is seen that when value of n exceed certain limit, values of P_{nA} (or P_{nB}) value become negligible small. Let us define this limit by introducing criterion ε. If ε- value attained to the limit, the longer block is formed with negligible probability:

$$\left(R_{A_m} - \sum_{k=1}^{k=n} p_{kA} \right) \leq \varepsilon \qquad (13)$$

The value of $\varepsilon=0,005$ was chosen as a criterion for completion of the calculation. This value for a certainty not exceed the precision of the integral intensities determination of triads at NMR-^{29}Si- spectra.

To present the distribution of blocks on their sizes more clearly it is more convenient to use the share of A- or B- units included at [-A_x-] or [-B_x-] blocks (instead of the probability of corresponding x- units block appearance). The dependence of shares (D) of A- and B- units expended on formation of corresponding A- or B-type blocks are exibited in fig. 5-8. Analysis of plots of D versus x have revealed some regularity of shaping of microstructure at copolymer chains formed.

As it was expected, in all cases the increase of relative content of one or other monomer is followed by increase of the size of corresponding blocks (Fig. 5). For the systems in which the relative content of one monomer is much less than other, it is characteristic higher share of short-chain blocks (consisting of 1 - 3 siloxane units) of minor component as compared to blocks consisting from units of major monomer (Fig. 6). Data presented on Fig. 7 allows to follow the changes of microstructure of the copolymers as function of time.

Proposed method for evaluation of distributions for dissimilar blocks by their size allows to reveal more fine differences, for example caused by different ways of preparation of copolymers under the same ratios of starting monomers (Fig. 8).

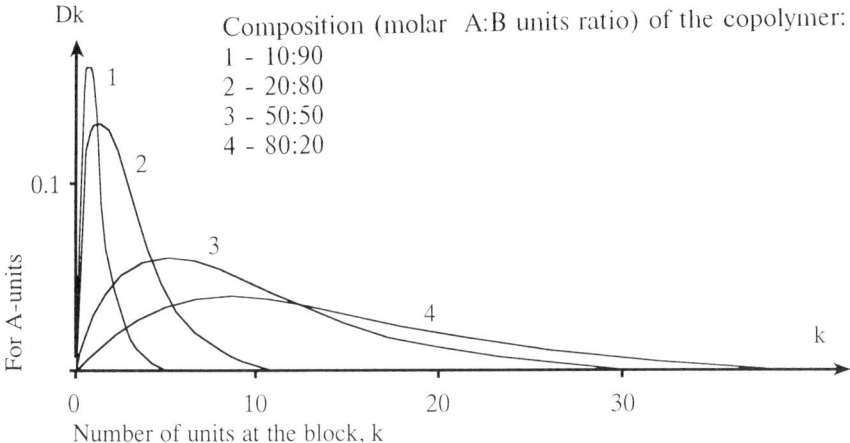

Fig. 5. Distribution on the size of microblock at copolymer chains. Change of size of A-microblocks depending on A:B units ratio at polymer chains. Key: DK is share of A- (or B-) units, expended to a formation of microblocks.

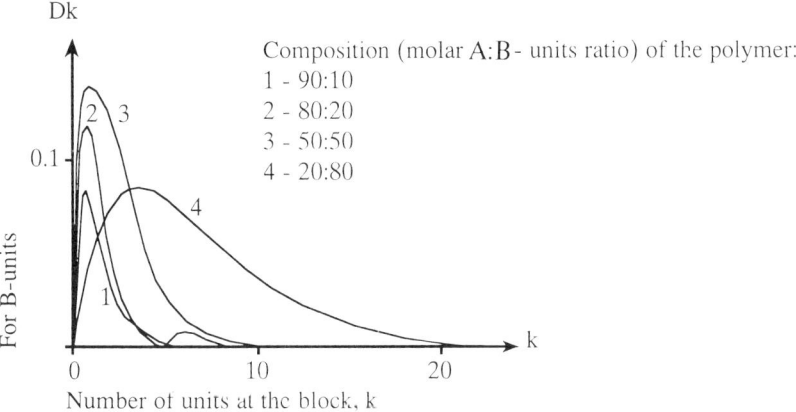

Fig. 6. Distribution on the size of microblock at copolymer chains. Change of size of B-microblocks depending on A:B units ratio at polymer chains. Key: DK is share of A- (or B-) units, expended to a formation of microblocks.

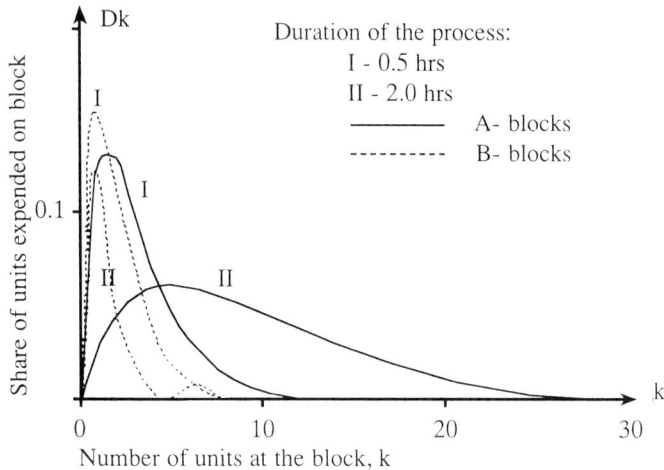

Fig.7: Changes of distribution of microblocks during polymerization process

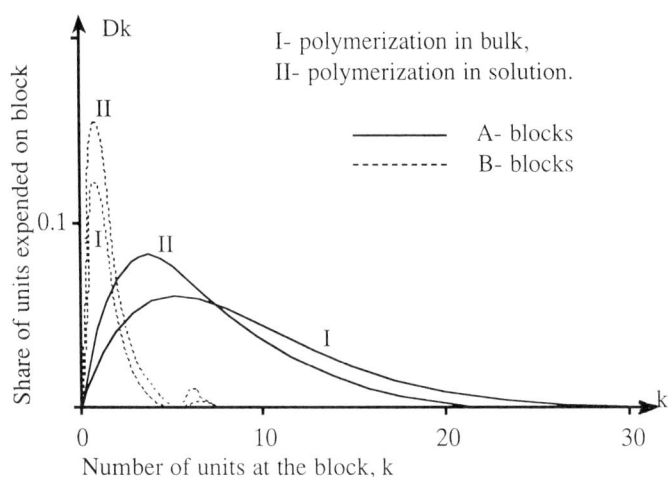

Fig.8: Size microblocks distribution at different reaction conditions

In wider sense, employment of NMR ^{29}Si- spectroscopy data in combination with proposed computation technique will be available in other cases. For instance, it may be useful for investigation of ring open copolymerization of other cyclosiloxanes or under studies of influence of technological parameters on the properties of copolymer obtained.

The presence of Ph_2SiO - units in copolymers is accompanied by increasing of T_g proportionally to their molar concentration at polymer chains. The linear relationship between T_g and content of these units is described by empirical equations:

$$2.07 \cdot x - T_{0g} = T'_g \quad (\pm 5\%) \qquad (14)$$

where: T_{0g} and T'_g - values for a linear polydimethylsiloxane (T_{0g}=-121^{0}C) and copolymer, correspondingly; x - molar concentration of Ph_2SiO-units at copolymer chains.

So to produce a low-temperature elastomers by copolymerization of A_4 and B_4 it is necessary to limit contents of modifying units in copolymers on the low level (no more than 10 mol.%). Experiments show also that copolymers of the same composition obtained by different methods manifest different properties. Copolymers prepared in "equilibrium" process (method 2) posess the lower tendency to crystallization as compared with "none-equilibrium" copolymers of similar composition, obtained by method 1.

These results are correlated with data presented on Fig. 8. Accordingly to these data an average length of corresponding both A- and B- blocks differs at copolymers of much the same composition. As a rule, the size is somewhat smaller in the case of equlibrium copolymers prepared by method 2 as compared with method 1. It is noteworthy to note that distinctions observed are not a result of differences in molecular weight or nature of end groups. Probably, they are associated with variety of distribution of sequences in the polymer chains. In this connection the data reported by previous investigators *(8,9,13)* arouse an interest. For example, G. Babu et al. *(13)* has demonstrated detectable discrepancy in T_g for dimethyl-diphenylsiloxane copolymers of similar composition.

Conclusions.

The equilibrium anionic ring opening copolymerization of unstrained cylotetrasiloxanes A_4 and B_4 is accompanied by intensive chain transfer reactions from the very beginning of the process. That manifests itself in (i) a continious interconversion of the active centers from both monomers, (ii) appearance of the mixed cyclosiloxanes of A_mB_n type, and (iii) changes in microstructure of macromolecules formed. Thus commonly employed methods of determination of relative reactivity of the comonomers are unapplicable. The microstructure of copoly-dimethyl(diphenyl)siloxanes formed has been estimated with a computational tecnique based on the NMR ^{29}Si- data. By application of this approach a relationship between the reaction conditions and microstructure of copolymer chains has been demonstrated. It was shown that physical properties of copolymers may be influenced to some extent by variations of experimental conditions during copolymerization process.

Acknowledgements

This work was supported in part by RFBI, Grant 96-03-32608.

Literature Cited.

1. Warrick, E.L., Pierce, O.R., Polmanteer, K.E., Saam, J.C., *Rubber Chem. & Technology,* **1979**, *v. 52*, p. 432.
2. Noll, W. *Chemistry and Technology of Silicones*; Academic Press; New York, **1968**.
3. Shetz, M., *"Silikonovy kaucuk"*, Publisher: SNTL, Praha, CSR, **1971**, p. 35.
4. *"Chemistry & Tecnology of organosilicon elastomers"*, Ed. Reihsfeld, V.O.. 1973, Khimia, Leningrad , USSR, **1973**, p. 37 - 58.
5. Wright, P.V., in *"Ring Opening Polymerization"*, Win, K.J., Saegusa, T. Eds., Elsevier Applied Sci. Publishers, Ltd, , v. 2, p. 223.
6. Merker, R.L.; Scott, M.J. *J. Polym. Sci.* **1966**, *v. 41*, p. 297.
7. Andrianov, K.A.; Zavin B.G.; Sablina G.F. *Polym. Sci. USSR* (Engl. Transl.) **1978**, *v.14*, p.1234.
8. Yilgor, L., Riffle, J.S., McGrath, J.E. *In Reactive Oligomers*; Harris, F.W, and Spinelli,H.J, Eds.; ACS Symposium Series 282; Am. Chem. Soc.: Washington, DC, **1985**; p. 161.
9. Ziemelis, M., M., Lee, M., Saam, J.C., *Polymer Prepr., Am. Chem. Soc., Div. of Polym. Chem.,* **1990**, *v. 31* (1), p. 38.
10. Carmichael,J.B., and Heffel J., *J.Phys. Chem.*, **1965**, *v. 69*, p. 2218.
11. Ziemelis, M.J, Saam J.C. *Macromolecules*, **1989**, v. 22, p. 2111.
12. Rozenberg, B.A., Irzhack, V.I., Enykolopjan N.S., *"Interchain Change in Polymers"*, Khimia, Moscow, USSR, **1975**, p. 194.
13. Babu, G.N.; Christopher, S.S.; Newmark, R.A., *Macromolecules*, **1987**, *v. 20*, p. 2654.
14. Mayo, F.R.; Lewis, F.M. *J. Am. Chem. Soc.* **1944**, *v. 66*, p.1594.
15. Laita, Z.; Jelinek, M. Vysokomol.Soedin. **1963**, *v.5*, p.1268.
16. Reikhsfeld, V.O; Ivanova, A.G. Vysokomol. Soedin. **1962**, *v.4*, p. 30.
17. Jancke,H.; Engelhardt,G.; Kriegsman, H.; Keller, F. *Plaste und Kautsch.* **1979**, *v. 26*, p. 612.
18. Engelhardt,G.;Jancke, H. *Polym. Bull.* **1981**, *v. 5*, p. 577.
19. Ryzhova O. G., Molchanov, B.V., Denisova E.V., et al., *Vysokomolekularnye Soedin.,* **1988**, *v. 30 (A).*, p. 1294.
20. Brandt, P.J.A.; Subramanian, R.;Sormani, P.M.;Ward, T. C.; McGrath, J.E. *Polym. Prepr., Am. Chem. Soc., Div. Polym. Chem.* **1985**, *v.26 (2)*, p. 213.

Indexes

Author Index

Aksenov, V. I., 197
Antkowiak, T. A., 77
Arest-Yakubovich, A. A., 197
Auschra, Clemens, 208
Aznar, R., 142
Bee, T. G., 167
Bortolotti, M., 50
Brumbaugh, D. R., 77
Chorvath, I., 267
Couve, J., 142
Dave, R. S., 255
Didier, Y., 291
Dimitrov, Ivaylo, 236
Dobreva-Schué, R., 142
Dotcheva, Dobrinka, 236
Doytcheva, Maria, 236
Ermakova, I. I., 197
Fernández-Fassnacht, Enrique, 129
Ferreira, A. A., 90
Fetters, L. J., 36
Flores, Rodolfo, 129
Golberg, I. P., 197
Gurnari, A., 50
Hahnfeld, J. L., 167
Hattori, Iwakazu, 62
Hourston, D. J., 291
Hsieh, Henry L., 28
Huang, J. S., 36
Ishiguro, Minoru, 98
Ishii, Masao, 186
Ito, Hiroshi, 218
Jang, Sung H., 2
Kaufmann, Marita, 208
Kirkpatrick, D. E., 167
Knoll, Konrad, 112
Kruse, R. L., 255
Lawson, D. F., 77
Lee, Youngjoon, 2
Lemstra, P. J., 267
Lindner, P., 36
Lindsay, C. I., 291
Lizarraga, Gilda, 2
Lovisi, H. R., 90
Ma, J.-J., 159
Maeda, Mizuho, 186
Majumdar, B. D, 159
Martínez, Enrico N., 129
Martins, M. L. S., 90

Matsubara, Tetsuaki, 98
McGrail, P. T., 291
Meijer, H. E. H., 267
Mertens, M. D. M., 267
Moctezuma, Sergio A., 129
Morita, K., 77
Müller, Axel H. E., 208
Nakata, Hiromichi, 186
Nakayama, S., 77
Nestegard, M. K., 159
Nicol, P., 142
Nicolini, L. F., 90
Nießner, Norbert, 112
Nishikawa, Makoto, 186
Ozawa, Y., 77
Petrova, Elisaveta, 236
Pike, W. C., 167
Quirk, Roderic P., 2
Rabkina, A. Yu., 304
Richter, D., 36
Ronova, I. A., 304
Roos, Sebastian, 208
Ryakhovsky, V. S., 197
Sablina, G. F., 304
Saffles, D., 77
Schreffler, J. R., 77
Schué, F., 142
Sheridan, M. M., 159
Siol, Werner, 208
Southward, R. J., 291
Stamenova, Rayna, 236
Stayer, M. L., 77
Stebbins, L. R., 255
Stellbrink, J., 36
Strelkova, T. V., 304
Sung, J., 36
Tadaki, Toshihiro, 62
Takamatsu, Hideo, 186
Tsutsumi, Fumio, 62
Tsvetanov, Christo B., 236
Udipi, K., 255
van Geenen, A. A., 267
van Schijndel, R. J. G., 267
Viola, G. T., 50
Willner, L., 36
Zavin, B. G., 304
Zhuo, Qizhuo, 2
Zolotarev, V. L., 197

Subject Index

A

Acrylate-based macromonomers. *See* Comb-shaped copolymers
Acrylonitrile-butadiene-styrene (ABS), polymerization of homogenous polymer/monomer systems, 268
Adhesive applications. *See* Pressure sensitive adhesive (PSA) tapes; Styrene-isoprene-styrene (SIS) block copolymers
Aggregate behavior of living polymer chains. *See* Self-assembly of living polymer chains
Alkyl methacrylates, anionic polymerization, 19–20
Alkylene oxides, anionic polymerization, 20–21
Aminoaryl end-functionalized poly(methyl methacrylate). *See* Poly(methyl methacrylate) (PMMA) with reactive end groups
Anionic polymerization. *See* Principles of anionic polymerization
Asphalt modification
 asphalt composition, 132t
 characteristics of tapered block copolymer (TBC) modifier, 130–131, 132f
 preparation and testing of asphalt-TBC mixtures, 131–132
 storage modulus and tan δ of asphalt-TBC mixtures as function of frequency, 134, 136f
 viscosity and softening point temperatures of asphalt-TBC mixtures as function of TBC content and type, 134, 135f
Asymmetric star block copolymers
 adhesive formulation and performance, 164–165
 experimental materials for styrene-isoprene star block copolymers, 160
 high and low molecular weight styrene end blocks for desired adhesion, 164–165
 morphology and rheology in comparison to regular star block copolymers, 162–163
 polymerization conditions, 160
 size exclusion profiles of mixed molecular weight end blocks and asymmetric star, 160, 161f
 synthesis route, 160
 tensile behavior in comparison to regular and commercial star block copolymers, 164
 See also Star-branched oligomers; Star-branched polymers

B

Block copolymers
 difunctional initiation and two-step sequential monomer addition, 11
 schematic triblock copolymer synthesis by two-step monomer addition with dilithium initiator, 12
 synthesis methods by anionic polymerization, 6–11
 three-step sequential monomer addition, 7–9
 two-step sequential monomer addition and coupling, 9–11
Branched polystyrenes
 branch density prediction equation, 178
 calculation of molecular weight and branch density for randomly branched PS, 174, 178–179
 electron transfer and radical coupling products from polystyryl dianion with benzyl halide and dihaloxylene, 170, 172f
 model compound reactions, 168–174
 molecular weight calculations by Macosko–Miller approach, 178
 nucleophilic substitution products from polystyryl dianion with benzyl halide and dihaloxylene, 170, 171f
 polymer analysis and comparison with molecular weight and branch density predictions, 179–180
 possible reaction paths of coupling reactions, 168, 169f, 170
 proton abstraction products from polystyryl dianion with benzyl halide and dihaloxylene, 170, 171f
 randomly branched polystyrene (PS) synthetic procedure, 179
 reaction of polystyryl dianion with α,α'-dibromo-p-xylene and α,α'-dichloro-p-xylene in benzene, 170, 173f
 reaction of polystyryl dianion with α,α'-dichloro-p-xylene and benzyl chloride in benzene/THF, 170, 173f
 reactions of polystyryl dianion with benzyl chloride and benzyl bromide in benzene, 170, 172f
 scheme for synthesis of three-arm stars and randomly branched samples, 168, 169f
 selective precipitation for removal of uncoupled product from three-arm PS, 174, 176f

separation of branched from linear polymer, 180, 181f
solvent polarity effect on selective precipitation, 180, 182f, 183f
synthesis and purification of three-arm polystyrene, 174, 176f
synthesis of randomly branched polystyrene, 174, 178–180
synthetic route and GPC data for three-arm star polystyrene using α,α',α"-trichloromesitylene, 170, 174, 175f
1,3-Butadiene polymerization
addition effect of N,N,N',N'-tetramethylethylenediamine (TMEDA), 72, 74, 76
characterization methods, 63–64
N,N-dimethyltoluidine/n-butyllithium (TD/BuLi) procedure, 63
formation ratio of TD-Li and BuLi in mixed solvents cyclohexane/ tetrahydrofuran CHX/THF, 74f
GPC curves for polybutadiene (PBD) from TD/BuLi and BuLi initiation, 64, 66f, 67
^{1}H NMR spectra of PBD from TD/BuLi and BuLi initiation, 64, 65f
IR spectra of PBD from TD/BuLi and BuLi initiation, 64, 65f, 66f
model experiments for verifying formation rate of TD/BuLi initiator, 64
polarity of solvents in TD/BuLi reaction, 67
polymerization of 1,3-butadiene in cyclohexane/tetrahydrofuran (CHX/THF), 68
polymerization of 1,3-butadiene in several solvents, 64, 67
relation between 1,2-bond content in polymer and "functionality introduction ratio", 74, 75f
relation between CHX/THF solvent ratio and "functionality introduction ratio" with varying TMEDA addition, 74, 75f
relationship between "functionality introduction ratio" and solvent polarity, 68, 69f
relationship between polymer UV intensity and solvent polarity, 68, 69f
TD and BuLi reaction product analysis, 70–72
Butadiene/styrene copolymers
pure and tapered block applications, 30–31
random copolymer applications, 31
sec-Butyllithium, use in block copolymer synthesis, 7–9

C

ε-Caprolactam. See Polyamides; Poly(2,6-dimethyl-1,4-phenylene ether) (PPE)/polyamide-6 blends
Carbon black

hysteresis reduction in carbon black-filled products from cyclic reagents, 79
re-agglomeration prevention for reduction in hysteresis, 78
Chain-end functionalization
amination by N-(3-chloropropyl)dialkylamines, 17
carbonation of polymeric carbanions with carbon dioxide, 16–17
functionalized initiators, 17–19
hydroxylation by ethylene oxide, 17
protected hydroxyl initiators, 18–19
telechelic, functionalized block, and star-branched polymers, 18
termination with electrophilic reagents, 16–17
Comb-shaped copolymers
applications, 208–209
determination of macromonomer (MM) reactivities in radical copolymerization, 211, 212f
functionality determination by liquid chromatography, 209, 210f
graft copolymer from methacryloyl-terminated PMMA MM and n-butyl acrylate, 210
industrial scale synthesis of PMMA MM by free radical polymerization, 209
microphase separation in bulk by differential scanning calorimetry (DSC), 211–213
MM technique state-of-the-art, 209
properties and practical applications, 213–216
synthesis of PMMA macromonomers by group-transfer polymerization, 209
thermoplastic elastomers properties, 213–215
viscosity index improvers for motor oils, 215–216
Coupling agents, difunctional linking agents for block copolymer synthesis, 9–11
Cyclosiloxanes. See Poly(methylphenylsiloxane)s

D

Diene monomers
anionic propagation kinetics, 4
kinetic scheme for diene polymerization, 5
Diene polymerization with homogeneous lithium amide initiators
base-catalyzed elimination of amino head-groups, 84
calibration curves for UV-visible determination of tertiary amines, 86f
experimental methods, 78–79
general procedure for polymer-bound amine determination, 78–79

323

hysteresis reduction in carbon black-filled products from cyclic reagents, 79
N-lithiopyrrolidinide•2THF (LPY•2THF) initiator, 79–86
LPY•2THF generation scheme, 81
model SEC traces of polymers with and without end-linking by $SnCl_4$, 81, 82f
solubilities and polymerization results for selected N-Li dialkylamides, 80t
summary of LPY•2THF-initiated random styrene-butadiene copolymer in mixed hexanes, 81t
three approaches to generation of homogeneous initiator reagents, 79

Diesters
coupling living poly(styrene-b-butadienyl)lithium with didecyl adipate, 155–156, 157f
coupling oligobutadienyllithium with dimethyl adipate, diethyl adipate, and dimethyl phthalate, 150, 152

Dimethylcyclosiloxanes. *See* Poly(methylphenylsiloxane)s

N,N-Dimethyltoluidine/n-butyllithium (TD/BuLi)
analysis of initiator formation, 70–72
effect of N,N,N',N'-tetramethylethylenediamine (TMEDA) addition, 72, 74, 76
^{1}H NMR spectrum of reaction products obtained in 75/25 cyclohexane/tetrahydrofuran, 73f
^{1}H NMR spectrum of reaction products obtained in 88/12 CHX/THF, 73f
^{1}H NMR spectrum of reaction products obtained in cyclohexane, 71f
^{1}H NMR spectrum of TD, 71f
polybutadiene initiator in mixed solvents, 68–69
polybutadiene initiator in several solvents, 63–67

Diphenylcyclosiloxanes. *See* Poly(methylphenylsiloxane)s

Divinylbenzene, star-branched polymers by linking reactions, 14–16

E

Epoxidized soybean oil, coupling oligobutadienyllithium for star-branched oligomers, 152–153, 154f
Epoxy resins, rubber-toughened epoxies, 267–268
Ethyl acetate, coupling oligobutadienyllithium for star-branched oligomers, 150, 151f
Ethylene oxide polymers

three polymeric classes, 236
See also Poly(ethylene glycol) (PEG); Poly(ethylene oxide) networks (PEON); Poly(ethylene oxide) (PEO)
Extrusion (reactive) process for polyamides, 256, 258, 259f

F

Free radical polymerization, industrial scale synthesis of end-functional PMMA macromonomer using chain transfer agents, 209

G

Graft copolymers. *See* Comb-shaped copolymers
Group transfer polymerization, PMMA macromonomer synthesis, 209

H

High impact polystyrene (HIPS)
characteristics of tapered block copolymer (TBC) modifier, 130–131, 132f
impact modification by TBC, 129–130
notched Izod impact, tensile and flexural modulus of HIPS-TBC blends as function of TBC content and type, 135–136, 137f, 138f
polymerization of homogenous polymer/monomer systems, 268
preparation and testing of HIPS-TBC mixtures, 133–134

Hysteresis
characterizations and cured, compounded hysteresis results for LPY•2THF-initiated random styrene-butadiene copolymer, 83t
reduction by prevention of carbon black re-agglomeration, 78
rolling resistance, 62–63

I

Impact modifiers
hydrogenated styrene-isoprene block copolymers for various polymers, 187, 190
tapered block styrene-butadiene copolymers for polystyrene, 129–130
Industrial applications
control variables in anionic polymerizations, 28–29

high cis-polyisoprene, 28–29
nonfunctional liquid polybutadienes, 30
polybutadiene rubber, 29–30
pure block and tapered butadiene/styrene copolymers, 30–31
random butadiene/styrene copolymers (solution SBRs), 30–31
random copolymers in tire-tread applications, 31
selective hydrogenation of styrene-diene copolymers, 31
styrene-butadiene block copolymers with >60% styrene, 32
styrenic thermoplastic elastomers, 31–32
telechelic polybutadienes, 30
Initiators
bifunctional initiators for Styroflex thermoplastic elastomer, 113, 114f
sec-butyllithium use in block copolymer synthesis, 7–9
difunctional initiation for block copolymer synthesis, 11
N,N-dimethyltoluidine/n-butyllithium for 1,3-butadiene in mixed solvents, 68–69
N,N-dimethyltoluidine/n-butyllithium for 1,3-butadiene in several solvents, 63–67
N-lithiopyrrolidinide•2THF (LPY•2THF) for dienes, 79–86
lithium dihydrocarbon amides for dienes, 77–78
schematic of anionic polymerization of styrene, 3–5
sodium alkoxide of tetrahydrofurfuryl alcohol (THF-ONa), 204
sodium-containing initiators for 1,2-polybutadiene production, 204–205
Isoprene polymers
applications and properties of hydrogenated styrene-isoprene block copolymers (S-EP-S), 187, 190, 191t
applications and properties of polystyrene vinyl-bonded polyisoprene block copolymers (S-VI-S), 190, 192, 196
electron micrographs of PP/S-EP-S and PP/S-VI-S blends, 192, 194, 195
hydrogenation of SBS (S-EB-S), 190
liquid isoprene rubber (L-IR) applications and properties, 187, 188f, 189f
microstructure control by vinyl-bonded units in polyisoprene, 192, 194f
polymers from high purity monomer, 186–187, 188f
polypropylene/hydrogenated vinyl-bonded polyisoprene blends, 192, 193t, 196
property comparison of SEBS and SEPS, 190, 191t

vibration damping effect of S–VI–S, 192, 193t
Ziegler versus anionic polymerization, 186–187

L

Lactam polymerization
anionic ring-opening polymerization, 21
polymerization schematic, 22
See also Polyamides
N-Lithiopyrrolidinide•2THF (LPY•2THF) initiator system
analysis of head-group pyrrolidine functionality, 84t
characteristics of LPY•2THF-initiated random styrene-butadiene copolymer, 81t
characterizations and cured, compounded hysteresis results, 83t
polymerization progress by ^{13}C NMR, 84, 85f
tetrahydrofuran (THF) for solubility, 79–80
Lithium-based initiators
difunctional initiation for block copolymer synthesis, 11
functionalized initiators for chain-end functional groups, 17–19
kinetics of initiation in aromatic and aliphatic solvents, 4
styrene polymerization with sec-butyllithium, 3–5
See also Initiators
Lithographic resist polymers
acid-catalyzed depolymerization of poly4HOMST films, 232f
anionic cyclopolymerization of 4-trimethylsilylphthalaldehyde (SPA), 226, 227f
bilayer (all-dry) lithography with polySPA, 227f
t-butoxycarbonyl (tBOC) resist imaging chemistry, 228f
elucidation of acid-catalyzed depolymerization mechanism of poly(4-hydroxy-α-methylstyrene) (poly4HOMST), 230–233
investigation of end group effect in chemically amplified resists, 227–233
lithographic imaging process schematic, 219f
poly(4-t-butoxycarbonyloxystyrene) (polyBOCST) synthesis via living anionic polymerization, 229f
poly(isopropenyl t-butyl ketone) (polyIPTBK), 222–225
polymer end group influence on chemically-amplified resist sensitivity, 227–230
poly(methyl α-trifluoromethylacrylate) (MTFMA), 220–222

325

polyphthalaldehydes by anionic
    cyclopolymerization, 226
preparation by radical polymerization, 218–219
proposed depolymerization from terminal
    polymer chain end, 232, 233f
resist polymers accessible only by anionic
    polymerization, 220–226
sensitivity curves of tBOC resists from
    polyBOCST and Ph$_3$SSbF$_6$, 230f
syntheses of poly4HOMST via cationic and
    living anionic polymerization, 231f

M

Metallocene polyethylene (PE)
    competitive analysis of fresh meat packaging
        films, 120, 121t
    mechanical property comparison with Styroflex,
        118, 119f, 120t
    permeability comparisons with Styroflex, 120,
        122f
    processability comparisons with Styroflex, 118,
        120t
Methacrylates, anionic polymerization, 19–20
Molecular weight, control variable in anionic
    polymerization, 5
Molecular weight distribution, control variable in
    anionic polymerization, 5–6
Morero's method, microstructure of polymer by
    IR spectroscopy, 63

N

Nitroaryl end-functionalized poly(methyl
    methacrylate). See Poly(methyl methacrylate)
    (PMMA) with reactive end groups
Nylon-6. See Polyamides; Poly(2,6-dimethyl-1,4-
    phenylene ether) (PPE)/polyamide-6 blends

O

On-line process monitoring
    batch copolymerization of random styrene-
        butadiene copolymer (SBR) with <1%
        styrene block, 55, 60f
    batch copolymerization of SBR with 10%
        styrene block, 55, 59f
    chromophore detection, 52
    decreasing precision at high temperature, 61
    evaluation of coupling efficiency, 55, 56f
    linear relationship between styrene
        concentration and absorbance, 55, 57f
    monitoring active chain ends at total
        conversion, 61
    on-line UV spectroscopy, 50–51
    pilot plant with process photometer, 51–52
    potential data from on-line UV spectroscopy,
        51
    production quality, 50
    relationship between molecular weight of
        diblock tapered copolymers and absorption
        signal at total conversion, 52, 55, 56f
    relationship between molecular weight of SBS
        copolymers and absorbances during
        synthesis, 52, 54f
    run-away reaction prevention, 51
    sequential SBS copolymerization by UV
        monitoring, 55, 58f
    strategies for pure triblock (SBS) copolymer, 55
    typical polymerization reaction, 52, 53f
    versatility of UV spectroscopy, 51
Order-disorder transition, determination for
    Styroflex thermoplastic elastomer, 115, 117f

P

PMMA. See Poly(methyl methacrylate) (PMMA)
    with reactive end groups
Poisson molecular weight distribution,
    requirements for formation, 6
Polyamide blends. See Poly(2,6-dimethyl-1,4-
    phenylene ether) (PPE)/polyamide-6 blends
Polyamides
    anionic graft copolymerization of caprolactam
        with hydroxy functional core shell rubber,
        265f
    anionic ring opening polymerization of lactams
        by activated monomer mechanism, 255–256
    dynamic mechanical properties of nylon-6 and
        85/15 copolymer of nylon-6 and
        polyetherimide, 260f
    fiber glass/nylon-6 reactive pultrusion setup,
        261, 263f
    graft copolymerization of caprolactam with
        polyetherimide, 258, 259f
    in situ copolymerization of caprolactam with
        terminally functional elastomeric prepolymer,
        262, 263f
    nylon block copolymer properties from RIM
        process, 264t
    nylon-6 from caprolactam by reactive extrusion,
        256, 258
    properties of particulate rubber-modified nylon-
        6 RIM, 262, 264t
    reaction path for caprolactam polymerization,
        257f

reactive extrusion process, 256, 258
reactive extrusion set-up for caprolactam polymerization, 259f
reactive injection molding (RIM) process, 262, 264–265
reactive thermoplastic pultrusion process, 258, 261
RIM of nylon block copolymers, 262, 263f
1,2-Polybutadiene (1,2-PBD)
  advantages of anionic polymerization process, 197
  control of molecular architecture and molecular weight distribution, 199–201
  GPC traces of typical commercial 1,2-PBD, 201f
  high molecular weight 1,2-PBD, 198–201
  industrial applications, 197
  low molecular weight liquid 1,2-PBD, 203–204
  properties and application areas for high molecular weight 1,2-PBD, 201–203
  properties of commercial 1,2-PBD rubbers, 203t
  properties of commercial liquid 1,2-PBDs, 203t
  properties of 1,2-PBD from sodium-containing initiators, 205t
  prospects for process improvements, 204–205
  sodium alkoxide of tetrahydrofurfuryl alcohol (THF-ONa) promising initiator, 204–205
  varying 1,2-unit content by electron donors, 198f
Polybutadienes
  branched versus linear polybutadiene rubber, 29–30
  carboxy or hydroxy telechelic polybutadiene applications, 30
  nonfunctional liquid polybutadienes applications, 30
  See 1,3-Butadiene polymerization
Polybutadienyllithium, thermal stability at high concentrations in ethylbenzene, 145, 146f
Poly(2,6-dimethyl-1,4-phenylene ether) (PPE)/polyamide-6 blends
  binary phase diagram of PPE/ε-caprolactam solutions, 273, 275–276
  calorimetric investigation conditions, 270
  cloud point measurement method, 270
  considerations for morphology development, 276–278
  effect of residual ε-caprolactam and oligomers on thermal properties, 286–287
  experimental materials, 269
  liquid-liquid demixing preceding crystallization, 275
  mechanism of activated anionic polymerization of ε-caprolactam, 276
  monomer conversion measurement by water extractable content (WEC) method, 271
  morphology blends with polyamide-6 as continuous phase, 278, 280f, 281f, 282f
  morphology blends with PPE as continuous phase, 278, 283f, 284f, 285f, 286
  morphology by scanning electron microscopy (SEM) and transmission electron microscopy (TEM), 271
  morphology of PPE/polyamide-6 blends, 278–286
  phase behavior of PPE/ε-caprolactam for < 50 wt% solutions, 272–273, 273
  phase diagram of PPE/ε-caprolactam solutions, 274f
  phase separation and phase inversion dependence on PPE concentration, 277–278
  preparation method for polyamide-6, 270
  preparation methods for PPE/polyamide-6 blends based on PPE concentration, 270–271
  preparation of PPE/ε-caprolactam solutions, 269
  reactive solvent approach with ε-caprolactam for PPE/nylon-6 blend, 267–269
  schematic ternary phase diagram PPE/ε-caprolactam/polyamide-6, 279f
  typical DSC scans for PPE/ε-caprolactam solutions, 274f
  upper critical solution temperature behavior by cloud point curve, 275
  WEC results for conversion of ε-caprolactam, 286–287
Polyetherimide
  dynamic mechanical properties of nylon-6 and 85/15 copolymer of nylon-6 and polyetherimide, 260f
  graft copolymerization of caprolactam with polyetherimide, 258, 259f
Poly(ethylene glycol) (PEG)
  anionic polymerization in solution, 236–237
  main factors for effective polymerization, 237
Poly(ethylene oxide) networks (PEON)
  applications, 247–248
  crosslink density throughout sample thickness, 249–250
  effect of benzophenone concentration on crosslinking efficiency, 250, 251t
  effect of PEO molecular weight on crosslinking efficiency, 251, 252t
  effect of photoinitiator type on crosslinking efficiency, 249t

formation of PEO/poly(vinyl acetate) networks, 252*t*, 253
irradiation temperature effect on crosslinking efficiency, 251
irradiation time maximum for crosslinking density, 250
photochemical crosslinking with and without benzophenone initiation, 247–249
sample preparation for UV irradiation experiments, 248

Poly(ethylene oxide) (PEO)
anionic suspension polymerization, 237–247
degree of crystallinity during course of polymerization, 246*t*
electron micrographs of initial catalyst particles and during first polymerization stage, 243*f*
ethylene oxide (EO) polymerization rate by calcium catalysts, 239
initiator efficiency determination, 241
kinetic investigations, 239–242
main characteristics of EO suspension polymerization for different catalyst systems, 239*t*
molecular weight, dispersity, and PEO yield as function of time, 240*t*
morphological investigation of forming polymer particles, 242–247
nonionic surfactant or amphipatic polymer for particle stabilization, 247
optical micrographs of PEO particle growth during polymerization, 245*f*
particle size and shape dependence on monomer concentration and temperature, 244–247
particle size distribution during course of polymerization, 244*f*
polymer yield versus time for constant monomer concentration, 240*f*
propagation rate constants for EO polymerization in solution and suspension, 242*t*
thermal properties at various reaction times, 246*t*
X-ray microanalysis of initiator metal content in popular PEO, 238*t*

Polyisoprene
commercial production, 29
high *cis*-configuration, 28–29
high *cis*-1,4-polymer by lithium metal-initiated polymerization, 2
See also Isoprene polymers

Poly(isopropenyl *t*-butyl ketone) (PolyIPTBK)
anionic polymerization of IPTBK with *t*-butyllithium, 222–225
Norrish type I photolysis, 222–223
quantum yield of main chain scission, 225

steric hindrance effects with Grignard and radical polymerization, 223, 224*f*

Polymer blends
poly(ethylene-*co*-vinylacetate) in methyl methacrylate for PMMA/EVA blend, 268
polymerization of homogeneous polymer/monomer solutions, 267–268
See also Poly(2,6-dimethyl-1,4-phenylene ether) (PPE)/polyamide-6 blends

Poly(methyl methacrylate) (PMMA) with reactive end groups
challenges of primary amino-ended PMMA preparation, 291–292
degree of end-functionality of nitroaryl-ended PMMA by ^{1}H NMR, 294, 295*f*, 296*f*
experimental materials, 292
^{1}H NMR spectrum of aminobenzoyl-ended PMMA, 301*f*
MALDI mass spectrum of aminobenzoyl-ended PMMA, 302*f*
matrix assisted laser desorption/ionization (MALDI) mass spectrometry of nitrobenzyl and nitrobenzoyl-ended PMMA, 294, 297*f*, 298*f*
nitroaryl-ended PMMA molecular weight and functionality data, 293
nitroaryl-ended PMMA synthetic strategy, 295*f*
primary aminoaryl-ended PMMA by reduction of nitroaryl-terminated PMMA, 300, 301*f*
procedure for nitroaryl-ended PMMA synthesis, 292
procedure for reduction of nitroaryl-ended PMMA, 292
product characterization methods, 292
reaction conditions and product functionalities for nitroaryl-ended PMMA reductions, 300*t*
resonance stabilization of PMMA anion, 299*f*
synthetic route by living anionic polymerization, 291–292
thermal stability of nitroaryl-ended PMMA, 300

Poly(methylphenylsiloxane)s
bulk and solution copolymerization methods, 305
composition of copolymers by ^{1}H and ^{29}Si NMR spectroscopy, 308, 309*f*
copolymerization control methods, 305–306
cyclic monomer addition in ring-opening polymerization, 306
cyclic trimer and oligomer yield for varying monomer composition, 311
distribution of microblock size based on the comonomer ratio, 312, 313*f*
effect of increased phenyl-containing monomer content on polymerization rate and product properties, 306, 307*f*

elucidation of reaction condition-structure relationship at different stages of copolymerization, 311–312
experimental materials, 305
^{1}H and ^{29}Si NMR spectroscopic methods, 306
increasing glass transition temperatures proportional to Ph$_2$SiO-units, 315
microblock distribution changes during polymerization process, 312, 314f
microblock size distributions for bulk and solution polymerizations, 312, 314f
octamethylcyclotetrasiloxane [Me$_2$SiO]$_4$ and octaphenylcyclotetrasiloxane [Ph$_2$SiO]$_4$ monomers, 305
ratio of monomers and copolymerization constant values, 306, 308
relative reactivity of comonomers by Mayo–Lewis correlations, 308, 310f
results of end-group enrichment by phenyl-containing siloxane monomer, 308t
Poly(methyl α-trifluoromethylacrylate) (PolyMTFMA)
addition-elimination side reactions with typical anionic initiators, 220, 221f
anionic polymerization with amines or salts in presence of 18-crown-6, 220–222
radical copolymerization with methyl methacrylate (MMA), 220, 221f
Polyphthalaldehydes, anionic cyclopolymerization of 4-trimethylsilylphthalaldehyde (SPA), 226, 227f
Polypropylene (PP), electron micrographs of blends with styrene-isoprene block copolymers, 192, 194, 195
Polystyrene
anionic propagation kinetics, 4
blends of Styroflex and Styrolux, 124, 128
sec-butyllithium-initiated polymerization of styrene schematic, 3
general purpose polystyrene architecture, 167
properties of Styroflex/Styrolux/polystyrene blends, 127t
stress-strain diagram of Styroflex/Styrolux blends, 126f
See also Branched polystyrenes; High impact polystyrene (HIPS); Styroflex styrenic thermoplastic elastomer; Styrolux polystyrene with SB block structure
Polystyryllithium
estimation of chain transfer to ethylbenzene, 143, 144f
thermal stability at high concentrations in ethylbenzene, 143–145, 146f
Poly(vinyl acetate), formation of poly(ethylene oxide)/poly(vinyl acetate) networks, 252t, 253

Poly(vinyl chloride) (PVC)
competitive analysis of fresh meat packaging films, 120, 121t
mechanical property comparison with Styroflex, 118, 120t
permeability comparison with Styroflex, 120, 122f
processability comparison with Styroflex, 118, 120t
Pressure sensitive adhesive (PSA) tapes
carton sealability by formulation, 105, 109
carton sealability test method, 100
formulation and performance of asymmetric star block copolymers, 164–165
holding power for low and high diblock content, 109
holding power or shear adhesion method, 100
hydrogenated styrene-isoprene block copolymers (S–EP–S) applications and properties, 187, 190, 191t
measurements of PSA properties, 99–100
peel adhesion of SIS copolymers with varying SI content, 105
peel adhesion or adhesive strength method, 99
PSA properties of investigated SIS copolymers, 102t
styrene-isoprene-styrene (SIS) block copolymers, 98
tack measurement method, 99
wettability to rough surface, 102, 105
See also Styrene-isoprene-styrene (SIS) block copolymers
Principles of anionic polymerization
absence of termination and chain transfer, 2–3
anionic polymerization of alkyl methacrylates, 19–20
anionic polymerization of alkylene oxides, 20–21
anionic polymerization of lactams, 21, 22
block copolymer synthesis by difunctional initiation and two-step sequential monomer addition, 11, 12
block copolymer synthesis by three-step sequential monomer addition, 7–9
block copolymer synthesis by two-step sequential monomer addition and coupling, 9–11
chain-end functionalization by termination with electrophilic reagents, 16–17
control variables, 5–6
functionalized initiators for chain-end functionalization, 17–19
initiation kinetics dependence on solvent type, 4
kinetic scheme for diene polymerization, 5

mechanism of alkyllithium-initiated
polymerization, 3–5
molecular weight control, 5
molecular weight distribution (MWD), 5–6
propagation kinetics for dienes, 4–5
requirements for polymer formation with
Poisson molecular weight distribution, 6
schematic of *sec*-butyllithium-initiated
polymerization of styrene, 3
star-branched polymers by divinylbenzene
linking reactions, 14–16
star-branched polymers by linking reactions
with silyl halides, 13–14
synthesis of block copolymers, 6–11
synthesis of chain-end functionalized polymers,
16–19
synthesis of star-branched polymers, 11–16
Pultrusion
moisture content reaction concern, 261
nylon-6 pultrustion process, 261, 263*f*
polyamides in composite manufacturing
process, 258, 261
typical thermoplastic matrix pultrusion process,
261

R

Reactive end-functionalized polymers. *See*
Poly(methyl methacrylate) (PMMA) with
reactive end groups
Reactive extrusion process for polyamides, 256,
258, 259*f*
Reactive injection molding (RIM)
nylon block copolymer preparation, 262
nylon-6 toughened polymers, 262, 264*t*, 265*f*
Reactive solvents
ε-caprolactam with poly(2,6-dimethyl-1,4-
phenylene ether) (PPE), 268–269
poly(ethylene-*co*-vinylacetate) in methyl
methacrylate for PMMA/EVA blend, 268
polymers in monomer as solvents, 267–269
See also Poly(2,6-dimethyl-1,4-phenylene
ether) (PPE)/polyamide-6 blends
Reactive thermoplastic pultrusion. *See* Pultrusion
Resist polymers. *See* Lithographic resist
polymers
RIM. *See* Reaction injection molding (RIM)
Ring-opening polymerization of cyclosiloxanes.
See Poly(methylphenylsiloxane)s
Road pavement
tapered block copolymers for asphalt
modification, 129–130
See also Asphalt modification
Rolling resistance

fuel economy of cars, 62
hysteresis characteristics, 62–63

S

SANS (small angle neutron scattering), head-
group aggregation behavior, 41–47
SBS copolymers. *See* Styrene-butadiene-styrene
(SBS) triblock copolymers
Self-assembly of living polymer chains
aggregate mechanism in non-linear curve fitting
program, 41
butadienyllithium head-group on polystyrene
chain (SBLi) at low concentration in Kratky
plot, 43, 45, 46*f*
coherent scattering intensity I(Q) versus
scattering vector Q for head-group, 42–43,
44*f*
data fit by fractal mathematical model, 43, 44*f*,
45, 46*f*
entropic and enthalpic contributions for
aggregation, 45, 47
equilibria for conjectured mechanism with
dienes, 37–38
experimental protocol, 41–42
generic propagation rate expression for
aggregate-based mechanism, 36
high-resolution small angle X-ray scattering
(SAXS) and rheological measurements, 41
I(Q) versus Q for SBLi head-group at two
concentrations , 45, 46*f*
large- and small-scale association capacities of
living anionic lipophobic head-groups, 47
large scale structure formation by head-groups,
47
mathematical model describing mass fractal, 43
mechanism involving self-assembled
aggregates, 36
scattering behavior of powdered Pyrex versus
anomalous scattering, 36–37, 39*f*
small angle neutron scattering for reevaluation
of head-group aggregation, 41, 47
static light scattering measurements, 36–37
supramolecular mass fractal structure of
aggregate, 43
viscosity behavior of oligomeric
butadienyllithium as function of measured
degree of polymerization DP_n of terminated
polymer, 38, 39*f*
viscosity measurements for concentrated styryl-
and dienyllithium head-groups, 38, 39*f*
viscosity ratio of aggregate to dimer versus
parent chain DP_n, 38, 40*f*, 41
Siloxane polymers

cyclic organosilicon monomers, 305
ionic ring-opening polymerization of cyclosiloxanes, 304
See also Poly(methylphenylsiloxane)s
Silyl halides, star-branched polymers by linking reactions, 13–14
SIS copolymers. *See* Styrene-isoprene-styrene (SIS) block copolymers
SLS (static light scattering), head-group aggregation behavior, 36–38
Small angle neutron scattering (SANS), head-group aggregation behavior, 41–47
Sodium-based initiators, polymerization and properties of 1,2-polybutadiene, 204–205
Star-branched oligomers
 coupling with diesters dimethyl adipate, diethyl adipate, and dimethyl phthalate, 150, 152
 coupling with epoxidized soybean oil (ESO), 152–153, 154f
 coupling with monoester ethyl acetate, 150, 151f
 GPC curves for oligomer before and after coupling with ethyl acetate, dimethyl phthalate, and ESO, 151f
 ^{1}H NMR of oligobutadienyllithium, 149f
 oligobutadienyllithium characterization, 150t
 oligobutadienyllithium synthesis in toluene at room temperature, 148
 synthesis by reaction of oligobutadienyllithium with diesters and ESO, 148, 149f, 150–154
 See also Asymmetric star block copolymers; Star-branched polymers
Star-branched polymers
 anionic polymerization, 11–13
 linking reactions with divinylbenzene, 14–16
 linking reactions with silyl halides, 13–14
 synthesis by coupling living poly(styrene-*b*-butadienyl)lithium with diesters, 155–156, 157f
 See also Asymmetric star block copolymers; Branched polystyrenes
Static light scattering (SLS), head-group aggregation behavior, 36–38
Styrene-butadiene (SB) copolymers
 applications for >60% styrene, 32
 effect of residual SB on performance of SBS triblock, 90–92, 95
 pure and tapered block applications, 30–31
 Styroflex experimental SB copolymer, 112
 tapered block copolymers for asphalt and plastic modification, 129–130
 See also Styroflex styrenic thermoplastic elastomer; Styrolux polystyrene with SB block structure; Tapered block copolymers (TBC) of styrene-butadiene (SB)

Styrene-butadiene-styrene (SBS) triblock copolymers
 abrasion resistance of commercial SBS copolymers by sequential addition and coupling processes, 93f
 applications, 90
 DIN abrasion versus SB diblock content, 94f
 experimental copolymer and SB diblock preparation and testing, 91
 gel permeation chromatograms of commercial SBS copolymers by sequential addition and coupling processes, 93f
 hardness versus SB diblock content, 94f
 impact of SB diblocks on physical properties, 92, 95
 melt flow index versus SB diblock content at 180°C, 95f
 potential synthetic routes, 90–91
 residual SB diblocks, 90–91
 tan δ versus SB diblock content, 95, 97f
 viscoelastic behavior of SBS/SB blends, 95, 96f
Styrene-isoprene (SI) diblock copolymers
 asymmetric star block copolymers, 159
 effect of SI content in SIS triblock copolymers, 98, 109, 111
 See also Asymmetric star block copolymers
Styrene-isoprene-styrene (SIS) block copolymers
 carton sealability test method, 100, 101f, 103f
 carton sealability improvements with high diblock content, 105, 109
 experimental samples by traditional coupling method, 98–99
 force balance in measurement of carton sealability, 110f
 holding power or shear adhesion test method, 100
 holding power versus diblock content, 109
 loss tangent and tensile properties with varying diblock content, 105, 106f, 107f
 measurement methods of pressure-sensitive adhesive (PSA) properties, 99–100
 measurements of rheological properties, 100
 molecular characteristics of three copolymers, 99
 peel adhesion test method, 99
 peel adhesion versus peel rate by formulation, 105, 107f, 108f
 PSA properties of studied copolymers, 100, 102
 PSA tape application, 98
 storage modulus and peel adhesion values by formulation, 103f, 104f, 106f
 stress relaxation of PSA compounds, 109, 110f
 styrene-isoprene diblock content, 98
 tack measurement method, 99
 wettability to rough surface, 102, 105

Styrenic thermoplastic elastomers
  industrial applications, 31–32
  See Styroflex styrenic thermoplastic elastomer
Styroflex styrenic thermoplastic elastomer
  basic synthetic routes, 113, 114*f*
  bifunctional initiators, 113, 114*f*
  competitive analysis of commercial fresh meat packaging films, 120–121
  determination of order-disorder transition temperature (ODT), 115, 117*f*
  differential scanning calorimetry analysis, 116*f*
  dynamic mechanical thermoanalysis, 116*f*
  hysteresis of film materials, 119*f*
  mechanical properties typical for thermoplastic elastomer, 118
  morphology by transmission electron microscopy (TEM), 113, 114*f*, 115
  processibility by common methods, 118, 120*t*
  properties of Styroflex versus plasticized PVC and metallocene PE as extruded blown film, 120*t*
  rheological behavior of different symmetrical triblock copolymers, 115, 116*f*
  stress-strain diagram of compression-molded flexible materials, 119*f*
  structure and synthesis, 112–113
  Styroflex/Styrolux blend system, 124, 125*f*, 126*f*
  Styroflex/Styrolux/polystyrene blend system, 124, 126*f*, 127*t*, 128
  synergistic blends with other styrene polymers, 121–128, 122*f*
  synthesis, structure, and morphology of Styrolux, 121, 123*f*, 124
  thermal properties, 115
  toughness predominant feature, 118, 120*t*
  water and oxygen permeability of packaging films, 120, 122*f*
Styrolux polystyrene with SB block structure
  blend system with Styroflex, 124
  blend system with Styroflex and polystyrene, 124, 128
  mechanical and optical properties of Styroflex/Styrolux blend, 125*f*
  morphologies of pure Styrolux, Styroflex, and two blends, 125*f*
  properties of Styroflex/Styrolux/polystyrene blends, 127*t*
  stress-strain diagrams for series of blends with Styroflex and polystyrene, 126*f*
  synthesis, structure, and morphology, 121, 123*f*, 124
  See also Styroflex styrenic thermoplastic elastomer

T

Tape applications. *See* Pressure sensitive adhesive (PSA) tapes
Tapered block copolymers (TBC) of styrene-butadiene (SB)
  asphalt and plastic modification applications, 129–130
  composition and applications, 30–31
  molecular weight, Tg, and styrene content for TBCs in study, 130–131, 132*f*
  notched Izod impact, tensile and flexural modulus of HIPS-TBC blends as function of TBC content and type, 135–136, 137*f*, 138*f*
  preparation and testing of high impact polystyrene-TBC blends, 133–134
  preparation and testing of TBC mixtures for asphalt applications, 131–132, 133*f*
  storage modulus and tan δ of asphalt-TBC mixtures as function of frequency, 134–135, 136*f*
  viscosity and softening point temperatures of asphalt-TBC mixtures as function of TBC content and type, 134, 135*f*
  See also Asphalt modification; High impact polystyrene (HIPS)
Telechelic polybutadienes, industrial applications, 30
Telechelic polymers, α,ω-difunctional polymers using functionalized initiators, 18–19
N,N,N',N'-Tetramethylethylenediamine (TMEDA), addition to initiator reaction of N,N-dimethyltoluidine and n-butyllithium, 72, 74–76
Thermal stability of polymer-lithium systems
  decreasing polymer-lithium concentrations as function of heating time, 147*f*
  lithium hydride elimination, 145
  polybutadienyllithium at high concentration in ethylbenzene, 145, 146*f*, 147*f*
  polymer and copolymer synthesis in ethylbenzene at high monomer concentration, 145, 148, 149*f*
  polystyryllithium at high concentration in ethylbenzene, 143–145, 146*f*
Thermoplastic elastomers
  comb-shaped copolymers from acrylate-based macromonomers, 213–215
  olefins, urethanes, copolyester ethers and others, 32
  styrenic thermoplastic elastomers, 31–32
  See also Styroflex styrenic thermoplastic elastomer
Tire industry

hysteresis characteristics, 62–63
reduction in hysteresis by carbon black re-agglomeration prevention, 78
rolling resistance, 62
α,α′,α″,-Trichloromesitylene, coupling agent of three-arm star polystyrene, 174, 175f

U

UV spectroscopy
monitoring anionic polymerization on-line, 50–51
See also On-line process monitoring

W

Water extractable content (WEC)
monomer conversion method for polyamides, 271
nylon-6 results, 286

Z

Ziegler polymerization, isoprene polymers with high stereoregularity, 186–187

Bestsellers from ACS Books

The ACS Style Guide: A Manual for Authors and Editors (2nd Edition)
Edited by Janet S. Dodd
470 pp; clothbound ISBN 0–8412–3461–2; paperback ISBN 0–8412–3462–0

Writing the Laboratory Notebook
By Howard M. Kanare
145 pp; clothbound ISBN 0–8412–0906–5; paperback ISBN 0–8412–0933–2

Career Transitions for Chemists
By Dorothy P. Rodmann, Donald D. Bly, Frederick H. Owens, and Anne-Claire Anderson
240 pp; clothbound ISBN 0–8412–3052–8; paperback ISBN 0–8412–3038–2

Chemical Activities (student and teacher editions)
By Christie L. Borgford and Lee R. Summerlin
330 pp; spiralbound ISBN 0–8412–1417–4; teacher edition, ISBN 0–8412–1416–6

Chemical Demonstrations: A Sourcebook for Teachers, Volumes 1 and 2, Second Edition
Volume 1 by Lee R. Summerlin and James L. Ealy, Jr.
198 pp; spiralbound ISBN 0–8412–1481–6
Volume 2 by Lee R. Summerlin, Christie L. Borgford, and Julie B. Ealy
234 pp; spiralbound ISBN 0–8412–1535–9

The Internet: A Guide for Chemists
Edited by Steven M. Bachrach
360 pp; clothbound ISBN 0–8412–3223–7; paperback ISBN 0–8412–3224–5

Laboratory Waste Management: A Guidebook
ACS Task Force on Laboratory Waste Management
250 pp; clothbound ISBN 0–8412–2735–7; paperback ISBN 0–8412–2849–3

Reagent Chemicals, Eighth Edition
700 pp; clothbound ISBN 0–8412–2502–8

Good Laboratory Practice Standards: Applications for Field and Laboratory Studies
Edited by Willa Y. Garner, Maureen S. Barge, and James P. Ussary
571 pp; clothbound ISBN 0–8412–2192–8

For further information contact:
Order Department
Oxford University Press
2001 Evans Road
Cary, NC 27513
Phone: 1-800-445-9714 or 919-677-0977
Fax: 919-677-1303

Highlights from ACS Books

Desk Reference of Functional Polymers: Syntheses and Applications
Reza Arshady, Editor
832 pages, clothbound, ISBN 0–8412–3469–8

Chemical Engineering for Chemists
Richard G. Griskey
352 pages, clothbound, ISBN 0–8412–2215–0

Controlled Drug Delivery: Challenges and Strategies
Kinam Park, Editor
720 pages, clothbound, ISBN 0–8412–3470–1

Chemistry Today and Tomorrow: The Central, Useful, and Creative Science
Ronald Breslow
144 pages, paperbound, ISBN 0–8412–3460–4

Eilhard Mitscherlich: Prince of Prussian Chemistry
Hans-Werner Schutt
Co-published with the Chemical Heritage Foundation
256 pages, clothbound, ISBN 0–8412–3345–4

Chiral Separations: Applications and Technology
Satinder Ahuja, Editor
368 pages, clothbound, ISBN 0–8412–3407–8

Molecular Diversity and Combinatorial Chemistry: Libraries and Drug Discovery
Irwin M. Chaiken and Kim D. Janda, Editors
336 pages, clothbound, ISBN 0–8412–3450–7

A Lifetime of Synergy with Theory and Experiment
Andrew Streitwieser, Jr.
320 pages, clothbound, ISBN 0–8412–1836–6

Chemical Research Faculties, An International Directory
1,300 pages, clothbound, ISBN 0–8412–3301–2

For further information contact:
Order Department
Oxford University Press
2001 Evans Road
Cary, NC 27513
Phone: 1-800-445-9714 or 919-677-0977
Fax: 919-677-1303